Maintenance Costs and Life Cycle Cost Analysis

T0225337

Maintenance Costs and Life Cycle Cost Analysis

Diego Galar, Peter Sandborn,
and Uday Kumar

CRC Press is an imprint of the
Taylor & Francis Group, an **informa** business

CRC Press
Taylor & Francis Group,
6000 Broken Sound Parkway NW, Suite 300,
Boca Raton, FL 33487-2742

First issued in paperback 2020

ISBN-13: 978-0-367-57300-3 (pbk)
ISBN-13: 978-1-4987-6954-9 (hbk)

Library of Congress Cataloging-in-Publication Data

Names: Galar, Diego, author. | Sandborn, Peter A., 1959- author. | Kumar, Uday, author.
Title: Maintenance costs and life cycle cost analysis/Diego Galar, Peter Sandborn, Uday Kumar.
Description: Boca Raton : Taylor & Francis, a CRC title, part of the Taylor & Francis imprint, a member of the Taylor & Francis Group, the academic division of T&F Informa, plc, [2017] | Includes bibliographical references and index.
Identifiers: LCCN 2016059173 | ISBN 9781498769549 (hardback: alk. paper) | ISBN 9781498769556 (ebook)
Subjects: LCSH: Plant maintenance. | Life cycle costing.
Classification: LCC TS192.G35 2017 | DDC 658.2--dc23
LC record available at https://lccn.loc.gov/2016059173

Visit the Taylor & Francis Web site at
http://www.taylorandfrancis.com

and the CRC Press Web site at
http://www.crcpress.com

Contents

Preface

The purpose of maintenance is to reduce societal, business, and personal risks. However, the cost associated with maintenance actions should not exceed the value of the risks being eliminated and the associated benefits reflected in terms of increased productivity or reduced number of accidents, etc. In world-class organizations, in general, maintenance decisions are based on a sound cost-benefit-risk analysis. Maintenance costs analysis should include direct, consequential, and indirect costs; intangible costs; opportunity costs; and the cost of potential risks. Benefits should include all direct and indirect revenues and intangible benefits, such as increased productivity from improved employee safety, etc. The contribution of maintenance to the success of organizations has been increasingly recognized. The role of maintenance managers has been naturally extended into their involvement in strategic planning issues, such as production capacity or renewal of equipment planning. This includes the selection of technological solutions and maintenance cost quantification and optimization within life cycle cost (LCC) analysis, which, in turn, contributes to the study of the consequences for the investments to be made.

Maintenance itself, in all its areas of intervention, is very difficult to manage, given the variability of situations where it has to intervene and its evolution over time. Probably for this reason, maintenance management is the most recent variable to be considered when organizations try to increase their competitive advantages. Approaches like on-condition maintenance, predictive maintenance, reliability, virtual reality, certification attempts, and key performance indicators are only some of the many maintenance fields or, in other words, LCC fields that motivate organizations to work better and increase their market value.

The competitiveness of organizations is increasingly dependent on the optimization of the LCC of their physical assets and how well they achieve their required function. Depending on the sector, the cost of maintaining physical assets represents around 5%–20% of the added value achieved by maintenance; 5%–12% of the total capital invested; 1%–15% of gross sales; 3%–10% of the production costs. These are strong reasons to emphasize the importance of maintenance costs in the annual budget and within LCC.

Maintenance costs are a fact of everyday life. Nevertheless, when engineers who have good knowledge of mathematics and technical skills are presented with cost theory, there is a clear gap in their knowledge, especially when discussing cost–benefit analysis, etc.

In any case, in this book, we have carefully avoided the approach adopted by economists, that is, a financial type of presentation understandable only to economists. All of the theory is included but we have tried to present it in an easy-to-understand manner; we do not want to use an economist's approach to explain the concepts of cost modeling. We have relied heavily on the assumption that engineers have a solid mastery of technology and management but not financial mathematics.

In developing the contents of this book, we respected the fact that organizations may use the existing methodologies as per their convenience and tradition. Therefore, we present the issues and challenges of maintenance costs in such a way that engineers can choose the approach or methodology convenient for their purpose. Our goal is to provide a comprehensive and understandable treatment of the fundamentals. Once engineers have mastered these, we can lead them to deeper analysis and more accurate LCC estimation but with a substantial difference, that is, a deep understanding of maintenance costing.

Clearly, there are many views of how the fundamentals of maintenance cost should be organized. We have attempted to create coherent chapters and sections and to present topics in a logical and sequential order. The text necessarily starts with the importance of the maintenance function in an organization and moves to LCC considerations and to budgeting constraints. We have intentionally postponed a discussion of intangible costs and downtime costs to later in the book mainly because of the controversy among managers.

We conclude the book with a short description of a number of sectors where maintenance cost is critical. The goal is to train the reader in a deeper study and better understanding of these elements for decision-making in maintenance, specifically in the context of asset management. Considering the ever-increasing role of asset management, we believe this is an effective way to introduce the rigor of the subject to those who will use it.

The book is motivated by the persistent pattern of failure of maintenance engineers to explain the basics of maintenance cost and associated risks and benefits to senior managers in their organizations. This motivation was reinforced by the recent success of several publications putting this type of research into the spotlight, especially the new ISO 55000.

This book is intended for managers, engineers, researchers, and practitioners directly or indirectly involved in the area of maintenance. We hope the book will contribute towards a better understanding of maintenance cost and that this enhanced knowledge will be used to improve the maintenance process.

It was a challenge for us to create a reader-friendly book that could be useful for both practitioners and researchers in the maintenance field. But it evolved naturally throughout our many years of teaching, research, and consulting in the area of maintenance engineering and life cycle costing.

Acknowledgments

It is with immense gratitude that we acknowledge the support and help of our universities, Maryland and Lulea, for their institutional support during this time showing a clear interest in the topic. We must thank students and the academic and administrative staff at Luleå Tekniska Universitet and the University of Maryland for giving us the opportunity and support to write this manuscript and perform research on the topic. Thanks for providing the facilities to do it.

The list of the people who should be thanked for this book is too long to mention everyone, but we will do our best. First of all, we thank our families. Without their continuous support and encouragement, we would never have been able to achieve our goals.

Thanks to the people who directly or indirectly participated in this book; none of this would have been possible without them. Their dedication and perseverance made the work enjoyable. Especially remarkable was the contribution of Dr. Aditya Thadouri. His support in the preparation of the manuscript was indispensable to the success of the project.

We want to thank all our colleagues at the Division of Operation and Maintenance, University of Lulea, Sweden, and CALCE at the University of Maryland, for encouraging us to produce this book as a compilation of maintenance cost methods widely used in our field, especially those who worked with us for many long hours discussing the coherence of a manuscript that addresses such a controversial and multidisciplinary topic.

It was a challenge to create a reader-friendly book that would be useful for both practitioners and researchers in the maintenance field! That is why we must also thank all international conferences, forums, maintenance associations, journals, etc., who fed our need of information and discussion.

Thanks to all of you. Without your help, this book would not be a reality.

Acknowledgments

Authors

Diego Galar has an MSc in telecommunications and a PhD in manufacturing from the University of Saragossa, Zaragoza, Spain. He has been professor in several universities, including the University of Saragossa and the European University of Madrid. He also was a senior researcher in I3A, Institute for Engineering Research in Aragon; director of academic innovation; and subsequently pro-vice-chancellor of the university. In industry, he has been technological director and CBM manager. He has authored more than a hundred journal and conference papers, books, and technical reports in the field of maintenance. Currently, he is professor of condition monitoring in the Division of Operation and Maintenance at Luleå University of Technology (LTU), where he is coordinating several EU-FP7 projects related to different maintenance aspects and is also involved in the SKF UTC center located in Lulea focused on SMART bearings. He is also visiting professor at the University of Valencia, Polytechnic of Braganza (Portugal), Valley University (Mexico), Sunderland University (UK), and Northern Illinois University (USA).

Peter Sandborn is a professor in the CALCE Electronic Products and Systems Center at the University of Maryland, College Park, Maryland. Dr. Sandborn's group develops obsolescence forecasting algorithms and performs strategic design refresh planning and lifetime buy quantity optimization. Dr. Sandborn is the developer of the MOCA refresh planning tool. MOCA has been used by private and government organizations worldwide to perform optimized refresh planning for systems subject to technology obsolescence. Dr. Sandborn also performs research in several other life cycle cost modeling areas, including maintenance planning and return on investment analysis for the application of prognostics and health management (PHM) to systems, total cost of ownership of electronic parts, transition from tin-lead to lead-free electronics, and general technology trade-off analysis for electronic systems. Dr. Sandborn is an associate editor of the *IEEE Transactions on Electronics Packaging Manufacturing* and a member of the editorial board of the *International Journal of Performability Engineering*. He is a past conference chair and program chair of the ASME Design for Manufacturing and Life Cycle Conference. He is the author of more than 150 technical publications and several books on electronic packaging and electronic systems cost analysis and was the winner of the 2004 SOLE Proceedings and 2006 Eugene L. Grant awards. He has a BS in engineering physics from the University of Colorado, Boulder, in 1982, and an MS in electrical science and a PhD in electrical engineering, both from the University of Michigan, Ann Arbor, in 1983 and 1987, respectively.

Uday Kumar the chaired professor of operation and maintenance engineering, is director of Luleå Railway Research Center and scientific director of the Strategic Area of Research and Innovation—Sustainable Transport at Luleå University of Technology, Luleå, Sweden. Before joining Luleå University of Technology, Dr. Kumar was professor of Offshore Technology (Operation and Maintenance Engineering) at Stavanger University, Norway. Dr. Kumar has research interests in the subject area of reliability and maintainability engineering, maintenance modeling, condition monitoring, LCC and risk analysis, etc. He has published more than 300 papers in international journals and peer-reviewed conferences and has made contributions to many edited books. He has supervised more than 25 PhD theses related to the area of reliability and maintenance. Dr. Kumar has been a keynote and invited speaker at numerous congresses, conferences, seminars, industrial forums, workshops, and academic institutions. He is an elected member of the Swedish Royal Academy of Engineering Sciences.

1

Relevance of Maintenance Function in Asset Management

1.1 Purpose of Maintenance Function

Webster's dictionary defines maintenance as

- To maintain
- Keep in existing condition
- Preserve, protect
- Keep from failure or decline

EN 13306:2010 defines maintenance management as "all activities of the management that determine the maintenance objectives, strategies, and responsibilities, and implement them by means, such as maintenance planning, maintenance control and supervision, improvement of methods in the organization, including economical, environmental, and safety aspects.

Therefore, the overall goal of maintenance within an industry is to preserve, with reliability, the system (or product) operation at a level that meets the company's business needs, and to ensure

A specific duration of failure-free performance under stated conditions.

Maintenance management will take care of the resource allocation required to accomplish the maintenance goals; and maintenance costs definitely play a key role in this equation to achieve those goals at minimum cost and also to know how to calculate, register, and control those costs.

Everyone has an understanding of what maintenance is. If asked, they will mention things like fix, restore, replace, recondition, patch, rebuild, and rejuvenate. To some extent, there is a place for these concepts, but they miss the mark in understanding the totality of maintenance. Maintenance is the act of maintaining. The basis for maintaining is to keep, preserve, and protect; that is, to keep in an existing state or preserve from failure or decline.

There is a major difference between this definition and the words and functions normally mentioned by those considered "knowledgeable" about maintenance.

The maintenance department is responsible and accountable for maintenance. It is responsible for the way equipment runs and looks and for the costs incurred in achieving the required level of performance. This is not to say that operators have no responsibility for the use of equipment when it is in their hands—they do. But to split responsibility between maintenance and any other department is to leave no one accountable.

Therefore, the maintenance department is responsible for the frequency and level of maintenance, thus balancing responsibility and cost control. That said, the maintenance function should report to top management. This allows maintenance problems to be dealt with in the best interests of the plant or company as a whole. Maintenance efforts and costs must not be manipulated as a means for a department to achieve its desired results.

When the maintenance department or group is held responsible and accountable for maintenance, its relationship with other departments takes on new meaning. The maintenance department can't afford to have adversarial relationships with others. It must base its interdepartmental relationships on credibility and trust. This is essential for the successful operation of a maintenance management system (Smith and Mobley, 2007).

1.1.1 Maintenance Function

The process of maintaining physical assets covers all actions necessary for

- Retaining an asset in a specified condition
- Restoring an asset to a specified condition

Deterioration of the condition and capability of a manufacturing system starts as soon as the system is commissioned. In addition to normal wear and deterioration, other failures may occur, especially when the equipment is pushed beyond its design limits or suffers operational errors. Equipment downtime, quality problems, processing speed losses, safety hazards, and/or environmental pollution all occur because of deterioration. These outcomes can negatively impact the operating cost, profitability, customer satisfaction, and productivity. To ensure a plant or a company meets its production targets at an optimal cost, maintenance management has to carefully plan its maintenance objectives and strategies.

Good maintenance assumes that maintenance objectives and strategies are not determined in isolation, but are derived from such factors as company policy, manufacturing policy, and other potentially conflicting demands and constraints in the company. According to Muchiri et al. (2010), maintenance objectives are related to the attainment of production targets (through high availability) at a required quality level, and within the constraints of the system's condition and safety. Furthermore, maintenance resources are used to ensure the equipment is in good condition, the plant achieves its design life, the safety standards are met, and the energy use and raw material consumption are optimized, among other factors.

Some of the important factors influencing the maintenance objectives are shown in Figure 1.1: ensuring the plant functionality (availability, reliability, product quality, etc.); ensuring the plant achieves its design life; ensuring plant and environmental safety; ensuring cost-effectiveness in maintenance; and ensuring the effective use of resources (energy and raw materials). The maintenance objectives of a given company or plant will obviously influence the specific performance indicators used to measure success.

Once the maintenance objectives are outlined, the next step is to formulate the maintenance strategy) and action plan: what type of maintenance needs to be done, who should execute the maintenance tasks, when to do it, and how often it can be done. In general, maintenance decision-making can be broadly explained in terms of maintenance actions (basic tasks), maintenance policies, and maintenance concepts, etc. (Pintelon and Van Puyvelde, 2006). Maintenance policies are the rules or set of rules describing the triggering mechanism for the various maintenance actions. Examples of these policies are failure-based maintenance (FBM), use-based or time-based maintenance (UBM/TBM), condition-based maintenance

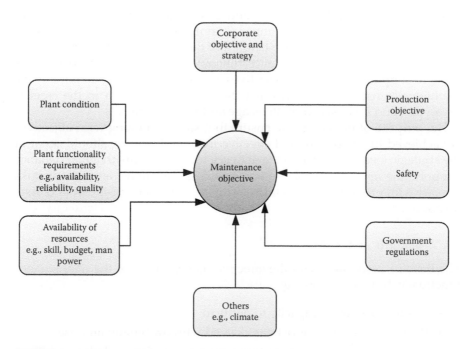

FIGURE 1.1
An illustration of factors influencing maintenance objectives.

(CBM), and design-out maintenance (DOM) (Coetzee, 1997; Madu, 2000; Waeyenbergh and Pintelon, 2002; Savsar, 2006).

Once the objectives and strategies have been established, the success of the maintenance function is dependent on the maintenance work management. The maintenance work management cycle, as outlined by Campbell (1995), consists of work identification, planning, scheduling, execution, and closing the job. Maintenance work is identified using the preventive, predictive, and failure-finding work orders generated by proactive maintenance. Alternatively, repair work occurs because of failure. At the heart of the maintenance function is work planning and scheduling; this defines what gets done and when. To complete the work cycle, effective work execution is vital to guarantee the achievement of the required equipment condition and performance.

Reviewing the maintenance objectives can yield insights into the complex environment within which the maintenance department functions. They are essential ingredients and indicators in the development of a maintenance performance measurement system, and they form a potential basis for performance evaluation (Muchiri et al., 2010).

In summary, the maintenance function provides a strategic link between a maintenance program and its corporate directions and core business. It supports the management in planning and implementing a maintenance program in alignment with its capital investment, and operational and disposal plans.

It requires a properly structured and professionally managed system to achieve this in a cost-effective manner. (Maintenance excludes general cleaning and refurbishment to a new standard of use (Assetic System Overview).)

The objective of the maintenance function is to ensure that assets continue to support the business objectives and service delivery requirements. Maintenance itself also ensures that the capital investment in the asset is preserved, and is consistent with its age and market value.

The purpose of planning is to ensure that short- and long-term objectives are achieved in an efficient and effective manner. Planning should occur at the strategic level, as well as at the delivery level, to ensure that maintenance supports the agency's corporate and business directions, and that it links directly with programs for capital investment, management-in-use and asset disposal.

1.1.2 Benefits and Risks

The benefits associated with the effective management of the maintenance function include the following:

- Assurance of asset capacity to perform
- Better management of risks associated with ownership and use
- Compliance with statutory requirements (e.g., for workplace health and safety)

- Economic service delivery
- Retention of asset value
- Rational demand for capital investment

The risks associated with a lack of effective maintenance management include the following:

- Loss of asset capacity and potential
- Loss of asset value
- Increased costs of service delivery
- Legal and other liabilities
- Poor image and community criticism
- Premature replacement
- Inappropriate maintenance standards and performance
- Unscheduled/unexpected major expenditure
- Wastage of resources
- Inappropriate funding of maintenance
- Stakeholder dissatisfaction
- Loss of productivity, and employee dissatisfaction

Maintenance requires a commitment of resources in an environment of competing demands. The risks of due to poor resource allocation for maintenance can be managed only through sound planning, informed decision-making, and professional management.

1.1.3 Maintenance Process: Goals to Achievements

Maintenance management has as its primary focus on the delivery of expected function from the asset and achievement of this objective encompasses the following three steps:

- Develop a structured maintenance framework to ensure appropriate maintenance policies, strategies, systems, and delivery mechanisms are in place to meet the service delivery needs of the managerial level and to facilitate the development of a maintenance plan/program.
- Develop a maintenance program/plan that encompasses asset life cycles and budget considerations and aligned to the corporate and physical asset strategies.
- Implement the maintenance program/plan in conjunction with capital investment, and operational and disposal plans.

1.1.3.1 Develop a Structured Maintenance Framework

A maintenance framework in complex organizations ensures consistency in the planning, implementation of, and reporting of maintenance. It is not a standard but it works as such from an internal point of view and can often be the precursor of standard deployments. In this case, the organization with a maintenance framework will be more successful in the deployment of ISO 55000. ISO 55000 provides an overview of the subject of asset management and the standard terms and definitions.*

The development of a structured maintenance framework includes the following actions:

- Define maintenance objectives
- Develop a maintenance policy
- Develop a maintenance strategy
- Implement a maintenance management framework
- Define maintenance objectives

As part of the corporate planning and physical asset planning processes, the roles and performance standards of assets should be defined and, where appropriate, performance indicators should be developed. The maintenance framework developed for the assets should enable the performance standards of the assets to be met efficiently and effectively. Keeping this in mind, maintenance objectives should be clearly defined. They should facilitate performance measurement and enable appropriate policies and strategies to be developed for their achievement (Maintenance Management Framework (I), 2012).

1.1.3.2 Develop a Maintenance Policy

A maintenance policy should be developed and established to help accounting people to implement a consistent maintenance approach across the spectrum of assets. It should be expressed in broad terms, owing to the wide range of assets, but should set a clear direction and enable the development of strategies.

It should also be consistent with the corporate and asset management policy developed for maintenance including issues related to the following:

- Compliance with governmental and corporate policies
- Compliance with legislative requirements
- Health, safety, and security

---

* http://www.assetmanagementstandards.com/.

- Risk management
- Asset preservation
- Standards of maintenance
- Roles and responsibilities
- Performance measurement
- Quality management

The policy should be endorsed at the appropriate level and incorporated in documentation relevant to the management and maintenance of physical assets (Maintenance Management Framework (I), 2012).

1.1.3.3 Develop a Maintenance Strategy

The maintenance policy sets the framework and direction for the development of a maintenance strategy to achieve the outcomes required. The strategy should address each aspect of the policy and establish how maintenance should be implemented for the various categories of assets (Maintenance Management Framework (I), 2012).

1.1.3.4 Implement a Maintenance Management Framework

The efficient and effective maintenance of assets requires both a strategic and an operational focus. To ensure that maintenance is complementary to the corporate and service delivery needs of the company, a comprehensive maintenance management framework must be set in place.

The key requirements of this management framework are the following:

- To identify management responsibilities
- To allocate maintenance resources
- To develop a maintenance information system
- To fund maintenance
- To measure maintenance performance
- To train maintenance personnel

These requirements should be reviewed on an ongoing basis to establish whether they can be improved or restructured to meet future needs.

Where responsibility for maintenance delivery has been assigned to external providers, contract conditions need to be suitably structured to ensure that the customer's interests are safeguarded, the responsibilities of purchaser and provider are clearly defined, and gaps in responsibility do not occur (Maintenance Management Framework (II), 2012).

1.1.3.5 Allocate Maintenance Resources

Subject to the specific needs of the various types of assets, the following factors should be considered in developing delivery strategies and in allocating resources:

- Maintenance tasks and activities
- Level and mix of professional and technical expertise
- Balance of in-house and outsourced expertise
- Procurement arrangements
- Funding constraints
- Management of resources
- Maintenance facilities

Asset managers should consider maintenance within the overall context of asset management and their core business to determine the scope, extent, and cost-effectiveness of undertaking maintenance themselves or outsourcing maintenance services.

1.1.4 Developing a Maintenance Program/Plan

Maintenance planning is the process through which a company can review the maintenance requirements for its assets and develop suitable strategies for addressing any gaps and future needs, in alignment with new capital investment and the rationalization of existing assets.

The plan should contain an overview of the physical asset portfolio and the key issues relating to maintenance, which include the following:

- The status of asset condition, performance, and availability for core service delivery
- Policies and service standards established for core services and their implications for maintenance
- Maintenance policies and standards and their implications for asset condition
- Financial, social, environmental, and other emerging issues affecting maintenance
- Future maintenance-planning strategies, including demand management
- Planned capital investment programs and other changes to the asset base, such as disposals and rehabilitation
- Risk management

1.1.4.1 Outline of a Maintenance Plan

The purpose of a program/plan is to review, in broad terms, the physical state and maintenance requirements of the assets and to establish a plan of action for future maintenance, supported by a sound financial strategy. The plan should clearly state the outcomes and present them in a structured format ready for implementation. It should also state the necessary endorsements from key stakeholders that will be required before the plan can be implemented.

The strategic direction of maintenance should be guided by the role of assets in supporting service delivery and the need to preserve the assets in appropriate conditions.

The plan should address long-term objectives as well as short- to medium-term objectives and should be aligned to strategies, including resourcing and implementation issues.

Fundamental to the planning process is the existence of a maintenance infrastructure which will support and provide the necessary information for operational and strategic planning.

The maintenance plan should contain an overview of the asset portfolio and describe the key issues relating to maintenance, which include the following:

- Policies and service standards established for the company's services and their implications for asset condition, performance, and capacity to support service delivery
- The current status of asset condition, performance, and capacity and how these have affected service delivery
- An overview of maintenance and how it has impacted assets
- Financial, social, environmental, and other issues that have affected maintenance
- Planned capital investment programs and other changes to the asset base, such as disposals and rehabilitation
- Key maintenance issues that need to be addressed

The asset profile should be reviewed to establish current and future trends and identify any changes that may impact future plans for maintenance. The review should include information from related planning activities such as capital investment planning and property rationalization/disposal planning to ensure consistency and that all factors affecting the asset base have been properly considered.

The review should seek to establish specific attributes of the asset that has affected maintenance of those assets, their performance in support of service delivery, and implications for future maintenance. An essential element is to establish the existing and anticipated future maintenance liabilities associated with the assets and to develop strategies to manage them.

The following should be considered in a review of asset types and their impact on maintenance:

- Type of asset
- Geographic distribution
- Design
- Operating environment
- Environmental issues

Issues arising from the analysis should be considered in terms of their long-term as well as their short-term impact on maintenance. Significant trends that have affected or are likely to influence maintenance planning should be identified. These should then be translated into maintenance strategies.

The age, condition, and complexity profiles of the asset base need to be analyzed and presented. The analysis should seek to establish any gaps between the condition profile necessary to support service delivery and the current profile. It should also identify whether such gaps have been influenced by the level of maintenance provided and in which particular aspects. For example, it may be hard to obtain spare parts for technically obsolete assets while high-tech assets in remote areas may suffer from the lack of maintenance expertise. Age profiles, together with other related information, will facilitate planning in relation to future maintenance, major repairs, replacements, and disposal.

The performance of assets should be reviewed to analyze whether their functionality and capacity to meet operational requirements have been significantly affected by the quality of maintenance provided. Based on the key influencing factors strategies to improve maintenance can then be developed. A review of maintenance and operational requirements may also identify opportunities for improving productivity and operational efficiencies (e.g., energy consumption).

1.1.5 Fund Maintenance: The Beginning of Maintenance Costs

Maintenance process and organization should be appropriately funded to avoid deferment or non-performance of essential maintenance actions which result in deterioration of asset performance.

The maintenance program should be reviewed in terms of key areas of expenditure by maintenance activity (e.g., preventive, condition-based, statutory, routine breakdown, and incidence maintenance) and asset category or element to identify areas where factors such as maintenance practices, usage, age, condition, complexity, locality, and environment have influenced trends in maintenance costs.

Deferred maintenance backlogs should be analyzed against various parameters to establish benchmarks for planning and future reference. As part of the maintenance funding projections, companies should consider how any backlog is to be managed and funded.

Projected maintenance works such as major repairs, overhauls, and replacement should be identified, quantified, and programmed over a 3- to 5-year horizon. The funding commitments required should be clearly identified and a cash flow analysis prepared in current dollar terms. Included in this category are emerging issues such as the impact of new legislation, and environmental, heritage, and health/safety concerns.

Anticipated maintenance commitments such as new assets coming out of warranty and requiring maintenance servicing (e.g., elevators or air-conditioning) should be identified, quantified, and programmed accordingly. Capital investment plans should be scanned to identify new assets likely to be built/acquired/leased which will require additional maintenance funding.

Major maintenance projects and improvements to be undertaken should be reported, outlining their costs/benefits. Such projects are usually undertaken with a view to reducing the future costs of maintenance and in most instances will require an initial capital investment.

Drawing on the review of assets and their maintenance projections, a financial plan can established for the forward estimates period defining funding requirements for maintenance and priorities for budget establishment. Longer-term funding implications should also be identified as projections for future planning considerations.

The financial plan should provide a comprehensive picture of maintenance funding requirements and how they are to be met. It should seek to identify where shortfalls in previous funding have occurred and how they affect funding requirements in the current forward estimates period. Nonrecurrent funding required for short-term commitments should be identified and presented as well.

In summary, the following issues should be considered in maintenance costing:

- Appropriate maintenance activities to be funded as maintenance
- Funding of ongoing maintenance
- Funding of major corrective maintenance
- Funding of deferred maintenance
- Funding of major replacements
- Funding of work brought on by new legislation, service requirements, and new assets
- Contingencies for major unpredictable failures

It is essential that methodologies for determining appropriate funding levels be developed and tested against asset condition, maintenance cost histories, and industry benchmarking.

Proper management procedures for the administration of maintenance budgets should also be established, together with procedures for budget planning, allocation, monitoring, control, and reporting.

1.1.6 Developing a Basis for Maintenance Performance Measurement

A system of performance measurement should be developed and implemented to track the performance not only of maintenance but also of service providers and other suppliers. In addition, a system of audits should be incorporated to test performance.

Performance measurement relies on accurate and reliable information being provided by maintenance information systems.

Performance measurement against set standards of asset condition, asset performance, and other financial benchmarks should be developed and applied. Specific performance indicators relevant to the particular class of asset or business process can be developed to evaluate the effectiveness and efficiency of maintenance task delivery.

These may include response times, mean time between failure (MTBF), and others. The focus should be on outcomes but with due regard for inputs and outputs.

In fact, to develop a structured approach to measuring the performance of the maintenance function, it is imperative to have a well-formulated maintenance strategy based on corporate and manufacturing strategies. In addition, the approach should be incorporated within a coherent theory of maintenance processes known to be critical factors in manufacturing and business success.

Good management of the maintenance process (efforts) is important if the desired results and maintenance objectives are to be achieved. The key steps in the maintenance process are work identification, work planning, work scheduling, and work execution (Campbell, 1995):

- Work identification refers to identifying the right work to be performed at the right time by the maintenance staff based on the maintenance objectives. It identifies and controls failure modes affecting the equipment's ability to perform its intended function at the required performance level. Activities are evaluated based on the consequences of failure of equipment performance so that maintenance resources are used most effectively. This, in turn, ensures that maintenance activities contribute to the performance results.

- Work planning develops procedures and work orders for the maintenance activities identified. This includes identification of resource requirements, safety precautions, and the instructions required to carry out the job.

- Scheduling evaluates the availability of all resources required for the work and the time frame for executing it. Scheduling also considers the impact of maintenance work on the production schedule.
- Work execution ensures that the scheduled activities are carried out within the allocated time and with the effective use of resources. This process ensures the maintenance work is done effectively.

To manage the maintenance process, performance indicators need to be defined for each step. Since maintenance processes determine the maintenance outcomes and results, the indicators related to the maintenance process are called leading indicators.

Once the maintenance processes are completed, the maintenance results for a given period need to be monitored. The results are measured in terms of the equipment's condition and performance, together with maintenance costs and the effective use of maintenance resources. Careful analysis of maintenance results supports the identification of performance gaps and the continuous improvement of equipment performance. Performance analysis involves comparison of the achieved results with the targets, comparison with the historical data, analysis of the trends, and review of the maintenance activities' costs. The indicators related to the maintenance results are called lagging indicators, as they are only known after a given period or after events have happened.

In summary, the monitoring and review of maintenance and its management are critical to ensure that maintenance is both efficient and effective; a system of performance measurement should therefore be developed and implemented to evaluate the maintenance management process, service delivery, and outcomes.

An asset management organization for continuous improvement of practices should also be established as seen in Figure 1.2.

The performance of the maintenance management process will be determined on the basis of the following:

- Degree of planning and asset user input into planning
- Production of asset maintenance plans
- Establishment of asset maintenance standards
- Appropriateness and effectiveness of maintenance strategies
- Development and achievement of maintenance schedules
- Procurement of services in accordance with policies
- Processes for the control of work
- Use of computerized maintenance management systems
- Quality and accuracy of maintenance data
- Establishment of reporting structures and production of management reports

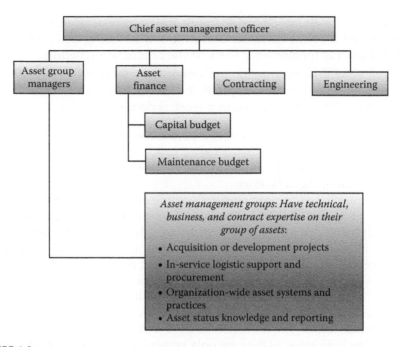

FIGURE 1.2
Asset management group structure. (Adapted from Hastings, N.A.J., *Physical Asset Management*, XXIX, 2010, 370 p., Hardcover. http://www.springer.com/978-1-84882-750-9.)

Finally, the outcomes of the expected maintenance will be measured in terms of efficiency and effectiveness as follows:

- Delivery of program on time and within budget
- Level of asset downtime
- Level of disruption to business operations
- Level of planned vs. unplanned maintenance
- Condition of assets
- Degree of asset user satisfaction
- Quality of work
- Quality of response
- Cost of service
- Level of deferred maintenance

Once initial performance indicators have been established, maintenance practices should be periodically evaluated against these figures. Adjustment, modification, and correction of the practices should then follow as part of a continuous improvement process. This process is important in maintenance management to ensure that the maintenance is responding to the specific needs of users and the asset.

Continuous improvement involves the identification of both strengths and weaknesses in the planning and implementation of maintenance strategies. Opportunities for improving the maintenance planning, delivery, and management processes will ensure the delivery of more effective and efficient maintenance.

1.1.7 Maintenance World of Tomorrow

In today's globally competitive marketplace, there is intense pressure for manufacturing industries to continuously reduce and eliminate costly unscheduled downtime and unexpected breakdowns. With the advent of the Internet and tether-free technologies, companies must make dramatic changes by transforming traditional "fail and fix (FAF)" maintenance practices into "predict and prevent (PAP)" e-maintenance methodologies. e-Maintenance addresses the fundamental needs of predictive intelligence tools to monitor degradation, rather than detecting the faults in a networked environment, to optimize asset utilization (Lee et al., 2006).

1.1.8 e-Maintenance

The term e-maintenance has been in use since early 2000 and is now a common term in the maintenance-related literature. However, it is inconsistently defined in maintenance theory and practice. For instance, according to Baldwin (2004b), the "e" in e-maintenance has several possible meanings:

e – Maintenance = Excellent maintenance

 = Efficient maintenance (do more with fewer people and less money)

  + Effective maintenance (improve reliability, availability,

   maintainability, and safety [RAMS] metrics)

  + Enterprise maintenance (contribute directly to enterprise performance)

The concept of e-maintenance generally refers to the integration of information and communication technologies (ICT) within the maintenance strategy and/or the plan to meet the needs emerging from innovative ways to support production (e-manufacturing) or business (e-business) (Li et al., 2005).

Should we consider e-maintenance a maintenance strategy (i.e., a management method), a maintenance plan (i.e., a structured set of tasks), a maintenance type (such as condition-based maintenance (CBM), corrective, etc.), or a maintenance support (i.e., resources, services to carry out maintenance) (Muller et al., 2008)? The following paragraphs provide some clues:

- *e-Maintenance as a maintenance strategy*: e-Maintenance can be defined as a maintenance strategy when tasks are managed electronically by the use of computer using real-time equipment data (i.e., mobile devices, remote sensing, condition monitoring, knowledge engineering, telecommunications, and Internet technologies) (Tsang, 2002). This definition can be refined as follows: "e-Maintenance is an asset information management solution that integrates and synchronizes the various maintenance and reliability applications to gather and deliver asset information where it is needed when it is needed" (Baldwin, 2004a).

- *e-Maintenance as a maintenance plan*: e-Maintenance can be seen as a maintenance plan when it meets the needs of the future e-automated manufacturing world in the exploration of proactive maintenance, collaborative maintenance, remote maintenance and service support, provision of real-time information access, and integration of production with maintenance (Ucar and Qiu, 2005). The implementation of an e-maintenance plan requires a proactive e-maintenance scheme and an interdisciplinary approach that includes monitoring, diagnosis, prognosis, decision, and control processes (Muller et al., 2006).

- *e-Maintenance as a maintenance type*: Generally speaking, e-maintenance represents the gradual replacement of traditional maintenance by more predictive/proactive maintenance.

- *e-Maintenance as maintenance support*: Last but by no means least, e-maintenance can be used as maintenance support. Because e-maintenance is a combination of web service technology and agent technology, some authors say intelligent and cooperative features can be created for the systems in an industrial automation system (Zhang et al., 2003). Other authors define e-maintenance as a distributed artificial intelligence environment, which includes information-processing capability, decision support and communication tools, and collaboration between maintenance processes and expert systems (Crespo and Gupta, 2006).

1.2 Reliability and Maintenance

The reliability of an item/system is the probability that it performs a specified function under specified operational and environmental conditions for a specified period of time. Quantitatively, reliability is the probability of success (non-failure).

Since the Industrial Revolution, maintenance has been a continuous challenge. Although impressive progress has been made in maintaining equipment in the field, it remains challenging because of various factors, including complexity, cost, and competition. Each year billions of dollars are spent on equipment maintenance worldwide. There is a definite need for effective asset management and maintenance practices that can positively influence success factors such as quality, safety, price, speed of innovation, reliable delivery, and profitability (Dhillon, 1992).

The reliability of assets has become an important issue during the design process because of our dependence on the satisfactory functioning of increasingly complex systems. Normally, the required reliability of a system is spelled out in the design specifications, and during the design phase, every effort is made to fulfill this requirement. Some specific factors that play, directly or indirectly, an instrumental role in increasing the importance of considering reliability in the design phase are high acquisition costs, complexity, safety, reliability- and quality-related lawsuits, public pressures, and global competition. Given the diversity of these issues, maintainability, maintenance, and reliability professionals obviously need to collaborate during product design, but to be effective, they must have some understanding of each other's discipline. Once this happens, many work-related difficulties will be reduced to a tolerable level or disappear altogether, resulting in more reliable and maintainable systems (Dhillon, 2006).

1.2.1 Reliability Engineering

The primary role of the reliability engineer is to identify and manage asset reliability risks that could adversely affect plant or business operations. This broad primary role can be divided into three smaller, more manageable roles: loss elimination, risk management, and life cycle asset management (LCAM).

1. *Loss elimination*: One of the fundamental roles of the reliability engineer is to track production losses and any abnormally high maintenance costs and find ways to reduce them. These losses are prioritized to focus efforts on the largest/most critical opportunities. The reliability engineer (in partnership with the operations team) develops a plan to eliminate or reduce the losses through root cause analysis, obtains approval of the plan, and facilitates its implementation.

2. *Risk management*: Another role of the reliability engineer is to manage risks to achieve an organization's strategic objectives in the areas of environmental health and safety, asset capability, quality, and production. Some tools used by a reliability engineer to identify and reduce risk include the following:

- Preliminary hazard analysis (PHA)
- Failure modes and effects analysis (FMEA)
- Criticality analysis (CA)

- Simplified failure modes and effects analysis (SFMEA)
- Maintainability information (MI)
- Fault tree analysis (FTA)
- Event tree analysis (ETA)

3. *Life cycle asset management*: Studies show that as much as 95% of the total cost of ownership (TCO) or life cycle cost (LCC) of an asset is determined before it is put into use. This suggests the need for the reliability engineer to be involved in the design and installation stages of new assets and the modification of existing assets (LCE Life Cycle Engineering Company).

1.2.1.1 Important Capabilities of Reliability Engineers

A qualified reliability engineers should have the capability to

1. Display current knowledge of predictive, analytical, and compliance technologies and apply these techniques to add value to a company.
2. Adapt and apply concepts such as TPM and RCM.
3. Develop and implement a proactive M&R plan(s) to eliminate maintenance requirements, minimize the use and costs of reactive maintenance (RM), maximize the benefits of preventive maintenance (PM) and predictive maintenance (PdM), and achieve increasing levels of integrated asset management.
4. Lead or technically support multidisciplinary teams.
5. During design, advise other engineers on reliability (prediction) on their systems and tactics to improve reliability in areas such as redundancy, parts derating, failure mode and effects analysis, and so on.
6. During design, participate in trade-off studies of performance, cost, and reliability; reliability estimates are a key input to LCC.
7. During development, continue to update reliability predictions and prepare reliability test plans.
8. During preproduction, verify reliability of system and subsystems through various types of testing.

1.2.2 Maintenance Engineering

Although humans have felt the need to maintain their equipment since the Stone Age, the beginning of modern engineering maintenance started with the development of the steam engine by James Watt (1736–1819) in 1769 in Great Britain (Dhillon, 2006).

Maintenance should start with proper planning. A risk assessment should be carried out and workers should be involved in this process. A maintenance plan is a useful tool. To create this tool, a list of premises and equipment to be maintained should be compiled. The plan should include details of the maintenance to be carried out on each item and when it should be done. Record keeping is important, and details of the work carried out should be archived.

The following are some of the tasks that are the responsibility of the maintenance engineer:

1. Ensures equipment is properly designed, selected, and installed based on a life cycle philosophy. Today, many companies purchase equipment based on low bids. Quite simply, if they are not performing the tasks listed for the maintenance foreman and maintenance planner, the company will lack the data it needs to purchase equipment based on a life cycle philosophy. Without the proper data, the purchasing and accounting departments will purchase items that cost the least. This may not be the best long-term decision, so collecting maintenance-cost data is crucial.

2. Ensures equipment is performing effectively and efficiently. This task is different from tracking uptime. It means ensuring that when the equipment is running, it is running at the designed speed and capacity. When focusing only on maintenance, many companies set goals in terms of uptime, without realizing the equipment may be running at only 50% or 60% capacity. Thus, understanding design capacity and speed is much more important than measuring uptime.

3. Establishes and monitors programs for critical equipment analysis and condition-monitoring techniques. The maintenance engineer is responsible for ensuring the appropriate monitoring techniques are used to determine equipment condition. This information is given to the planner so an effective overhaul schedule can be determined. These techniques should also help eliminate unplanned maintenance downtime.

4. Reviews deficiencies noted during corrective maintenance. The maintenance engineer and the planner periodically review equipment maintenance records. If they observe continual problems with equipment, and the problems are not included in the preventive or predictive maintenance program, the maintenance engineer must find solutions.

5. Provides technical guidance for computerized maintenance management systems (CMMs). The maintenance engineer reviews the data in the CMMS (SAP, Maximo, etc.). He or she makes recommendations about the types of data and the amount of data being collected. The maintenance engineer may also note problem causes and suggest action codes to track maintenance activities.

6. Maintains stock items, surplus items, and rental equipment and advises on their use and disposition. The maintenance engineer reviews spare parts policies for plant equipment to ensure the right parts are in stock and in the right quantity.

7. Promotes equipment standardization. The maintenance engineer will help to ensure the company is purchasing standardized equipment. Equipment standardization increases the number of sources for spare parts, reduces the number of spare parts required, reduces the amount of training necessary, and, overall, reduces the maintenance budget. Standardization requires data from the CMMS. If the organization is not collecting data through the maintenance foreman and maintenance planner, the maintenance engineer will not have the data required to implement equipment standardization.

8. Consults with maintenance workers on technical problems. The maintenance engineer consults at a technical level with maintenance workers on equipment- or work-related problems. This consultation may be about advanced troubleshooting or even equipment redesign.

9. Monitors new tools and technology. The maintenance engineer is responsible for staying abreast of all the tools and technology available in the maintenance marketplace. This means he or she is responsible for reading books and magazines, attending conferences, and interfacing with other maintenance engineers to gather these data.

10. Monitors shop qualifications and quality standards for outside contractors. The maintenance engineer is responsible for ensuring all outside contractors are qualified and the work performed is of the proper quality. The maintenance engineer must do a factory inspection if equipment is refurbished at a contractor's site.

11. Develops standards for major maintenance overhauls and outages. The maintenance engineer is responsible for examining outage and overhaul plans for completeness and accuracy. He or she makes appropriate recommendations to the planner for adjustments in the plans and/or schedules.

12. Performs a cost-effective benefit review of the maintenance program. Periodically, the maintenance engineer reviews maintenance programs for areas of responsibility and determines whether the work should be performed by operators, maintenance workers, or outside contractors. In addition, he or she reviews what work needs to be done, what work can be eliminated, and what new work needs to be identified and added to the maintenance plan. This is called job plan optimization.

13. Provides technical guidance for the preventive and predictive maintenance programs. The maintenance engineer periodically reviews

the preventive and predictive maintenance programs to ensure the proper tools and technologies are being applied. This review is typically done in conjunction with the maintenance planner.

14. Performs benchmarking by monitoring competitors' activities in maintenance management. The information about competitors' maintenance programs may come from conferences, magazines articles, or peer-to-peer interfacing and should be reviewed for ideas on potential improvements in the company's maintenance program.

15. Serves as the focal point for monitoring performance indicators for maintenance management. The maintenance engineer is responsible for developing performance indicators for maintenance and reviewing these with the maintenance manager.

16. Optimizes maintenance strategies. The maintenance engineer examines maintenance strategies and ensures they are cost-effective. No maintenance strategy lasts forever in a plant; conditions change, so strategies must be reviewed once a year.

17. Analyzes equipment and operating data. The maintenance engineer ensures equipment is operating as close to design parameters as possible. This means there is no wasted production because of less-than-optimal equipment capacity. If you don't measure, you don't know your plant performance (Ntozelizwe, 2015).

1.3 Changing Role of Maintenance in Asset Management

An asset can be generalized as something owned and/or maintained that is expected to provide benefits to the user for a certain period from its date of purchase, construction, or maintenance date.

"Asset management" is now used to describe the integrated, whole-life, risk-based management of industrial infrastructure that evolved principally in the United Kingdom's North Sea oil and gas industry during the late 1980s and early 1990s. The term has long been used in the financial sector to describe maximizing the value of an investment portfolio. In the latter context, a quantitative treatment of risk and the deliberate management of trade-offs between short-term performance and longer-term security are well established. The financial services sector has some limitations in its applicability elsewhere and shows short-term versus long-term accounting distortions, but the body of knowledge and the baselines of skills, tools, and procedures are fully developed and scientifically applied (Woodhouse).

ISO 55000 defines asset management as the "coordinated activity of an organization to realize value from assets." In turn, assets are defined as follows: "An asset is an item, thing or entity that has potential or actual value to an organization." This is deliberately broader than a definition applying to only physical assets (note: qualifying notes to these definitions appear in ISO 55000).

Asset management involves the balancing of costs, opportunities, and risks against the desired performance of assets, to achieve the organizational objectives. This balancing might need to be considered over different time frames.

Asset management also enables an organization to examine the need for, and performance of, assets and asset systems at different levels. Additionally, it enables the application of analytical approaches to manage an asset over the different stages of its life cycle, starting with the conception of the need for the asset, through to its disposal, and includes the managing of any potential post-disposal liabilities.

Asset management is the art and science of making the right decisions and optimizing the delivery of value. A common objective is to minimize the whole-life cost of assets, but there may be other critical factors such as risk or business continuity to be considered objectively in this decision-making (Woodhouse, 1997).

The development of an asset management strategy is the key to managing assets throughout the life cycle from acquisition to disposal. It is part of an overall framework that includes an asset management policy and strategy, a total asset management plan, and individual asset management plans. This allows a rational approach to capital acquisition and renewal and the adoption of a risk management approach to asset management and business-driven maintenance.

A number of steps must be undertaken if asset management is to be a valuable tool and become the basis for an asset management strategy. The steps occur in logical sequence and failure to undertake any step may result in poor decision making and suboptimal performance of the asset and service delivery or adverse cost impacts.

1.3.1 Asset Life Cycle

The asset life cycle approach is a developmental or holistic approach, enabling the issues that impact an asset to be reviewed at different stages of its life. With the correct use of this approach, available information is synthesized, financial implications are understood and considered, and longer-term or whole-life asset cost and management implications are realized rather than simply the short-term costs or implications of the creation/acquisition of the asset alone (Kriedemann, 2016).

1.3.2 Asset Planning

The goal of asset management is to meet the required level of service in the most optimal way. In this process, asset planning is fundamental to the effective management of assets, as it is the first phase in the asset's life cycle. It is also an area where the greatest impact on the life cycle costs for an asset can be optimized.

Asset planning involves confirming the service level that is required by the customer and ensuring that the proposed asset is the most effective solution to meet the customer's needs.

Assets should be created/acquired /upgraded only where there is sufficient demand to warrant this. The process for identifying this need will vary depending on the asset. It may involve the physical demand placed on the asset (e.g., flow rates that require the upgrade or creation of a new asset to meet the demand and anticipated demand based on projections). Or there may be a need to seek views about requirements for assets such as requested facilities that may be needed. Valid statistical techniques are required to support data gathering to ensure the need identified is not biased by the sampling technique or lobbying by concerned stakeholders.

Whatever process is used for identifying demand, it must be documented and allow rigorous evaluation.

Demand and customer expectations will provide the basis for the level of service required from the assets. Much work has already been done in identifying levels of service. These sources should be used when identifying levels of service.

As part of the planning approach, alternatives to asset creation must be considered. This may be the augmentation of existing assets to meet the new service demand (Kriedemann, 2016).

1.3.3 Asset Creation/Acquisition

Asset creation/acquisition is justified when the outlay for the provision or improvement to an asset can reasonably be expected to provide benefits beyond the year of outlay. The main reasons for creating an asset are to satisfy or, where necessary, improve the level of service, meet new demands from users, or to provide a commercial return.

Traditionally, organizations have focused on preliminary costs of asset creation or acquisition only. However, the inclusion of new assets into the asset portfolio increases the demand on the maintenance budget, if not immediately then at some future time. It is therefore necessary to evaluate those full life cycle costs as well (see Chapter on LCC).

All assets should be managed from a life cycle perspective. Costs should include the environmental and social costs associated with the asset, not just the financial costs. These "costs" should be documented as it may not be possible to financially quantify them.

Project management is required to manage asset creation, acquisition, and upgrading. Good management results in clear outcomes, time frames, and costs, forming the basis for the budget decision-making required to fund a project. The degree of detail required will vary depending upon the complexity. More complex projects will require a more detailed analysis to allow asset-related decisions to be made (Kriedemann, 2016).

1.3.4 Asset Operations

Asset operations refer to the day-to-day running and availability of the asset, considering working hours, energy management, programming of down-time, and so on. It refers primarily to dynamic assets such as motor vehicles, plant, machinery, and equipment but can provide information on strategies for all assets. For dynamic assets, the operational and upkeep costs (including recurrent maintenance expenditure) represent a significant proportion of the total life cycle costs. Therefore, the day-to-day efficiency with which operations are carried out is important in optimizing the overall life cycle cost of the asset.

The operational costs must be factored into the total asset costs and included in the annual budget. Implementing efficiency in asset operations is the responsibility of those in charge of assets. Benchmarking against similar assets internally and with other service providers will assist in this process (Kriedemann, 2016).

1.3.5 Asset Maintenance

Asset maintenance should be based on a cost–benefit approach to ensure that the asset is meeting service requirements without excessive maintenance expenses. The level of maintenance priority should be reflected by the asset's ability to deliver on the organizations strategic priorities and customer service standards. The effects of failure due to inadequate maintenance must be considered when prioritizing maintenance budgets and programs among all asset classes.

1.3.6 Asset Condition/Performance

All management decisions on maintenance, rehabilitation, renewal, and service delivery revolve around the condition and performance of assets. Having a clear knowledge of these aspects will limit the company's exposure to business risks and possible loss of service potential caused by premature failure of assets. It will also allow the timely application of maintenance actions.

By undertaking regular condition- and performance-monitoring exercises, life cycle strategies including maintenance and/or rehabilitation options can be determined, updated, and refined. Ultimately, asset replacement programs can be more accurately predicted and planned.

If failure is imminent, asset managers will likely have had sufficient time to look at options other than replacement. Additionally, they will be able to manage the failure and minimize some of the consequences. The performance of assets is closely aligned to the level of service provided to customers and this can generally be measured in terms of reliability, availability, capacity, and meeting customer demands and needs.

Key performance indicators are required for each asset class. Reporting indicator values on a regular basis will improve the decision-making for the asset; it will be more transparent and visible to the com- munity of interested stakeholders. Indicators should cover the following aspects of assets:

- Capacity or utilization
- Levels of service
- Condition or mortality
- Costs

To assist in implementing condition-based assessment or performance, the company must establish systems and techniques to enable reliable comparisons to be made among similar asset classes. This may involve detailed testing methods or visual assessment by trained and competent staff. The assessment processes and standards will be documented as part of the management plans for individual assets.

Benchmarking using the assets of similar companies is a potential way to move toward better practices in asset management (Kriedemann, 2016).

1.3.7 Asset Replacement

Decisions taken to rehabilitate/replace assets are often associated with potentially large investments by the company. This is generally because these assets have become uneconomical to own and operate in their existing state. The measure that indicates rehabilitation or replacement is their degree of failure to deliver the services as required.

Asset renewal and rehabilitation refer to capital activities associated with assets that have been identified through a strategic planning process as requiring a change to ensure effective service delivery.

Arguably, the most important aspect here is the ability to understand the different failure modes that each asset may suffer, and then predict when each of these is likely to take place. Therefore, risk management processes that also target criticality should form part of the preliminary study of all assets.

The business risk assessment will allow companies to assess the relative merits and therefore optimize the decision-making process of the various strategies available (Kriedemann, 2016).

1.3.8 Asset Disposal/Rationalization

Asset disposal terminates the use and control of an asset, but may generate the need for a replacement to support the continuing delivery of a service. Decisions to dispose of an asset require thorough examination and economic appraisal.

Like acquisition decisions, disposals should be undertaken within an integrated planning framework that takes account of service delivery needs, corporate objectives, financial and budgetary constraints, and the overall resource allocation objectives.

Performance monitoring forms an important part of the decision-making process. Where assets are underutilized or underperforming, a decision must be made about continuing to fund the asset.

When an asset is underutilized, then, prior to any decision to acquire another similar asset, the potential for using the existing underutilized asset should be considered (Kriedemann, 2016).

1.3.9 Financial Management

Accurate recording, valuing, and reporting procedures are needed so that decisions to modify, rehabilitate, find alternative use for, or dispose of an asset can be soundly based. To be accountable for the operation, maintenance, and financial performance of the assets and services that they are delivering, companies must be able to make informed decisions about those assets. (Kriedemann, 2016).

All assets should be valued at their "fair cost." This means all assets will require revaluations on a regular basis.

Valuations are important for the following purposes:

- Facilitating appropriate pricing and funding levels
- Insurance
- Benchmarking
- Asset management decision making

Determining depreciation (used in conjunction with estimated remaining life).

1.3.10 Asset Management Definition and Function

Initially, asset management was seen by many as another name for maintenance management. This short-sighted view cost many organizations their competitive edge. Asset management was also confused with pension funding and equity investment, which was even more misleading. However, the term here refers to "physical" or "engineering" asset management. The scope of the subject is obviously very broad, but some of the more important areas are covered in this book.

As international industrial competition is on the rise because of globalization, business survival is increasingly threatened. To ensure survival in the short, medium, and long terms, profitability from assets needs to be maximized. This implies ensuring productivity and profitability improvement, while including the improvement of safety, product quality, and production efficiency, not to mention the reduction of maintenance costs.

If organizations are structured into functions, for example, maintenance, operations, and engineering, with functional heads of departments, "silo thinking" can result. Budgets will be planned and controlled with little consideration of the performance of the whole organization. Politics can also lead to "silo thinking." The maintenance manager needs to work in cross-functional teams to make decisions that are in the best interest of the organization, while maintaining the maintenance work control program to meet budgeted performance. It is the former activity that normally needs development as it requires a paradigm shift from the past (Huggett, 2005).

Asset management activities permeate many levels of an organization and are not confined to a central group. For this reason, we use the term asset management function as a flexible descriptor for the activities involved and employ the term asset managers for those involved significantly, but not necessarily exclusively, in asset management activities.

The purpose of the asset management function is to provide resources and expertise to support the acquisition, in-service support, and disposal of the physical assets required by an organization. A central asset management function will be needed at the company level, providing inputs to asset planning, playing a role in major acquisitions and developments, and providing the systems needed to support assets throughout their life. Asset management is distinct from operations and does not usually involve the direct design or building of the assets themselves.

Asset management is also normally distinct from maintenance, but the technical service functions that support maintenance are part of asset management. The exact terminology and reporting structures may vary from organization to organization (Hastings, 2010).

1.3.11 Structure of Asset Management

Asset management activities and responsibilities impact a wide range of roles within an organization and are not confined to a specific department. However, in a large organization, effective asset management will benefit from the existence of recognized asset management personnel with expertise in specific areas. These may be formed into distinct groups, the title of which will depend on company history and structure.

An example is the Defence Materiel Organization (DMO) in the Netherlands; the DMO has divisions managing land, sea, and air assets. Within each division are asset management groups called system program offices (SPOs). Each SPO manages a particular group of prime assets, such as a type of ship or aircraft, and all the associated subsidiary assets. In air assets more generally, groups will correspond to the main aircraft types, with additional groups for ground-handling equipment, sales systems, and customer support systems.

In general, an asset management group consists of asset managers with suitable technical backgrounds, and personnel in accounting and finance, legal, contracting, procurement, and engineering roles. The financial, legal, and engineering staff will be collocated with asset management groups from their professional area. For particular projects, teams will be formed with personnel and skills consistent with the content and size of the project. The asset management groups constitute the basis upon which teams can be formed (see Figure 1.2). Asset management groups play key roles in acquisition and development decisions, acquisition and development projects, and the creation and management of organization-wide systems for equipment support for new and existing assets (Hastings, 2010).

1.3.12 Asset Management Strategy

An asset management strategy is a broad-level frame work incorporating actions plans set by senior management as a guide to how an organization intends to achieve its aims. Figure 1.3 illustrates an overall life cycle model and related asset strategy factors.

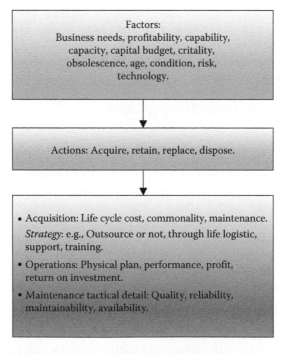

FIGURE 1.3
An asset strategy model. (Adapted from Hastings, N.A.J., *Physical Asset Management*, XXIX, 2010, 370 p., Hardcover. http://www.springer.com/978-1-84882-750-9.)

1.3.13 Changing Role of Maintenance Management in Asset Management

Table 1.1 lists some common issues facing management in the first column, and provides thoughts on how these issues might be addressed in the second column. All are discussed in more detail in the sections that follow.

1.3.14 Improve Asset Maintenance Strategy or Renew the Asset?

Should we continue improving our maintenance strategy or simply renew the asset? Which of these two choices is the most cost-effective? We don't want to spend money on improving the maintenance strategy when efficiency and safety could be improved by renewing the asset—but is it not more cost-effective to improve the maintenance strategy to increase the reliability and availability of the present asset?

The choice between these two alternatives needs to be addressed jointly by maintenance, operations, safety, quality, and finance, to ensure a common vision, strategy, and cost-effective investment.

Examples of terminologies to address these types of issues are "whole-life costing" and "life cycle costing." They both necessitate consideration of the whole life cycle of an asset and the changes over the life of the asset. Both assess the following:

- Maintenance costs—both planned and unplanned
- Downtime for maintenance—both planned and unplanned
- Efficiency of the asset—productivity and output

TABLE 1.1

Common Issues Facing Management When Starting an Asset Management Continuous Improvement Program

Our competitor's assets are more efficient; we need to do more PM.	But is it not time to renew the asset with new, more efficient assets?
Our engineering projects and modifications are often given lower priority than those of other departments.	But in the economic interests of the company, priority needs to be given to investments with the best payback.
This motor has been in stock for 3 years—get rid of it.	But it often pays to keep an insurance spare even if it never moves.
We can't make decisions because we don't have enough data—we need a bigger computer to store more data.	But many good decisions can be made with weak data—we need to develop skills to work with the best data available.
Asset management is another name for maintenance management.	But optimal asset efficiency requires other decision inputs: finance, process, safety, quality and production engineering, as well as maintenance.

Source: Huggett, J., Asset Management—The changing role of maintenance management, Principal consultant, The Woodhouse Partnership Ltd. IMC Conference, 2005, www.twpl.com.

- Operating costs—energy and so on
- Quality of operation or service—product and service rejects, among others
- Safety to staff and the public

The only sensible way to make a decision about continuing with the existing asset or changing it for a new one, and then choosing which new one, is to turn each of these considerations into monetary values (Huggett, 2005).

"Whole-life costing" and "life cycle costing" enable assets to be compared to select the most cost-effective long-term investment. However, they do not address "optimal total life cycle costing," a concept illustrated in Figure 1.4. The figure compares two assets (A and B) from different manufacturers to see which is more cost-effective in the long term.

As the figure shows, renewing asset A (green line) at 15 years is the most cost-effective option, and the best option for asset B (blue line) is to renew it at 25 years. If we decided the asset is to function for 40–50 years before renewing it, there will be no real difference between asset A and B. Overall, asset B is more durable, costs more, and has better availability and reliability over its life cycle. Its optimal life cycle is 25 years. However, asset A is the more cost-effective choice if it is renewed on a 15-year cycle.

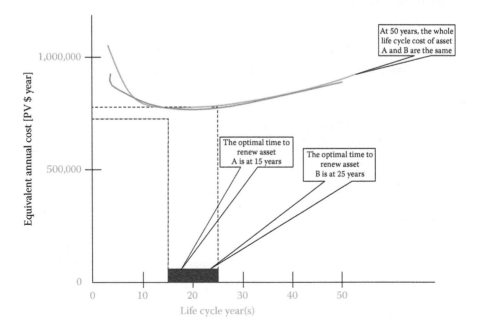

FIGURE 1.4
Comparison of cost-effectiveness of two assets. (Adapted from Huggett, J., Asset Management—The changing role of maintenance management, Principal consultant, The Woodhouse Partnership Ltd. IMC Conference, 2005, www.twpl.com.)

If management decides on a new asset, as illustrated in Figure 1.5, the optimal time to replace it in the future according to the estimated projections of the asset management team of the organization is around year 15. However, without this type of modeling, they could decide to keep the asset past its optimal replacement date, and perhaps replace it at 30–50 years. The asset would outwardly appear to be functioning perfectly well, but if their competitors had chosen asset A and decided to replace their assets on a 15-year cycle, they could well be 5%–10% more cost-effective. This is often enough to make the difference between being a market leader and going out of business (Huggett, 2005).

This is not the end of the story, however. The scenario above needs to consider the replacement of an "old" asset with a new one, as illustrated in Figure 1.6.

In the figure, the new asset is the best long-term choice. However, it is still cost-effective to continue with the present asset for 4–5 years before replacing it with the new one. That could be enough time to improve the maintenance and operating strategy of the present asset to gain short-term benefits. The decision to purchase and install the new asset could be reviewed in 2–3 years' time, as costs might have changed, and this could alter the course of the strategy (Huggett, 2005).

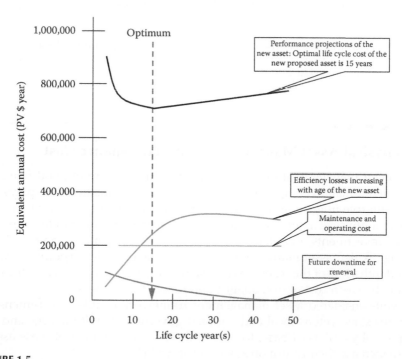

FIGURE 1.5
Asset A: Optional replacement time. (Adapted from Huggett, J., Asset Management—The changing role of maintenance management, Principal consultant, The Woodhouse Partnership Ltd. IMC Conference, 2005, www.twpl.com.)

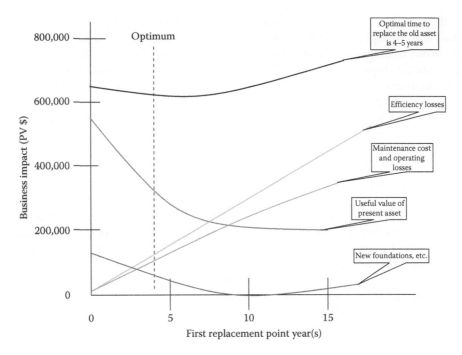

FIGURE 1.6
Optional replacement time of existing asset with new asset. (Adapted from Huggett, J., Asset Management—The changing role of maintenance management, Principal consultant, The Woodhouse Partnership Ltd. IMC Conference, 2005, www.twpl.com.)

1.4 Physical Asset Management and Maintenance Cost

Competent management of physical assets is at the core of good financial and operational oversight for asset-intensive organizations. It plays an important role in generating more accurate financial and operating reports. Proficient physical asset management is an essential element in the supervision of investments, maintenance, replacements, retirements, and decision making. Skilled asset management professionals provide corporations with an in-depth view of the resources invested in the physical assets and the correlation between cost of use, quality, and risk of these assets.

A well-considered asset management framework enables asset-intensive companies, as well as local and regional governments with a large and/or complex physical asset base, to strengthen the management of these assets and to increase their return on assets employed (ROAe).

Good asset management provides the information needed to improve decisions on physical assets, information technology (IT) infrastructure, internal controls, and financing needs. The resulting insights may allow asset owners to better quantify and anticipate the impact of future decisions and/or regulations on ROAe. Good asset management contributes to the identification of cost and investment opportunities, such as fine-tuning the maintenance budget. In addition, it underpins the calculation of different kinds of valuation exercises for the entire asset base.

In short, asset management involves processes of planning and monitoring physical assets during their useful lives. This means, managing assets effectively requires a high level of management interest. The main objective of asset management is to achieve the least cost for the acquisition, use, maintenance, and disposal of assets while accommodating the organization's overall objectives. Put otherwise, physical asset management (PAM) is the planning, acquisition, maintenance, and disposal of physical assets with due regard for economy, effectiveness, and efficiency, as well as full compliance with all applicable regulations and policy directives (Guidelines for Physical Asset Management).

1.4.1 What Are Assets?

In the context of PAM, an asset is a noncurrent asset, that is, a long-term asset, normally financed by the capital budget, with future economic benefits. The benefits arising from a noncurrent asset last longer than one financial year (Guidelines for Physical Asset Management).

1.4.1.1 Asset Life Cycle and Strategy

1.4.1.1.1 Asset Life Cycle

With limited resources available, it is important for asset managers to understand that asset consumption is a real and significant cost of program delivery. The application of life cycle costing techniques and the establishment of appropriate accountability frameworks are crucial.

The extended life of an asset has important implications for program managers. An acquisition decision based on the lowest purchase price, but which ignores potential operating costs, may result in a higher overall cost over the asset's life (Guidelines for Physical Asset Management).

1.4.1.1.2 Phases of Asset Life Cycle

It is important to understand the phases of an asset's life cycle and the impact of each phase on total program costs and outputs. The physical life of an asset needs to be distinguished from its useful life to an organization. The useful life is the period over which the benefits from its use are expected to be derived. The fact that assets have a life cycle distinguishes them from other program resource inputs. Typically, those responsible for acquisition decisions (and costs) in an organization differ from those responsible for operating and maintaining assets, and both groups often differ from those responsible for their disposal. Problems may arise as a consequence of this fragmentation of management over the asset's life cycle.

The physical life cycle of an asset has four distinct phases:

- Planning for acquisition of assets
- Acquisition of assets
- Maintenance of assets
- Disposal of assets

Understanding the phases of an asset's life cycle and the attendant costs is an important first step in managing assets on a whole-life basis. The use of life cycle costing techniques allows a full evaluation of the total cost of owning and maintaining an asset prior to its acquisition. This creates the opportunity to determine the most cost-effective program delivery solution. Estimating life cycle costs before acquisition also establishes a standard which becomes the basis for monitoring and controlling costs after acquisition.

Life cycle costs consist of capital and recurrent costs (Guidelines for Physical Asset Management).

1.4.1.1.3 Capital Costs

Capital costs are the costs of acquiring an asset. These include not only the purchase price but all associated fees and charges, and the delivery and installation costs incurred in putting the asset into operational use. They also include planning costs such as those for feasibility studies and tendering. One significant capital cost not routinely recognized by budget-dependent companies is the "finance" cost of the funds "locked up" in the value of the asset. An exception to this is the situation where companies borrow against future appropriations as part of the running cost arrangements and are charged interest. However, this reflects finance costs at the margins and generally does not extend over the life of the asset. The purchasing organization incurs a "finance" cost for capital funds either directly, as the interest expense on borrowing and the expected return on equity (stock), or indirectly, as interest foregone on funds that would otherwise have been available to the organization. When evaluating non-asset solutions and alternative acquisition strategies, it is important for the company to recognize this "cost."

1.4.1.1.4 Recurring Costs

Recurring costs include energy, maintenance, and cleaning costs. They may include employee costs where specialist staff is dedicated to the operation of the asset. Planned refurbishment and enhancements over the asset's life, while "capital" in nature, may be considered recurrent costs for planning purposes. Disposal costs should be included as well, particularly if they are expected to be significant. This may be the case when the asset, processes associated with it, or its outputs produce undesirable effects requiring rectification or remedial work. Environmental concerns are a good example of this.

The relative significance of capital and recurring costs as a proportion of total life cycle costs will depend on the nature of the asset. The cost of operating and maintaining an asset over its useful life can often be greater than its acquisition cost. In such cases, the use of full life cycle costing in evaluating alternatives is imperative to ensure overall program costs are recognized and minimized (Guidelines for Physical Asset Management).

1.4.1.1.5 Basic Principles of Effective Physical Asset Management

The following are the basic principles of effective physical asset management:

- Asset management decisions are integrated with strategic planning.
- Asset planning decisions are based on an evaluation of alternatives which consider life cycle costs, benefits, and risks of ownership.
- Accountability is established for asset condition, use, and performance.
- Disposal decisions are based on analysis of the methods which achieve the best available net return.
- An effective control structure is established for asset operation and maintenance.

1.4.1.1.6 Relative Importance of Assets

Principles of asset management apply to all assets—they do not, however, apply equally. The characteristics of the assets will dictate the extent and degree to which a particular principle is applied. One gauge of the relative importance of each management principle to particular groups of assets is the cost at each stage of their lives. For example, furniture and fixtures (typically high-volume, low-value items) provide an essential service and their contribution to an organization needs to be recognized. By their nature, they are typically low-maintenance items; it may suffice simply to monitor their condition in lieu of a costly preventive maintenance plan. However, if they constitute a relatively large percentage of the total value of total assets held, their acquisition and replacement planning assume greater importance (Guidelines for Physical Asset Management).

1.4.1.1.7 Strategy

A three-pronged strategy is proposed to achieve the objectives of physical asset management outlined in the preceding paragraphs:

1. Formulating guidelines for all stages of physical asset management, namely, planning for acquisition of assets, acquisition of assets, maintenance of assets, and disposal of assets.
2. Setting up action committees for implementation. Ideas can be generated through brainstorming, and these can be incorporated in the guidelines.

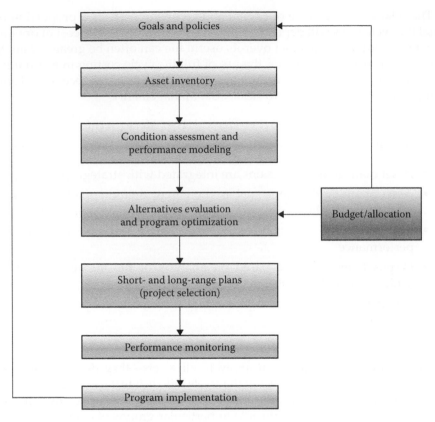

FIGURE 1.7
Steps of physical asset management. (Adapted from Guidelines for Physical Asset Management, Management Audit Bureau, Ministry of Finance and Economic Development.)

3. Development of software for an online asset register, with modules for preventive maintenance, actual maintenance, and disposal of assets.

Figure 1.7 shows a flow diagram with the various steps of physical asset management (Guidelines for Physical Asset Management).

1.4.2 Maintenance and Physical Asset Management

As illustrated in Figure 1.8, the maintenance function has a strong impact on the short- and long-term profitability of an organization. Its interaction with the other functions should obviously be organized in a systematic way. This task is an important part of physical asset management (EN 16646 Maintenance, 2016).

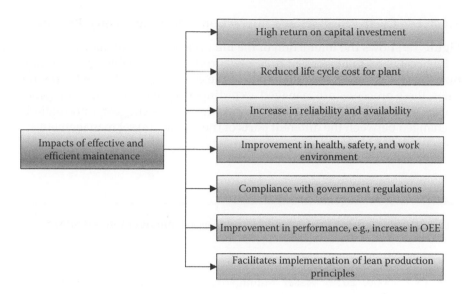

FIGURE 1.8
Impact of maintenance activities. *Note:* OEE, overall equipment efficiency. (Adapted from EN 16646 Maintenance, Maintenance within physical asset management, 2016.)

1.4.2.1 Getting Help in the Development of Asset Management

Supporting tools are available to help develop physical asset management activities and improve operations. Several new standards suggest ways to develop a management system for physical assets, while paying attention to organization-specific requirements.

Standards also help to develop cooperation between different organizational functions and life cycle stages and define clear roles for the maintenance function within the organization's asset management processes (EN 16646 Maintenance, 2016).

1.4.2.2 Life Cycle Stages

The following are the core life cycle processes:

- Acquire appropriate physical assets, if they exist in the market, or create them if they do not exist in the market at acceptable economic rates.
- Operate the assets to optimize their value to the organization.
- Maintain the assets to optimize their value to the organization.
- Modernize (upgrade) the assets to obtain greater value over the life cycle of the asset. *Note*: modernization contains the same life cycle phases as the whole asset system.
- Decommission and/or dispose of the assets when the end of useful life is reached.

1.4.3 Life Cycle Processes and Interaction with Maintenance Process

Figure 1.9 shows the interaction of the life cycle with maintenance.

Physical asset management processes produce sustainable value for organizations based on the requirements of the organization. As the figure indicates, these processes are managed according to a physical asset management policy, strategy, and plans. As the connecting arrows suggest, information is exchanged between the different parts of the physical asset management system (EN 16646 Maintenance, 2016).

1.4.3.1 Asset Management: Standards

At the moment, there are five well-known asset management standards:

- BSI PAS 55: Technical specification
- ISO 55000: 2014 "Asset Management—Overview, Principles and Terminology"
- ISO 55001: 2014 "Asset Management. Management Systems. Requirements"

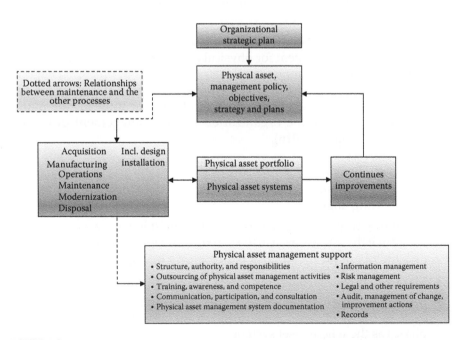

FIGURE 1.9
Interaction of life cycle with maintenance. (Adapted from EN 16646 Maintenance, Maintenance within physical asset management, 2016.)

- ISO 55002: 2014 "Asset Management. Management Systems. Guidelines for the Application of ISO 55001"
- EN 16646: 2014 "Maintenance within Physical Asset Management" (EN 16646 Maintenance, 2016)

1.4.4 Develop a Maintenance Information System

A comprehensive and appropriately structured asset register should form the hub of any asset management system. Information stored in asset registers varies for different asset types and is determined by user requirements. However, the asset register should be adequate to support, among others, systems for maintenance cost recording, work scheduling, and financial planning. It should include the following information:

- Asset description, value, and location
- Asset attributes
- Asset age, service life, and replacement requirements
- Statutory requirements (for registration, testing, etc.)
- Community-sensitive issues (e.g., heritage factors, contamination risks, etc.)
- Ownership and control details

Complex physical assets require significant technical information (in the form of manuals, drawings, specifications, and other records) to support maintenance. Disciplined management systems should be established both centrally and at the site level to ensure that maintenance personnel have timely access to information. Maintenance tasks also generate documents and records, some of which may be required by legislation to be kept or submitted to the appropriate authority. Maintenance procedures should address the issue of records management and ownership.

Computerized maintenance management systems (CMMS) facilitate the management of maintenance activities, including work programming, cost control, reporting, logistics, and operational planning. Consideration should be given to integrated systems which provide linkages to other asset management and financial management modules.

Information to support analysis and planning is essential to maintenance since it must anticipate planning needs as well as operational requirements.

1.4.5 Cost Avoidance for Physical Assets

Physical asset life cycle costs can be considerable and are often not transparent to budget managers. Hidden operational costs relating to physical assets cause significant problems for budget forecasting and managing cash flow.

When new capital assets are purchased, it is critical that asset life cycle costs are identified, so they can be challenged, compared to alternative purchases and included in future budget forecasts.

To meet the twenty-first-century challenges of reducing and controlling operating costs, it is essential that all businesses have embedded processes to control and monitor their physical asset life cycle costs.

Once transparent life cycle costs are in place, savings can be realized for many years to come. Future changes can be business-resilient, save money, and be aligned to long-term business aims (Asset Systems Support Management [ASSM]).

1.4.5.1 Business Benefits of Asset Management

Detailed physical asset life cycle cost planning will deliver significant savings to asset-related operating budgets. Unfortunately, different budget centers often operate independently. To make the best business decisions, there should be transparency across centers to enable cross-referencing of information (ASSM).

1.4.5.2 Proactive Use of Maintenance Budgets

Maintenance budgets are often fixed and stagnant. Without regular reviews based on accurate information, costs can begin to rise. By implementing an effective asset management system containing accurate, good-quality data, budgets can be maintained and reviewed regularly to make better business-focused decisions, adding value to the bottom line.

Keeping information on a single asset transparent allows reports to be quickly generated to inform business decisions. Organizing information on maintenance, calibration, validation, statutes, regulations, and reliability, along with capital, utilization, depreciation, and revenue costs, in one system, paints the whole picture for a single asset.

By having all asset-related information in one system, a company can make good business decisions that will lead to controlled and reduced costs, indicating improved business performance (ASSM).

1.4.5.3 Improved Ability to Manage Current Resources and New Capital Assets

The key challenge when managing assets and resources is a lack of clear, transparent information. To make the best business decisions, it is important to have access to all asset-related information, including life cycle costs when setting annual budgets for new purchases (ASSM).

1.4.5.3.1 *Managing Life Cycle Costs*

The following are some examples of asset life cycle cost considerations:

- Energy consumption
- Cleaning
- Fixed asset maintenance
- Equipment maintenance
- Statutory inspections and regulatory compliance

Developing embedded processes and control measures is critical for managing asset life cycle costs. These measures include the following:

- Early-stage planning of all new facilities and capital purchases, including a consideration of life cycle costs
- Identifying key performance requirements to ensure commercial effectiveness and minimized asset life cycle costs
- Creating asset registers at the earliest opportunity for new facilities and companies
- Conducting asset audits every 2–3 years
- Recording annual costs for each asset in the asset register, for example, operating costs and forecasts of maintenance and repair costs
- Identifying support costs for each asset, including maintenance, routine repairs, calibrations, and regulatory requirements
- Implementing support and maintenance regimes that meet user requirements, that is, keeping equipment fit for its intended purpose, including reliability, utilization, and regulatory requirements
- Identifying the costs of physical depreciation and developing a schedule for its management. This may include the following:
  - Building maintenance
  - Refurbishment
  - Damage correction at the end of a lease, or the cost of returning to original condition
- Considering how assets will be disposed of at the end of their life cycle and any costs incurred. This may include the following:
  - Legislation to be complied with for disposal
  - Recycling capability
  - Specialist material disposal costs
  - Hazardous material disposal costs (ASSM)

1.4.5.4 *Informed and Accurate Financial Planning and Reporting*

Financial planning related to assets is often disjointed, as the information required by accountants for finance reporting is very different from the asset life cycle costs that operations groups require. Asset reports are required by a large number of business groups, including accounting, procurement, quality, health and safety, maintenance, business operations, budget managers, and equipment users. In fact, almost every group needs some information on assets to carry out its business functions. For this to work well, both the asset and its associated information must be transparent (ASSM).

1.5 Focusing on the Bottom Line

The maintenance manager normally has a number of projects that he or she proposes to the organization on an annual basis. They include modifications to the plant and plant processes, design changes, abnormal maintenance, major maintenance work, and so on. Some projects may fall into the maintenance budget, but others may represent additional expense. The latter need to be considered for funding along with projects proposed by other departments, such as operations, safety, and quality.

Figure 1.10 illustrates how projects can be compared and prioritized to give the best payback or profitability to the organization. Top management has the authority to invest in certain projects for reasons that transcend profitability, but clear identification of important and profitable projects makes a good case for investment, particularly when allocation of funds is restricted (Huggett, 2005). In the case study illustrated by the figure, 5 of the top 15 proposed projects come from maintenance. Those not accepted this year can be proposed again next year (Huggett, 2005).

In the figure's case study, each project has three estimates: worst case, base case, and best case. Some projects have a wide variation in inputs but when ranked against some of the less cost-effective projects, they are more profitable even in the worst cases. One of the reasons often given for not quantifying inputs to decisions is the lack of time to do so. Before getting bogged down with "analysis paralysis," it is wise to get the best information currently available and, using engineering judgment, to estimate the best and worst cases. This will quickly enable a manager to evaluate whether a case is worthwhile.

Lack of data and decisions based on incomplete information make a bad recipe for bottom-line approaches. The term "bottom line" is generally used in business finance to refer to profitability. More specifically, bottom-line profit is the net income of a company after all costs of goods

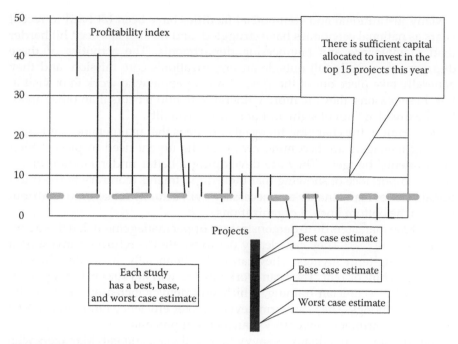

FIGURE 1.10
Projects ranked by profitability (similar to outputs from internal rate of return [IRR] and discounted payback in years). *Note:* The case study in this figure illustrates one of the most important skills in asset management: how to work with poor or nonexistent data. (Adapted from Huggett, J., Asset Management—The changing role of maintenance management, Principal consultant, The Woodhouse Partnership Ltd. IMC Conference, 2005, www.twpl.com.)

sold, fixed costs, and irregular items are accounted for. A bottom-line focus means the company puts earning profit for owners or shareholders as its top priority.

Virtually all for-profit businesses have some level of motivation to ensure bottom-line profit. They weigh tangible business goals heavily when making decisions.

A company not taking a bottom-line approach might place significant emphasis on social or environmental responsibilities, potentially at the expense of optimized profit (Huggett, 2005).

1.5.1 Maintenance and the Bottom Line

How financially savvy are you? A decade ago, that question might have been a mildly interesting topic of conversation among maintenance and engineering managers. Today, it is a hot-button issue that managers ignore at their peril.

Many institutional and commercial facilities have been hit hard in recent years as national economies have struggled, and no area has been hit harder than maintenance and engineering departments. The activities of these departments usually fall outside an organization's core mission, and they generally take place out of the view of most people who work in or visit it. For these reasons, they are more vulnerable to budget cuts than other areas. It's the dreaded "out of sight, out of mind" mentality.

One result of the changing financial climate is that maintenance and engineering managers are becoming aware of the urgent need to protect their departments' budgets. They are developing a better understanding of the critical importance of showing that their departments (contrary to the traditional view of maintenance as a necessary evil) actually make solid contributions to the bottom lines of organizations.

What can managers do to demonstrate to upper management that these contributions are real? They can clearly demonstrate the return on investment and the long-term savings. They can make financially smart decisions on whether to buy or rent. They can work hard to win the support of top management for technician training, which will deliver increased productivity and labor savings. They can wring even greater efficiency (and lower costs) from the department's inventory management process.

The common denominator is savvy financial management. Managers who lack the skills and experience to make smart financial decisions and who cannot speak the language of the organization's chief financial officer (CFO) in explaining their decisions are likely to find themselves and their departments on the short end of key decisions (Hounsell, 2012).

Today, with industry so focused on the bottom line, the cost of downtime has a huge impact on profitability. If equipment starts to wear, it is possible to turn out products with unacceptable quality and not know it for a long time. Eventually, machine wear will seriously affect not only productivity but also product quality.

World-class companies have already taken a game-changing approach, implementing a new model changing maintenance systems into smart service and asset management solutions. Such models reduce downtime and provide the ability to look ahead at the quality of products by closely watching equipment performance and machine wear. Rather than resorting to traditional reactive maintenance, that is, "fail and fix," companies can opt for "predict and prevent" maintenance (Krar, 2013).

A company must take a step back and review how it manages equipment performance. If equipment continues to fail after performing preventive maintenance or overhauls, a change is clearly needed. The focus must be on ensuring the reliability of assets. As a starting point, everyone should understand the definition of reliability and what it means to the success of the company. "Reliability" should be the collective buzzword (Smith and Mobley, 2007).

1.5.2 Maintenance beyond the Bottom Line

Is maintenance being taken seriously? Some industries continue to look only at the bottom line. They are not embracing new asset management solutions that offer to equip maintenance professionals with predictive maintenance technologies to help increase efficiency, reduce downtime, and prepare for economic uncertainty.

In challenging times, when budgets for new capital expenditure are tight, effectively maintaining existing assets is paramount, but large chunks of industry fail to understand that maintenance should be a given. Maintenance is as important as, say, quality control. Too many businesses are simply looking at the bottom line, failing to see the bigger picture. One thing is certain, however: equipment failure means downtime, and downtime costs money, and lots of it. No one would voluntarily throw money into a black hole but that is precisely what companies are doing if they define maintenance as a cost and not as an investment.

Reducing costs, improving efficiency, and eliminating costly equipment failures are vital. Many innovative technologies, products, systems, and services are now available to help companies achieve these goals, but some cannot accept the reality of what a downturn means to their business. They refuse to look horizontally to reduce costs through better maintenance.

In short, the industry as a whole is not taking maintenance as seriously as it should. Companies should be taking any opportunity available to them to look at new ways to develop their maintenance strategy. If they fail to act on issues as basic as maintenance, it will inevitably mean the demise of some, simply because they didn't have the courage or foresight to look beyond the bottom line (Blutstein, 2012).

1.5.3 Managing Availability for Improved Bottom-Line Results

Over the last several years, managers up to the CEO have come to recognize equipment uptime as a key part of any successful operating strategy. Recognition of this need has generated the use of equipment availability as one of the key performance indicators of a maintenance organization. Unfortunately, performance goals are often set based on "gut feel," or by benchmarking with similar branch plants or facilities within the organization or with similar organizations within the same industry.

Both methods involve high levels of uncertainty and can lead to overspending on maintenance and overtaxing maintenance resources. The uncertainty of "gut feel" speaks for itself, but benchmarking can also involve high levels of uncertainty, for example, not knowing the exact guidelines other plants, facilities, or organizations use to record unavailability. Four plants within the same company may have four different sets of guidelines for recording downtime. Correctly, managing availability is the solution to the problem (Keeter, 2003).

1.5.3.1 Availability Types

For engineering asset context, availability is defined in three forms (see Figure 1.11):

- Inherent availability is also called designed availability. It assumes that spare parts and manpower are 100% available with no delays and is expressed as ratio of mean operating time divided by sum of mean operating time and mean repair time.
- Achievable availability (Aa)—the expected level of availability for the performance of corrective and preventive maintenance. Aa is determined by the design of the equipment and the facility within which that asset is expected to work. It assumes that spare parts and manpower are 100% available with no delays.
- Operational availability (Ao)—the bottom line of availability. Ao is the actual level of availability of an asset realized in the day-to-day operation of the facility. It reflects maintenance resource levels and organizational effectiveness.

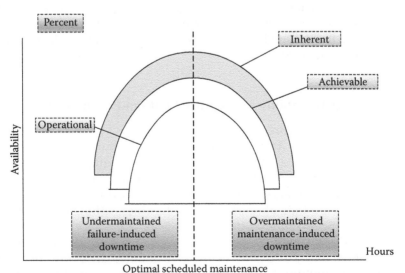

FIGURE 1.11

Three types of availability. (Adapted from ARMS Reliability Engineers, Managing availability for improved bottom-line results, USA, LLC—© 2002 by Bill Keeter, http://reliabilityweb.com.)

It is important to understand the distinctions among the three types:

- Aa considers availability when shutdowns are planned.
- Ai refers to expected performance between planned shutdowns.
- Ao isolates the effectiveness and efficiency of maintenance operations.

These distinctions lead to the following definitions:

- The shape and location of the Aa curve is determined by the design of the company or plant.
- An operation has a given Aa, based on whether scheduled or unscheduled maintenance strategies are selected for a failure. A goal of availability-based maintenance operations is to find the peak of the curve and operate at that level.
- Ao is the bottom line of performance, that is, the performance experienced as the company or plant operates at a given production level.
- The vertical location of Ao is controlled by decisions for resource levels and the organizational effectiveness of maintenance operations. By definition, Ao cannot be larger than Aa.

If Aa is not known, it will not be possible to develop world-class maintenance operations (Keeter, 2003).

1.5.3.2 Factors Determining Availability

Availability is a function of reliability and maintainability—in other words, how often equipment is likely to fail and how long it will take to get the equipment back to full production capability. Reliability, maintainability, and, therefore, availability are determined by the interaction of the design, production, and maintenance functions. By extension, availability is largely determined by how well designers, operators, and maintainers work together (Keeter, 2003).

1.5.3.3 Optimizing Availability

First of all, Aa is to be optimized because no organization can attain a level greater than Aa (see Figure 1.12) (Keeter, 2003).

All equipment fails based on its design even when operated and maintained perfectly, but many types of failures are really due to the materials (rather than the design). Every maintenance activity, whether scheduled or unscheduled, is representative of an equipment failure. Scheduled or time-based maintenance seeks to correct failures before they can affect equipment performance. Unscheduled maintenance is corrective maintenance performed as the result of breakdown or the detection of incipient failure (Table 1.2) (Keeter, 2003).

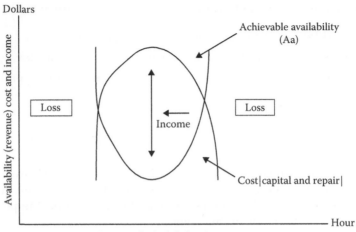

FIGURE 1.12
Optimal availability/cost. (Adapted from ARMS Reliability Engineers, Managing availability for improved bottom-line results, USA, LLC—© 2002 by Bill Keeter, http://reliabilityweb.com.)

1.5.3.4 Design of Achievable Availability

Aa is the result of the factors in the listed below and is depicted graphically in Figure 1.12 (Keeter, 2003). Company or plant design determines the shape and location of the Aa curve. Therefore, this design establishes the possible Aa (Keeter, 2003).

- Maintenance strategies determine the company's or plant's location on the Aa curve. Therefore, these strategies establish the actual Aa.
- The extreme right of the Aa curve represents 0% scheduled maintenance. There are no surprises here, because all maintenance is performed during a scheduled maintenance period. Availability is well below optimum. This extreme can be compared to coming in to the pits during every lap of a race to ensure you have no breakdowns on the race course. It would work, but you would never win the race.
- Trading off scheduled maintenance for unscheduled maintenance results in a climb back up the availability curve to the left. A nearly linear increase in availability occurs until we reach the point where unscheduled maintenance due to breakdowns takes away from availability gains. Operating farther to the left places the equipment under more stress and increases organizational chaos.
- After reaching the left of the peak of Aa, further reductions in scheduled maintenance become poor strategies.

TABLE 1.2

Top-Level Factors Affecting Availability

Reliability	Maintainability
Is increased as the frequency of outages is reduced and time between failures or shutdowns is increased.	Is increased as the duration of plant, subsystem, or equipment downtime is reduced.
Factors driven by design decisions	
Operating environment	Accessibility to the work point
Equipment-rated capacity	Features and design that determine the ease of maintenance
Maintenance while the system, subsystem, or item of equipment continues to function	Plant ingress and egress
Installed spare components within an equipment item	Work environments
Redundant equipment and subsystem	
Simplicity of design and presence of weak points	
Factors driven by maintenance decisions	
Preventive maintenance based on failure-trend data analysis	How maintenance tasks are detailed, developed, and presented to the maintenance technician
Trend diagnoses and inspection of equipment conditions to anticipate maintenance needs	Quality of the system of maintenance procedures
Quality of maintenance tasks (including inspections)	The probability of human, material, and facility resources being available to maintenance tasks
Skills applied to maintenance tasks	Training programs
	Management, supervision, and organizational effectiveness
	Durability of handling, support, and test equipment
Factors driven by operations decisions	
Use of equipment relative to its rated capacity	Organizational effectiveness as a factor in the troubleshooting process
How spares are incorporated in normal process operation	Organizational effectiveness and procedures to ready equipment for maintenance and startup
Shutdown and startup procedures	

Source: ARMS Reliability Engineers, Managing availability for improved bottom-line results, USA, LLC—© 2002 by Bill Keeter, http://reliabilityweb.com.

The cost curve represents strategic decisions to invest large amounts of capital up front to increase Aa through design or to spend operating dollars to increase Aa through more intensive maintenance strategies. These decisions are driven by many factors, such as the need to get a product to market quickly, the availability of capital, or the operating mentality of the company.

The availability/cost curve relationship highlights the fact that availability is a proxy for revenues. At some point of either extreme of the cost curve or the availability curve, the cost of availability will exceed the income it allows. Without availability management, operating beyond those intersections can occur without management's awareness; normal accounting practices and other maintenance performance indicators cannot easily reveal it.

The difference between Aa and Ao is the inclusion of maintenance support. Aa assumes resources are 100% available and no administrative delays occur in their application. Therefore, maximum Ao theoretically approaches Aa. In reality, every human endeavor has a natural upper limit that prevents Ao from reaching Aa.

The shape and location of the Ao curve are determined by the level of maintenance operation resources and organizational effectiveness. Both have upper bounds above which additional spending will not yield better results. At that point, Aa must be increased to give Ao room to move upward. Aa can be increased by new maintenance strategies, provided the plant or company is not operating at the peak of the Aa curve; capital investment is required to move the Aa curve upward if it is.

This is an important point. Without availability engineering and management, it is easy to unknowingly spend beyond the point of maximum return. This may occur when performance falls short of management's desired productive capacity. Management tries to achieve gains by putting increased stress on maintenance support, but the Ao curve has already been unwittingly forced against the Aa curve. The result is throwing good money after bad. Spending is in the loss zone to the right of the intersection of the Aa and cost curves (Keeter, 2003).

1.5.3.5 Determining Achievable Availability for an Existing Facility

Few physical asset managers have the luxury of being an integral part of the design phase. Therefore, this section is dedicated to analyzing the current state of a company or plant facility to determine its Aa. Determining Aa is a four-step process:

1. Build a reliability block diagram (RBD) of critical systems.
   - Use publicly available reliability data for failures of these systems. Using plant or company data can skew the results, depending on organizational effectiveness.

- Use plant or company data or work estimation techniques to determine mean time to repair.
2. Determine logistical delays created by hard design.
   - To/from shops
   - To/from stores
   - Accessing equipment
3. Add in scheduled maintenance downtime for the chosen preventive maintenance strategy.
4. Perform availability simulations.

The scope of the analysis is determined by resources, time, and the desired quality of the result (Keeter, 2003).

1.5.3.6 Building the Reliability Block Diagram (RBD)

The RBD is a graphical representation of the systems, subsystems, and components arranged in a way that reflects equipment design configuration. The RBD is the cornerstone of the availability model because it shows how failure affects process uptime (Keeter, 2003).

1.5.3.7 Refining the RBD

All complex machines are built from the same basic machine elements of couplings, bearings, gears, motors, belts, and so on. The RBD is refined by breaking down the top-level RBD into several sub-RBDs that represent each top-level system.

1.5.3.8 Obtaining Failure and Repair Data

After the RBDs are built, failure and repair data must be obtained for use in availability simulations. Obtaining these data is a time-consuming task. The desired degree of certainty dictates the level of effort required for this stage of building the model. It is important to remember this is not an exact science. Perfection is not required.

There are many sources for failure data, as shown in Tables 1.3 and 1.4. Binomial and Weibull distributions are typically used for modeling purposes. Most availability simulators accept either type.

Obtaining repair data is a much more difficult task. Repair data are typically not available in tabular form. Repair times are dependent on the configuration of the equipment and the company or plant. Equipment with parts located in tight spots requires much longer repair times than equipment with plenty of space in which to work (Keeter, 2003).

TABLE 1.3

Sources of Failure Data

Source	Data Available
Reliability Analysis Center	Electronic parts reliability data (EPRD) Nonelectronic parts reliability data (EPRD) Available in print and software versions
Paul Barringer's web site: www.barringer1.com	Weibull data for many components plus links to other available data and reliability web sites
Improving Machinery Reliability (Practical Machinery Management for Process Plants, Volume 1) Heinz P. Bloch. ISBN: 087201455X	Table of equipment failure data plus practical information on improving equipment and system reliability
Plant date	

Source: ARMS Reliability Engineers, Managing availability for improved bottom-line results, USA, LLC—© 2002 by Bill Keeter, http://reliabilityweb.com. Provide a good standard against which to judge actual achieved repair times

TABLE 1.4

Sources of Failure Data

Method	Advantages	Disadvantages
Analyzing plant data	Usually does not require special training. Data are usually available in plants that have mature maintenance reliability programs.	Data may be unreliable. Data are affected by organizational effectiveness.
Works estimation systems	Eliminate organizational effectiveness as a factor. Provide a good standard against which to judge actual achieved repair times. Provides detailed work steps and procedures.	Require training on the system used. Require much time to analyze the equipment and break repairs into tasks.

Source: ARMS Reliability Engineers, Managing availability for improved bottom-line results, USA, LLC—© 2002 by Bill Keeter, http://reliabilityweb.com. Provide a good standard against which to judge actual achieved repair times

1.5.3.9 *Closing the Gaps*

There are gaps between Ai and Aa and Aa and Ao. Closing these gaps in a cost-effective way makes an organization more successful, but this requires a thorough understanding of the top-level factors that determine availability and finding ways to improve each of these factors. The goal is to select strategies that

TABLE 1.5

Improve Maintenance Operations

Optimize the preventive/predictive maintenance system	Analyze failure data to target specific equipment for specific tasks.
	Analyze equipment using reliability-centered maintenance techniques to determine the best PM/PdM tasks to perform.
Use predictive maintenance techniques.	Perform vibration analysis.
	• Monitor new installations for minimum vibration.
	• Monitor existing installations for overall vibration level, signatures, and trends.
	• Monitor repaired equipment to ensure that repairs were done properly.
	Use thermography.
	• Monitor electrical equipment.
	• Monitor like mechanical equipment for temperature differences.
	Perform lubricant analysis.
	• Analyze incoming lubricants.
	• Track trends.
Improve outage planning and scheduling.	Use project-planning techniques.
	• Gantt charts
	• CPM
	• Reverse planning
	Ensure good planning for every job.
	• Parts staged
	• Good written procedures
	• Adequate resources.

Source: ARMS Reliability Engineers, Managing availability for improved bottom-line results, USA, LLC—© 2002 by Bill Keeter, http://reliabilityweb.com.

- Minimize number and length of scheduled outages to drive Aa closer to Ai
- Minimize number and length of unscheduled outages to drive Ao closer to Aa
- Minimize number and length of scheduled outages

In general, the number and length of scheduled outages can be decreased by making operational and noncapital equipment improvements. For example, Table 1.5 suggests ways to improve maintenance, and Table 1.6 offers suggestions to improve equipment (Keeter, 2003).

1.5.3.10 *Minimizing Number and Length of Unscheduled Outages*

The number and length of unscheduled outages or breakdowns can be decreased by using precision maintenance techniques and making noncapital equipment modifications, as suggested in Table 1.7 (Keeter 2003).

TABLE 1.6

Improve Equipment

Modify equipment to allow for accomplishing PMs on optima.	Lubrication, belt checks, etc.
Modify equipment for easier access.	Quick-release guards
	Open space around equipment

Source: ARMS Reliability Engineers, Managing availability for improved bottom-line results, USA, LLC—© 2002 by Bill Keeter, http://reliabilityweb.com.

1.5.3.11 Improve Equipment

- Open space
- Better replacement parts (Keeter 2003)

1.5.3.12 Capital Improvements to Increase Availability

After Aa is optimized by improving maintenance operations and making noncapital equipment modifications, the next step is to increase availability by capital investment in the plant or by capital investment in the equipment to increase Ai (Tables 1.8 and 1.9) (Keeter, 2003).

1.5.3.13 Matching Availability Goals to Annual Business Needs

No business operates in a static environment. Availability goals that are appropriate today may not be appropriate next year, next month, next week, or even tomorrow. Equipment availability is as sensitive to the vagaries of the business climate as any other key performance measure. Business decisions that either enhance or impair availability are made every day at every level of the organization. Availability goals must be reviewed and managed the same as any other business goals because of their sensitivity to available capital and operating funds. Figure 1.13 shows typical inputs and outputs of the maintenance process.

The business situation determines whether inputs or outputs are optimized. During times of overcapacity, a company will be cost-constrained, so the input side of the process must be optimized to reduce costs. During times of under-capacity, business is output-constrained, and the output side of the process must be optimized to increase business output. The important point to remember is that, at best, business output will remain constant during periods when process inputs are being optimized. The longer inputs are constrained, the more negative will be the effect on the outputs.

By using availability modeling and simulation to consider availability, a company will be able to lay out equipment and set up its facility for optimum reliability and maintainability, install reliable equipment, and establish good maintenance procedures, giving it a good chance of having reliable and optimally available assets and a reliable and optimally available facility. Using availability modeling

TABLE 1.7

Improve Maintenance Operations

Use predictive maintenance techniques.	Vibration analysis. Thermography. Lubricant analysis. Failure trend data.
Manage spare parts.	Kanban. Poke-a-yoke. Failure trend analysis. Storage standards. Repair standards for repairables. Remote spares locations. PM procedures for spares.
Manage lubricants.	Lubricant list. Lubrication routes. Proper storage. Ensure clean lubricant of correct type in correct amount at correct spot.
Manage failures.	Use "canned procedures." Be prepared. Analyze for root cause. Physical root. Human root. Organizational root.
Train.	Maintenance personnel. • Precision installation and repair techniques. • Equipment monitoring techniques. • Failure analysis. • Importance of being proactive. • Cost of failures. • Business goals. Front-line supervision and operators. • Proper equipment operation and its relationship to availability. • Their role in achieving availability goals. • What to clean and how to clean it. • Equipment-monitoring techniques. • Importance of being proactive. • Cost of failures. • Business goals.

Source: ARMS Reliability Engineers, Managing availability for improved bottom-line results, USA, LLC—© 2002 by Bill Keeter, http://reliabilityweb.com.

and simulation to continually monitor operational needs, improve maintenance and operating procedures, and thoroughly understand the three subtypes of availability (achievable, inherent, and operational) and the current status of its plant or facility in relationship to Aa will promote long-term success by improving bottom-line results throughout the life of the plant (Keeter, 2003).

TABLE 1.8

Increasing Achievable Availability

Modify surroundings.	Add space around equipment for easier Access. Improve ingress and egress.
Modify shops.	Relocate closer to equipment. Improve shop and support equipment. • Better cranes • Better tools • Better technology • Improve layout.
Modify stores.	Relocate closer to equipment. Improve storage equipment. Improve stores' management software. Provide for controlled remote stores.

Source: ARMS Reliability Engineers, Managing availability for improved bottom-line results, USA, LLC—© 2002 by Bill Keeter, http://reliabilityweb.com.

TABLE 1.9

Increasing Inherent Availability

Upgrade equipment.	Match equipment to production need. Improve control system and embed troubleshooting tools.
Modify equipment.	Reduce part count. Modify guarding. Modify lubricant delivery system. Add redundancy.

Source: ARMS Reliability Engineers, Managing availability for improved bottom-line results, USA, LLC—© 2002 by Bill Keeter, http://reliabilityweb.com.

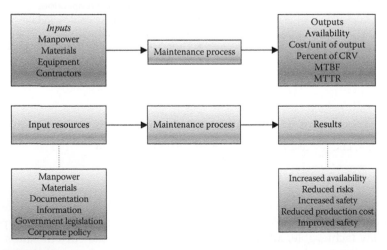

FIGURE 1.13
Maintenance function as a process within an asset management framework.

References

Ackermann, F., Eden, C., Brown, I., 2005. *The Practice of Making Strategy: A Step by Step Guide.* Sage Publication, London, U.K.

Assetic System Overview, asset management & protection. Asset Management for LG, NFP, Ports and Utilities. Assetic Pty Ltd. www.assetic.com.

ASSM—Asset Systems Support Management. Email: info@assetandsystems.co.uk or http://assetandsystems.co.uk/cost-savings-for-physical-assets/.

Baldwin, R. C., 2004a. (I). Enabling an E-maintenance infrastructure. *Maintenance Technology Magazine.* www.mt-online.com.

Baldwin, R. C., 2004b. (II). How do you spell e-maintenance? *Maintenance Technology Magazine,* www.mt-online.com.

Blutstein, A., 2012. Maintenance looking beyond the bottom line. Focus on Maintenance. Plant & Works Engineering (Ed.). https://www.eriks.co.uk/Document/695_Focus-on-Maintenance-and-Fenner-QD-NEO.pdf (accessed May 13, 2017).

Campbell, J. D., 1995. *Uptime: Strategies for Excellence in Maintenance Management.* Productivity Press, Portland, OR.

Coetzee, L. J., 1997. *Maintenance.* Maintenance Publishers, Centurion, South Africa.

Crespo, M. A., Gupta, J., 2006. Contemporary maintenance management: Process, framework and supporting pillars. *Omega* 34(3), 313–326.

Dhillon, B. S., 1992. Maintenance. In: Dhillion, B. S. (Ed.), *Reliability and Quality Control: Bibliography on General and Specialized Areas.* Beta Publishers Inc., Gloucester, Ontario, Canada.

Dhillon, B. S., 2006. *Maintainability, Maintenance and Reliability for Engineers.* Taylor & Francis, Boca Raton, FL.

EN 16646 Maintenance, 2016. Maintenance within physical asset management.

Gits, C. W., 1984. On the maintenance concept for a technical system: A framework for design. PhD thesis. Eindhoven University of Technology, Eindhoven, the Netherlands.

Gits, C. W., 1992. Design of maintenance concepts. *International Journal of Production Economics* 24, 217–226.

Hastings, N. A. J., 2010. *Physical Asset Management.* Springer, London, U.K., XXIX, p. 370 Hardcover. http://www.springer.com/978-1-84882-750-9.

Hounsell, D., 2012. Maintenance and the bottom line. FMD—Facility maintenance decisions. Editor March 2012. http://www.facilitiesnet.com/maintenanceoperations/article/Maintenance-and-the-Bottom-Line-Facility-Management-Maintenance-Operations-Feature--13053.

Huggett, J., 2005. Asset management—The changing role of maintenance management. Principal consultant. The Woodhouse Partnership Ltd. IMC Conference, 2005. www.twpl.com.

Keeter, B., 2003. Managing availability for improved bottom-line results. *Maintenance Technology* 16(4), 12–18.

Krar, S., 2013. The importance of maintenance. Changing from a FAIL and FIX approach to a PREDICT and PREVENT approach. https://www.automation-mag.com/images/stories/LWTech-files/94%20Intelligent%20Systems.pdf. Retrieved November 20, 2013.

Kriedemann, M., 2016. Strategic asset and service management plan. Manager infra-structure. Douglas Shire Council. Presented at Council workshop: June 14, 2016.

LCE Life Cycle Engineering Company. Article of reliability engineering. For more information to e-mail: info@LCE.com or https://www.lce.com/Whats-the-role-of-the-Reliability-Engineer-1227.html.

Lee, J., Ni, J., Djurdjanovic, D., Qiu, H., 2006. Intelligent prognostics tools and e-maintenance. NSF Center for Intelligent Maintenance System, University of Cincinnati, University of Michigan, USA. *Computers in Industry* 57(2006), 476–489.

Lee, J., Scott, L. W., 2006. Zero-breakdown machines and systems: Productivity needs for next-generation maintenance. In: *Engineering Asset Management. Proceedings of the 1st World Congress on Engineering Asset Management (WCEAM) 11–14 July 2006*. Springer, London, U.K., pp. 31–43.

Li, Y., Chun, L., Nee, A., Ching, Y., 2005. An agent-based platform for web enabled equipment predictive maintenance. In: *Proceedings of IAT'05 IEEE/WIC/ACM International Conference on Intelligent Agent Technology*, Compiegne, France, 2005.

Madu, C., 2000. Competing through maintenance strategies. *International Journal of Quality & Reliability Management* 17(9), 937–948.

Maintenance Management Framework (I), 2012. Building maintenance policy, stan-dards and strategy development. Queensland Department of Housing and Public Works, October 2012. www.hpw.qld.gov.au.

Maintenance Management Framework (II), 2012. Policy for the maintenance of Queensland Government buildings. Queensland Department of Housing and Public Works, October 2012. www.hpw.qld.gov.au.

Muchiri, P., Pintelon, L., Gelders, L., 2010. Development of maintenance function performance measurement framework and indicators. Centre for Industrial Management, Katholieke Universiteit Leuven, Heverlee, Belgium.

Muller, A., Marquez, A. C., Iung, B., 2008. On the concept of e-maintenance: Review and current research. *Reliability Engineering and System Safety* 93(2008), 1165–1187.

Muller, A., Suhner, M. C., Iung, B., 2006. Proactive maintenance for industrial system operation based on a formalised prognosis process. *Reliability Engineering System Safe* Article on line, domain: 10.1016/j.ress.2006.12.004.

NASA Reliability Centered Maintenance, 2008. *Guide for Facilities and Collateral Equipment*. National Aeronautics and Space Administration.

Ntozelizwe, B. M., 2015. Roles & responsibilities of a maintenance engineer. *LIZWE Engineers*, April 2015.

Pinjala, K. S., 2008. Dynamics of maintenance strategy. Doctoral report. Katholieke Universiteit Leuven, Leuven, Belgium.

Pinjala, K. S., Pintelon, L., Verreecke, A., 2006. An empirical investigation on the relationship between business and maintenance strategies. *International Journal of Production Economics* 104, 214–229.

Pintelon, L., Van Puyvelde, F., 2006. *Maintenance Decision Making*. Acco, Leuven, Belgium.

Savsar, M., 2006. Effects of maintenance policies on the productivity of flexible manu-facturing cells. *Omega International Journal of management Science* 34, 274–282.

Smith, R., Mobley, R. K., 2007. *Rules of Thumb for Maintenance and Reliability Engineers*. Butterworth-Heinemann, Oxford, U.K., 1st edn. http://reliabilityweb.com/articles/entry/the_maintenance_function.

Tsang, A., 2002. Strategic dimensions of maintenance management. *Journal of Quality in Maintenance Engineering* 8(1), 7–39.

Ucar, M., Qiu, R. G., 2005. E-maintenance in support of E-automated manufacturing systems. *Journal of the Chinese Institute Industrial Engineering* 22(1), 1–10.

Waeyenbergh, G., Pintelon, L., 2002. A framework for industrial maintenance concept development. *International Journal of Production Economics* 77, 299–313.

Woodhouse, J., 1997. What is asset management? *Maintenance and Asset Management* 12, 26–28.

Woodhouse, J., 2003. Asset management: An emerging science. *International Journal of COMADEM* 6(3), 4–10.

Zhang, W., Halang, W., Diedrich, C., 2003. An agent-based platform for service integration in e-maintenance. In: *Proceedings of ICIT 2003, IEEE International Conference on Industrial Technology*, vol. 1, Maribor, Slovenia, 2003, pp. 426–433.

2

Maintenance Costing in Traditional LCC Analysis

2.1 Traditional LCC

A product needs to move from a "technological-driven push" to a market-driven development to accommodate a "market pull" if it is to be successful (Van Baaren and Smit, 2000). The customer who buys the product frequently decides how it is going to look, be operated, and used. This means the design phase is essential, with input from the end-users.

For management to know its needs and give useful input, especially if the company wants to invest money in capital-intensive, new and emerging technologies and systems, life cycle costing (LCC) techniques can provide essential tools for engineering and economic analyses. An LCC analysis (LCCA) can provide a basis for evaluating the economic and engineering performance of proposed assets, including machines/equipment and production systems. LCC models are often projected as economic tools but they can also provide the basis for improvements in system effectiveness, making them useful for selecting equipment and production systems. LCC methodology may also be used to select alternative production schemes, determine modifications of existing systems or of machines/equipment, or make investments in new and improved technology, selecting machines/equipment from different suppliers.

The abbreviation LCC is used for both life cycle cost analysis and life cycle costing analysis, but the two terms are different. Life cycle cost refers to the total costs associated with the product or system over a defined life cycle, that is, all cost related to acquisition and utilization of a product over a defined period of the product's lifetime.

$$LCC = acquisition\ costs + operating\ costs + disposal\ costs \qquad (2.1)$$

Life cycle costing, however, refers to evaluation of alternative products, alternative system design configurations, alternative operational or maintenance solutions, and so on, and can be defined as "a systematic analytical process of evaluating various alternative courses of action with the objective of choosing the best way to employ scarce resources" (Fabrycky and Blanchard, 1991).

Life cycle costing is a tool for decision-making when several alternatives are under consideration, whereas life cycle cost evaluates the cumulative cost of a product throughout its whole life cycle. LCC can be very complex and requires large quantities of data. Life cycle costing is an economic decision tool and is a continuation of what was earlier called cost benefit analysis, but it takes a more systematic approach. In life cycle costing, the purpose is to analyze the differences between two or more alternative products in economic terms to select the best investment alternative. The analysis tries to identify the major cost drivers that contribute to the differences between the alternatives (Lund, 1998).

2.1.1 Life Cycle Phases

A product's life cycle can be divided into five distinct phases, as shown in Figure 2.1:

A. Need analysis and specification phase
B. Conceptual design phase
C. Detail design and development phase
D. Construction, production, and commissioning phase
E. Installation, system use, phase out, decommissioning, and disposal phase

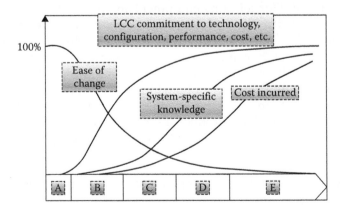

FIGURE 2.1
LCC committed, cost incurred, knowledge, and ease of change for various life cycle phases. (Adapted from Fabrycky, W.J. and Blanchard, B.S., *Life-Cycle Cost and Economic Analysis*, Prentice Hall, Inc., Englewood Cliffs, NJ, 1991.)

The initial design iterations start with customer specifications and a needs analysis, which is converted into design specifications in the conceptual phase. In this phase of a product's life cycle, an iterative creative process starts, as the designer tries to come up with several design alternatives, out of which one should be selected.

The specifications of the product need to be evaluated and decided before various concepts are considered. In the phase of design specification, there is an opportunity to quantify reliability and maintainability (R&M) considerations/characteristics, in particular, maintainability requirements, or workforce maintenance skills, such as whether a multiskilled workforce will be required, and/or whether there are acceptance or test criteria and reliability requirements. At this point, input from a LCC cost analysis, covering each of the phase of the product's life, is helpful. If the chosen design fulfils the needs and specifications, the consequent phases are preliminary design followed by final design. Little capital is invested in the needs phase of the design process. In fact, much of technology, configuration, performance, and cost details are committed in the conceptual design phase. The ease of change of a design decreases rapidly as the design progresses in time. The system-specific knowledge is low in the conceptual phase, but increases fast as the design progresses in time. Incurred cost is also low in the beginning of a design project, but starts to rise fast in the detailed design phase, raising the overall life cycle costs.

There is an opportunity in the early part of the design process to consider which environmental conditions the product will be subjected to through its lifetime. It is up to the purchaser to evaluate whether the product is fit for the operational condition in which it is to be used and to specify the requirements for the users' conditions (environmental, operational, social, etc.). The operator/buyer knows the environment the product is to be used in, while the producer of the machine/equipment knows the strengths and weaknesses of the product. It is therefore important that the two work together to develop a product that fits the purpose and the environment (Markeset and Kumar, 2000), and to do so before it is too late to make changes.

To sum up, as observed in Figure 2.1, the customer has the greatest influence on the product in the specification of needs phase. Later, the cost of changing the product rapidly increases, greatly increasing the overall life cycle costs and reducing the opportunity to make changes. When the acquisition phase is over and the utilization phase starts, the possibilities of influencing design are limited. Even though the purchaser is not always involved in the design and development phase, it is important to be aware of the commitment to life cycle cost. It is in the early phases of product development that a large percentage of life cycle cost is invested and at this point LCC analysis can be invaluable.

Life cycle costs are deterministic, including acquisition costs and disposal costs, or probabilistic, such as cost of failures, repairs, spares, and downtime.

Most of the probabilistic costs are directly related to the R&M characteristics of the system. A complete life cycle cost projection (LCCP) analysis may include accounting/financial elements (depreciation, present value of money/discount rates, etc.).

Given the specific interests of the various departments within a company, there may be problems projecting costs. For example:

- Engineering may avoid specifying cost-effective equipment needed to preclude costly failures so as to meet capital budgets.
- Purchasing may buy lower-grade equipment to get a favorable purchase price.
- Project engineering may build a plant with a view of successfully running it over 6 months rather than taking a long-term view of low-cost operation.
- Process engineering may want to operate equipment using the philosophy that all equipment is capable of operating at 150% of its rated condition without failure and other departments will clean up equipment abuse.
- Maintenance may defer required corrective/preventive actions to reduce budgets; if so, long-term costs will increase to ensure short-term management gains.
- Reliability engineering may be assigned improvement tasks with no budgets for accomplishing their goals.

Management is responsible for harmonizing potential conflicts. The glue binding these conflicting goals together is a teamwork approach to minimizing LCC. When properly used with good engineering judgment, LCC provides a rich set of information for making cost-effective, long-term decisions. LCC can be used as a management decision tool for making operating and support cost estimates, integrating the effects of availability, reliability, maintainability, capability, and system effectiveness into charts that are understandable.

2.1.2 Considerations of Life Cycle Cost

Many actions are taken during the service life or the life cycle of assets. Recall that the six stages of service life are planning, design, construction, operations and maintenance, renewal and revitalization, and disposal (Figure 2.2). The majority of these actions, particularly those in the first stage, have implications on the service life of the equipment and, thus, they influence the cost. The calculation of the life cycle cost (LCC) is based on the concept of the service life as this applies to a specific system.

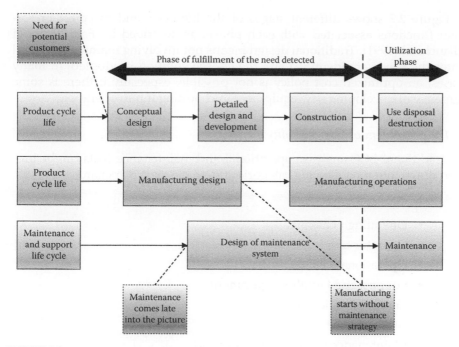

FIGURE 2.2
Life cycle of production and maintenance. (Adapted from Fabrycky, W.J. and Blanchard, B.S., *Life-Cycle Cost and Economic Analysis,* Prentice Hall, Inc., Englewood Cliffs, NJ, 1991.)

2.1.2.1 Problems of Traditional Design

Traditional engineering focuses on the acquisition phase of the life cycle, but as the analysis of the design phase in the previous section suggests, a product or system is properly coordinated and operated to be competitive through actions taken long before its conception.

Engineers must consider operational viability during the first stages of the development of a product, as well as its impact on the entire life cycle cost. This implies the use of a sequential method using relevant milestones in the LCC if the outcome desired is a cost-effective design.

A good design that fulfills the required function of the asset or asset system often has indirect and undesired effects in the form of operational problems. This results from the exclusive consideration of the required function, instead of approaching the most challenging task of the design phase, namely, that the design should satisfy RAMS parameters. There is plenty of specialized knowledge on this topic, and the problem can be solved. Fabrycky and Blanchard (1991) say the difficulty is the systematization of the integrated use of the known methodologies.

Economic considerations can be applied to engineering, as Thuesen and Fabrycky (2000) show, and should be integrated into the product's service life, production, and support or maintenance.

Figure 2.2 shows different stages of the life cycle and the most important functions associated with each phase, as described by Fabrycky and Blanchard (1991). Traditional design means not involving maintenance until after the needs are identified and the chosen solution is adopted. In many cases, an optimized cost policy is not possible, especially if there is some urgency in developing new equipment and putting it into operation.

2.1.2.2 Problem of Cost Visibility

In traditional LCC analysis, operational and maintenance costs can be broken down into the following categories (Fabrycky and Blanchard, 1991):

- Operations cost
    - Operating personnel
    - Operator training
    - Operational facilities
    - Support and handling equipment
    - Energy/utilities/fuel
- Maintenance cost
    - Maintenance personnel and support
    - Spare/repair part
    - Test and support equipment maintenance
    - Transition and handling
    - Maintenance training
    - Maintenance facilities
    - Technical data
    - System/product modifications
    - Disposal cost

The disposal cost may be substantial depending on the product and location (this is easily seen in the North Sea where the retirement of production platforms and equipment is a serious problem). Such costs are clearly visible, but a number of costs, not visible at a glance, have a huge impact on final LCC figures.

The combination of reduced spending power, budget limitations, and increased competition has generated interest in the total cost of equipment. With the current economic situation, it has become more complicated, with the addition of problems related to determining the real cost of the system. These include the following:

1. The total cost of the system may not be visible, particularly the costs associated with its operation and maintenance. The problem

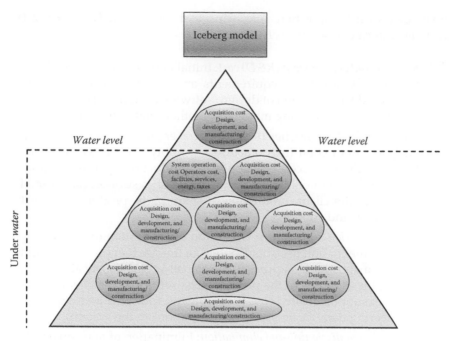

FIGURE 2.3
Iceberg effects in costs associated with service life. (Adapted from Blanchard, B.S. and Fabrycky, W.J., *Systems Engineering and Analysis*, 3rd edn., Prentice Hall International Series in Industrial and System Engineering, Englewood Cliffs, NJ, 1997.)

of cost visibility is a consequence of the "iceberg effect," as shown in Figure 2.3 and explained by Blanchard and Fabrycky (1997). To avoid the iceberg effect, cost analysis must tackle all aspects of the life cycle cost.

2. Often cost factors are applied incorrectly. Individual costs are wrongly identified or placed in the incorrect category, so that variable costs are treated as fixed (and vice versa).

3. Accounting procedures may not allow realistic and opportune evaluations of the total cost. In addition, it may be difficult (if not impossible) to determine the costs on a functional basis.

4. Budgetary practices may be inflexible with respect to switching funds from one category to another or from one year to the next to facilitate improvements in acquisition costs or to exploit the system.

2.1.2.3 Structure of Cost Breakdown

Generally, costs during the life cycle are divided into categories based on the organizational activity. These categories and their elements comprise a cost breakdown or decomposition. According to Thuesen and Fabrycky (2000), the main cost categories are as follows:

1. *Research and development (R&D) cost*: Initial planning, market analysis, product investigation, requirements analysis, engineering design, data and documentation of design, software design, tests and evaluations of the engineering models, and other related functions.
2. *Production and construction cost*: Industrial engineering and operation analysis, production (manufacture, assembly, and tests), construction of facilities, process development, production operations, quality control, and initial requirements of logistics support (e.g., initial support to the client, production of spare parts, production of test equipment, and support).
3. *Cost of operation and support*: Operation of the system or product by the customer or end user, distribution of the product (marketing and sales, transport and shipping), maintenance and logistical support during the life cycle of the product (e.g., service to the client, maintenance, support related to test equipment, transport and handling, engineering data, facilities, and modifications to the system).
4. *Cost of decommissioning and elimination*: Elimination of non-repairable elements during the service life, disposal of the system or product, recycling of material, and all applicable requirements of the logistical support for disposal, with special emphasis on environmental and health and safety issues.

The cost breakdown relates the organization's objectives and activities to its resources. It constitutes a logical subdivision of the cost for each area of functional activity, important system elements, and common or similar elements. Life cycle costs are summations of cost estimates from inception to disposal for both equipment and projects as determined by an analytical study and estimate of total costs experienced during their life. They are the total costs estimated to be incurred in the design, development, production, operation, maintenance, support, and final disposal of a system over its anticipated useful life span (Department of Energy (DOE), 1995).

The objective of life cycle cost analysis (LCCA) is to choose the most cost-effective approach from a series of alternatives to achieve the minimum long-term cost of ownership. LCCA helps engineers justify equipment and process selection based on total costs rather than the initial purchase price. Usually the costs of operation, maintenance, and disposal exceed all other costs many times over. The best balance among cost elements is achieved when the total LCC is minimized (Landers, 1996). As with most engineering tools, LCC provides the best results when art and science are merged with good judgment.

Procurement costs are widely used as the primary (and sometimes only) criterion for equipment or system selection. This criterion is simple to use but often results in bad financial decisions. Procurement costs tell only one part of the story—frequently the story is far too simple, with damaging, even disastrous, results on the overall financial well-being of the enterprise, for example, an extremely inexpensive piece of equipment may be attractive—but that equipment may break down frequently and fail far too soon, incurring huge expenses. As John Ruston says, "It's unwise to pay too much, but it's foolish to spend too little." This is the operating principle of LCC (Barringer et al., 1996).

Life cycle costs can be used in the following:

- *Affordability studies*: measure the impact of a system or project's LCC on long-term budgets and operating results.
- *Source selection studies*: compare LCCs of competing systems among various suppliers of goods and services.
- *Design trade-offs*: influence design aspects of plants and equipment that directly impact LCC.
- *Repair-level analysis*: quantify maintenance demands and costs rather than using rules of thumb, such as "maintenance costs ought to be less than the capital cost of the equipment."
- *Warranty and repair costs*: suppliers of goods and services, along with end users, can understand the cost of early failures in equipment selection and use.
- *Suppliers' sales strategies*: merge specific equipment grades with general operating experience and end user failure rates using LCC to sell for the best benefit rather than just selling on the basis of low initial cost.

The concept of life cycle cost was popular in the mid-1960s. Many original works on LCC are now out of print. More recent publications by Blanchard and Fabrycky are now useful sources for LCC information:

- *Maintainability* (Blanchard et al., 1995) for commercial issues
- *Logistics Engineering and Management* (Blanchard, 1992) for Department of Defense (DoD) issues
- *Systems Engineering and Analysis* (Blanchard and Fabrycky, 1990) for management and design issues
- *Life-Cycle Cost and Economic Analysis* (Fabrycky and Blanchard, 1991) for conceptual and theoretical issues

Technical societies use life cycle analysis. For example, the Society of Automotive Engineers includes life cycle costs in its *RMS Guidebook* (Society of Automotive Engineers (SAE), 1995) with a convenient summary of the principles. The Institute of Industrial Engineers includes a short section on life cycles and how they relate to life cycle costs in the *Handbook of Industrial Engineering* (Institute of Industrial Engineers (IEE), 1992). Notably, LCC has been the subject of a tutorial at the annual Reliability and Maintainability Symposium (Blanchard, 1991), sponsored by major technical societies.

In other words, LCC is experiencing resurgence. The concept has passed the test of time, and practitioners have learned to minimize its limitations. That said, the limitations can result in substantial setbacks if good judgment is not used. The following are some of the most frequently cited LCC limitations:

- LCC is not an exact science: everyone gets different answers and the answers are neither wrong nor right—only reasonable or unreasonable. There are no LCC experts simply because the subject is too broad.

- LCC outputs are only estimates and can never be more accurate than the inputs and the intervals used for the estimates; this is particularly true for cost–risk analysis.

- LCC estimates lack accuracy, and errors in accuracy are difficult to measure as the variances obtained by statistical methods are often large.

- LCC models operate with limited cost databases, and the data in the operating and support areas are difficult and expensive to obtain.

- LCC models must be calibrated to be highly useful.

- LCC models require large volumes of data; often only a few handfuls of data exist, and most of the available data are suspect.

- LCC requires a scenario for the following: how the money expenditure model will be constructed for the acquisition of equipment, how the model will age with use, how damage will occur, how learning curves for repairs and replacements will occur, how cost processors will function (design costs, labor costs, material costs, parts consumption, spare parts costs, shipping costs, scheduled and unscheduled maintenance costs) for each time period, how many years the model will survive, how many units will be produced/sold, and similar details required for building cost scenarios. Most details require extensive extrapolations, and obtaining facts is difficult.

- LCC models (by sellers) and cost-of-ownership (COO) models (by end users) have credibility gaps because different values are used in each model. Often credibility issues center on which is right and which is wrong (a win–lose issue) rather than harmonizing both models (for a win-win effort) using available data.

- LCC results are not good budgeting tools. They're effective only as comparison/trade-off tools. Producing good LCC results requires a project team approach because specialized expertise is needed.

- LCC should be an integral part of the design and support process to design for the lowest long-term cost of ownership. End users can use LCC for affordability studies, source selection studies of competing systems, warranty pricing, and cost-effectiveness studies. Suppliers find LCC useful for identifying cost drivers and ranking competing designs and support approaches.

- LCC is ubiquitous in military projects but is less frequently applied in commercial areas, mostly because there are so few practitioners of LCC. References to LCC in the DoD area include the following: MIL-HDBK-259 has LCC details; MIL-HDBK-276-1 and MIL-HDBK-276-2 are guides on importing data into specific software (Barringer et al., 1996).

Remember this adage when considering LCC limitations: In the land of the blind, a one-eyed man is king! We don't need perfect vision. We just need to see more clearly than our fiercest competitor, so we can improve our operating costs.

2.1.3 Cost Categories

There are various ways of classifying the cost components of an LCCA, depending on what role they play in the mechanics of the methodology. The most important categories in LCCA distinguish between investment-related and operational costs; initial and future costs; and single costs and annually recurring costs (Fuller and Petersen, 1996).

2.1.3.1 Investment Costs versus Operational Costs

Life cycle costs typically include both investment costs and operational costs. The distinction between investment and operation-related costs is most useful when computing supplementary economic measures such as the savings-to-investment ratio (SIR) and adjusted internal rate of return (AIRR). These measures evaluate savings in operation-related costs with respect to increases in capital investment costs. This distinction will not affect the LCC calculation itself; nor will it cause a project alternative to change from cost-effective to non-cost-effective or vice versa. However, it may change its ranking relative to other independent projects when allocating a limited capital investment budget.

All acquisition costs, including costs related to planning, design, purchase, and construction, are investment-related costs. Some LCC methodologies also require that residual values (resale, salvage, or disposal costs) and capital replacement costs be included as investment-related costs. Capital replacement costs are usually incurred when replacing major systems or components, paid from capital funds. Operating, maintenance, and repair (OM&R) costs, including energy and water costs, are operational costs.

Replacements related to maintenance or repair (e.g., replacing light bulbs or a circuit board) are usually considered to be OM&R costs, not capital replacement costs. OM&R costs are usually paid from an annual operating budget, not from capital funds (Fuller and Petersen, 1996).

2.1.3.2 Initial Investment Costs versus Future Costs

The costs incurred in the planning, design, construction, and/or acquisition phase of a project are classified as initial investment costs. They usually occur before an asset or system is put into service. Those costs that arise from the operation, maintenance, repair, replacement, and use of an asset or system during its service period are future costs. Residual values at the end of an asset or system's life, or at the end of the study period, are also future costs (Fuller and Petersen, 1996).

2.1.3.3 Single Costs versus Annually Recurring Costs

It is useful to establish two categories of project-related costs based on their frequency of occurrence, whether a one-time event or an ongoing process. This categorization determines the type of present-value factor to be used for discounting future cash flows to present value (Fuller and Petersen, 1996).

- Single costs (one-time costs) occur one or more times during the study period at non-annual intervals. Initial investment costs, replacement costs, residual values, maintenance costs scheduled at intervals longer than 1 year, and repair costs are usually treated as single costs.
- Annually recurring costs are amounts that occur regularly every year during the service period in approximately the same amount, or in an amount expected to change at some known rate. Energy costs, water costs, and routine annual maintenance costs fall into this category.

LCCA is a method of project evaluation in which all costs arising from owning, operating, maintaining, and ultimately disposing of a project are considered to be potentially important to the evaluation.

LCCA is particularly suitable for the evaluation of alternatives that satisfy a required level of asset or system performance, but that may have different initial investment costs; different operating, maintenance, and repair (OM&R) costs (including energy and water usage), and possibly different lifetimes. However, LCCA can be applied to any capital investment decision in which higher initial costs are traded for reduced future cost obligations (MacKay and Ronan, 2015).

LCCA provides a significantly better assessment of the long-term cost-effectiveness in asset management than alterative economic methods that focus only on first costs or on operation-related costs in the short run.

There are almost always several cost-effective maintenance alternatives for any given system. Many of these alternatives may be cost-effective, but (usually) only one can actually be used in a given application. In such cases, LCCA can be used to identify the most cost-effective alternative for that application. This is generally the alternative with the lowest life cycle cost and maximum availability performance, or whatever is expected.

LCCA can also be used to prioritize the allocation of funding to a number of maintenance investment projects within an asset when insufficient funding is available to implement all of them, that is, for the selection of overinvestments and further cost availed, as discussed in Chapter 5.

LCCA stands in direct contrast to the Payback method of economic analysis. The Payback method generally focuses on how quickly the initial investment can be recovered and, as such, is not a measure of long-term economic performance or profitability. The Payback method typically ignores all costs and savings occurring after the point in time at which payback is reached. It does not differentiate between maintenance alternatives, and it often uses an arbitrary payback threshold. Moreover, the Simple Payback method, which is commonly used, ignores the time value of money when comparing the future stream of savings against the initial investment cost.

LCCA is a powerful tool of economic analysis. As such, it requires more information than do analyses based on first-cost or short-term considerations. It also requires additional understanding of concepts such as discounted cash flow, constant versus current dollars, and price escalation rates. The alternative, however, is to ignore the long-term cost consequences of investment decisions, to reject profitable investment opportunities, and to accept higher-than-necessary utility costs (Fuller and Petersen, 1995).

This chapter considers incentives to use LCCA for maintenance, decision support system (DSS) evaluation, and several methodologies and cost approaches for this purpose. It provides guidance to asset managers on how to use LCCA to evaluate impacts of investments and expenses to ensure better projects with reduced future operating and maintenance costs.

2.2 Life Cycle Cost Analysis as a Project Follow-Up for Assets

An important component of a company's activities is prioritizing its capital improvement program to meet its most pressing needs. This prioritization occurs at the end of the capital project development process, which consists of project identification/initial validation, risk reduction, and LCCA,

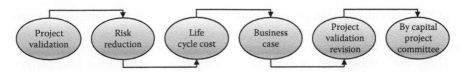

FIGURE 2.4
Capital project development process. (Adapted from Water Research Foundation (WERF), *Overview: What is Life Cycle Costing?* GWRC, GHD Consulting, Inc., Alexandra, VA, 2011a.)

all of which are used to establish the final business case for each project. As shown in Figure 2.4, the LCCA is part of the business case preparation (Water Research Foundation (WERF), 2011a).

LCCA allows a company to compare competing capital and O&M projects. Given the condition of the company's assets, the amount of capital available from the budget, and historical evidence, the project manager must decide which project alternatives will incur the least life cycle costs over the life cycle of the assets involved while delivering performance at or above a defined level. This analysis will enable the company to do the following:

- Make decisions about capital and O&M investments based on least life cycle costs
- Rank projects based on total cost of ownership (TCO)
- Combine the costing data with the project validation and risk reduction scores to prioritize the projects
- Make more informed decisions
- Submit better reports to key stakeholders

A thorough LCCA yields a higher level of confidence in the project decision and is part of the project validation calculation. The LCCA is combined with a risk reduction analysis and summarized in a business case, thereby providing a consistent approach to the review of projects.

The life cycle of an asset is defined as the time interval between the initial planning for its creation and its final disposal (Figure 2.5). This life cycle is characterized by these key stages:

- Initial concept definition
- Development of the detailed design requirements, specifications, and documentation
- Construction, manufacture, or purchase
- Warranty period and early stages of usage or occupation
- Prime period of usage and functional support, including O&M costs, as well as the associated series of upgrades and renewals
- Disposal and cleanup at the end of the asset's useful life

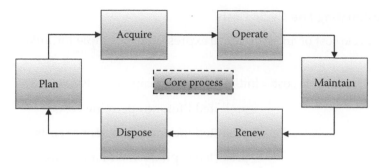

FIGURE 2.5
Life cycle of an asset. (Adapted from Water Research Foundation (WERF), *Overview: What is Life Cycle Costing?*, GWRC, GHD Consulting, Inc., Alexandra, VA, 2011a.)

As shown in Figure 2.5, certain day-to-day, periodic, and strategic activities occur for any asset. The asset life cycle begins with strategic planning, and moves on to creation, operation, maintenance, and rehabilitation, to the final decommissioning and disposal at the end of the asset's life. The life of an asset will be influenced by its ability to continue to provide a required level of service. Many assets reach the end of their effective life before they become nonfunctional (regulations change, the asset becomes noneconomic, the expected level of service increases, capacity requirements exceed design capability). Technological developments and changes in user requirements are key factors impacting the effective life of an asset (Water Research Foundation (WERF), 2011a).

2.2.1 Objectives of the Life Cycle Costing Methodology

Life cycle costing (*Note*: "life cycle costing" and "life cycle cost projections" can be used interchangeably) analysis can be carried out during any phase of an asset's life cycle. It can be used to provide input into decisions on asset design, manufacture, installation, operation, maintenance support, renewal/refurbishment, and disposal (Water Research Foundation (WERF), 2011a).

The following are the objectives of life cycle costing:

- Minimize the TCO given a desired level of sustained performance.
- Support management considerations affecting decisions during any life cycle phase.
- Identify the attributes of the asset which significantly influence the life cycle cost drivers so assets can be effectively managed.
- Identify the cash flow requirements for projects.

2.2.2 Estimating Life Cycle Costs

The life cycle cost of an asset can be expressed by a simple formula:

$$\text{Life cycle cost} = \text{Initial} \left(\text{projected}\right) \text{capital costs}$$

$$+ \text{projected lifetime operating costs}$$

$$+ \text{projected lifetime maintenance costs}$$

$$+ \text{projected capital rehabilitation costs}$$

$$+ \text{projected disposal costs}$$

$$- \text{projected residual value.}$$

Note the prominent role of projected costs versus historic (actual) costs in analyzing life cycle costs; because of its forward-looking "best guess" nature, life cycle costing is at least as much "systematic art" as it is analytical technique (Water Research Foundation (WERF), 2011a).

Operating, maintenance, and repair (OM&R) costs are often more difficult to estimate than other building expenditures. Since operating schedules and standards of maintenance vary from asset to asset, there is great variation in these costs, even for assets of the same type and age (Fuller, 2010).

OM&R costs generally begin with the service date and continue through the service period. Some OM&R costs are annually recurring costs which are either constant from year to year or change at some estimated rate per year. The present value of annual recurring costs over the entire service period can be estimated using appropriate factors. Others are single costs which may occur only once or at non-annual intervals throughout the service period. These must be discounted individually to present value (Fuller and Petersen, 1996).

2.2.2.1 Estimating OM&R Costs from Cost-Estimating Guides

Ongoing efforts to standardize OM&R costs have produced many helpful manuals and databases. Keep in mind that if OM&R costs are essentially the same for each of the project alternatives being considered, they do not have to be included in the LCCA.

Some of data estimation guides derive cost data from statistical cost-estimating relationships of historical data and report, for example, average owning and operational costs of assets. The popular CERL M&R Database derives data from time–motion studies, which estimate the time required to perform certain tasks (Fuller and Petersen, 1995).

2.2.2.2 Estimating OM&R Costs from Direct Quotes

A more direct method of estimating nonfuel OM&R costs is to obtain quotes from contractors and vendors. For cleaning services, for example, you can get quotes from contractors, based on prevalent practices in similar buildings. Maintenance and repair estimates for equipment can be based on manufacturers' recommended service and parts replacement schedules. You can establish these costs for the initial year by obtaining direct quotes from suppliers. For a constant-dollar analysis, the annual amount will be the same for the future years of the study period, unless, as is sometimes the case, OM&R costs are expected to rise as the system ages. In this latter case, the real (differential) escalation rate for that cost must be included in the analysis (Fuller and Petersen, 1996).

2.2.3 Impact of Analysis Timing on Minimizing Life Cycle Costs

A major portion of projected life cycle costs stems from the consequences of decisions made during the early phases of asset planning and conceptual design. These decisions comprise a large percentage of the life cycle costs for that asset. Figure 2.6 indicates the level of cost reduction that can be achieved at various stages of the project (Water Research Foundation (WERF), 2011a).

As the figure shows, the best opportunities to achieve significant cost reductions in life cycle costs occur during the early concept development and design phase of any project. At this time, significant changes can be made at the least cost. At later stages of the project, many costs are "locked in" and

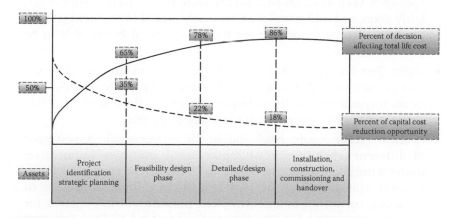

FIGURE 2.6
Opportunities for cost reduction. (Adapted from Water Research Foundation (WERF), *Overview: What is Life Cycle Costing?*, GWRC, GHD Consulting, Inc., Alexandra, VA, 2011a.)

not easily changed. To achieve the maximum benefit available during later stages, it is important to explore the following:

- A range of alternative solutions
- The cost drivers for each alternative
- The time period for which the asset will be required
- The level and frequency of usage
- The maintenance and/or operating arrangements and costs
- Quantification of future cash flows
- Quantification of risk (Water Research Foundation (WERF), 2011a)

2.2.4 Selecting Potential Project Alternatives for Comparison

The concept of the life cycle of an asset provides a framework to document and compare alternatives, including the following:

- *Do Nothing*: The Do Nothing option is literally not investing any money in any form of maintenance or renewal, including that recommended by the design engineer or vendor. This alternative is generally intended to set a conceptual baseline for asking the question: "What value does what we do now or what we plan to do add to extending the life/functionality/reliability of the asset over doing nothing at all?" Another way of looking at the core question posed here is "Why should we continue to do what we do or are anticipating doing?" There are rare occasions when Do Nothing is a valid approach, such as when an asset is due to be replaced or shut down fairly soon, and additional maintenance/capital investment is irrelevant to keeping it running for the short period before it is to be decommissioned.

- *Status Quo*: The Status Quo option is defined as maintaining the current operations and maintenance behavior—typically that defined by the manufacturer or the design engineer. It is the realistic baseline case against which other alternatives are compared.

- *Renewal (major repair, rehabilitation, or replacement)*: Assessment of different rehabilitation or replacement strategies requires an understanding of the costs and longevity of different asset intervention strategies. Each strategy is costed for the expected life of the asset, converted to an equivalent present worth, adjusted for varying alternative life lengths, and compared to find the least overall cost.

- *Non-Asset Solutions*: In certain circumstances, the non-asset solution, that is, providing the same level of service without a major additional investment, can be a viable alternative, for example, using pricing strategies to reduce the consumption of water.

- *Change Levels of Service*: Most life cycle costing assumes a constant level of service across options being compared. When this is not the case (quite common, in reality), comparisons across alternatives with different levels of service or different levels of benefit must introduce a projected benefits section for each alternative, in addition to the cost projections.

- *Disposal*: Disposal occurs when the asset is retired at the end of its useful life, sometimes when the function or level of service originally desired is no longer relevant.

It is unlikely that all seven alternatives will be feasible for each analysis; rather than waste money on obviously irrelevant options, it is wise to reduce the analyzed set to only those considered feasible (Water Research Foundation (WERF), 2011a).

2.2.5 Effect of Intervention

A single intervention option for the entire life cycle is not likely to be the best approach to maximizing the life of an asset. Multiple strategies and options will need to be studied to determine the optimal strategy or combination of strategies for maximum life extension.

Optimal decision-making uses LCCA as a core tool for determining the optimum intervention strategy and intervention timing (Water Research Foundation (WERF), 2011a).

2.2.6 Estimating Future Costs

Knowing with certainty the exact costs for the entire life cycle of an asset is, of course, not possible. Future costs can only be estimated with varying degrees of confidence, as they are usually subject to a level of uncertainty that arises from a variety of factors:

- The prediction of the utilization pattern of the asset over time
- The nature, scale, and trend of operating costs
- The need for and cost of maintenance activities
- The impact of inflation
- The opportunity cost of alternative investments
- The prediction of the length of the asset's useful life

The main goal in assessing life cycle costs is to generate a reasonable approximation of the costs (consistently derived over all feasible alternatives), not to find the perfect answer.

With the rehabilitation and/or replacement of assets during the life cycle, operations and maintenance costs will need to be adjusted appropriately. Both costs are likely to increase as the asset ages. The pattern of increase will vary by asset type and operational environment. In many cases, as the asset ages, it requires an increasing number of visits per year by the maintenance team, longer time to execute the work order, and more maintenance staff; these costs are both real and material and can be simply "modeled" in a spreadsheet.

The timing of the rates of increases in the flow of costs over time is instrumental in determining total life cycle costs and can substantially impact the outcome of the investment decision. It is therefore important to do the following:

- Be systematic, realistic, and detailed in estimating the future flow of real costs.
- Document the assumptions.
- Take inflation into account when discounting of future costs (Water Research Foundation (WERF), 2011a).

2.2.7 Managing Cash Flow

Using LCCA to determine the alternative with the lowest life cycle costs is important, but organizational cash flow issues may need to be considered as well. There will always be competing demands for the available cash resources of the organization at any given time. Management of cash flow is simple if the pattern is predictable over the long term. It is conceivable that the lowest cost solution might not be the best solution from the aggregate cash flow perspective.

Simply stated, LCCA provides a sound basis for projecting cash requirements to assist the chief financial officer in managing the cash flows of the organization (Water Research Foundation (WERF), 2011a).

2.2.8 Selecting a Discount Rate

As has long been recognized, money has what economists call a "time value." Generally speaking, this means that given the opportunity to receive a given sum of money today or the same amount at a future point in time, most people prefer to receive the money today because they can use the money now rather than defer its use of the money to a later time. The preference can be summed up as follows:

- Having the money in hand allows the recipient to satisfy wants now (e.g., consume, invest).
- The payee may not show up to hand over the money at the designated future date.
- Due to inflation, the money received at a future date may buy less.

When an organization has a choice of incurring a cost now or in the future, it generally considers the benefits of alternative uses that exist now for the available funds, otherwise called the "opportunity cost of capital." Future costs are regarded as less significant because they do not have the potential to grow, for example, through investment of current funds, over the intervening period.

For example, if a $100 purchase is to be made today, it is necessary to have $100 available now. However, if the purchase can occur in 3 years for $100, it would be possible to generate the required $100 by investing $75.13 at an interest rate of 10% per year for the next 3 years. If the $100 can be used in some other way by the organization, it may be able to generate more than 10% per year, which would make the deferral of this future cost even more attractive. This reduced value of future money relative to the present is called the "time value of money."

The systematic reduction of a sum of money available in the future is termed "discounting" and the result is termed a "future value." The mathematical effect is the reverse of the familiar compounding of interest, where money invested today is increased at a constant rate to reflect the accrual of interest. Note that a discount rate can be thought of as a negative interest rate.

When a cost that is expected to be paid at a future point in time is adjusted back to the present, the result is said to be the "present value of the future sum."

The formula for calculating the present value is

$$PV = \frac{FV}{(1+i)^n},$$ (2.2)

where
  PV is the value at time = 0
  FV is the future value at time period = n (i.e., the projected cost at a future year, month, quarter, etc.)
  i is the discount rate per time period
  n is the number of time periods

To derive the present value of a stream of costs (e.g., where costs associated with the life cycle are projected for each consecutive year in a stream of years out to a given year end and where those costs can vary from year to year), the present value for each year is first calculated; then the present values for each year are summed.

The life cycle cost projection uses this sum of consecutive years to calculate present values so that cost projections for a given cost component can vary from year to year; costs varying from year to year is the expected norm for most assets, as operations and maintenance factors are likely to vary over time in real dollars.

Setting a valid discount rate for analytic purposes is critical. For private companies, the discount rate is usually set equal to the company's borrowing rate for the period under analysis. More sophisticated approaches are available, but they are generally variations on the same theme.

When the asset analysis concerns renewal, a strong argument can be made that there is actually very little "opportunity cost" for the required funds if the function of the asset is to be sustained because failure to reinvest will inevitably lead to cessation of the function (due to failure of the existing asset). Using the money for other purposes instead of reinvestment is a *de facto* determination that the function is no longer needed, and this neutralizes the purpose of LCC as a renewal analysis.

Likewise, the risk associated with a reinvestment project is likely to be minimal unless substantially new technologies are replacing the existing ones.

These observations argue for a low discount rate, one generally independent of the traditional use of the current borrowing rate or the current investment rate.

When selecting a discount rate, the user should distinguish between "real" cost projections and "nominal" cost projections. Real costs are calculated using current money (i.e., not adjusted for inflation); nominal costs are adjusted for expected inflation.

Note that discounting is not the same as adjusting for inflation; they are two separate, albeit interrelated, forces. If the user desires to see discounted output that includes inflation, the inflation rate can be built into the discount rate using the following relationship:

$$\text{Nominal discount rate} = \big[(1 + \text{real discount rate})$$

$$* (1 + \text{expected inflation rate})\big] - 1, \qquad (2.3)$$

The user should carefully document whether the LCCA is in real or nominal money (similarly, whether the discount rate is real or nominal). Each has its purpose: if the user wishes only to analyze the interaction of "real" cost factors in determining future costs (e.g., to isolate the impact of increased energy consumption as an asset ages but independent of the likely increase in energy rates), he or she should forego inflation. If, however, the user wishes to project what will likely be needed in a given budget year as actual outlay, he or she should consider the nominal costs (Water Research Foundation (WERF), 2011b).

2.2.9 Time Value of Money

The value of money is time-dependent—what you have of value today is going to have a different value in the future, as shown in the following equation:

$$\text{Value today} = \frac{\text{Future value}}{(1 + \text{Discount rate})^{\text{time}}}$$

To compare the value of assets, the assets have to be compared on an equal basis. Future LCC costs and income have to be discounted to today's value. The payback, net present value, and internal rate of return are some of the more common discounting methods .

2.3 Trade-Off Tools for LCC

One helpful tool for easing LCC calculations involving probabilities is the effectiveness equation; it judges the chances of producing the intended results. The effectiveness equation has been described in several different formats (Raheja, 1991; Blanchard et al., 1995; Kececioglu, 1995; Pecht, 1995; Landers, 1996). In all cases, however, the issue is finding a system effectiveness value that gives the lowest long-term cost of ownership. The equation can be expressed as follows:

$$\text{System effectiveness} = \frac{\text{Effectiveness}}{\text{LCC}}$$

Cost is a measure of resource usage; cost estimates can never include all possible elements but will hopefully include the most important ones.

Effectiveness is a measure of value received; it rarely includes all value elements as many are too difficult to quantify. Effectiveness varies from 0 to 1 and can be expressed as follows:

$$\text{Effectiveness} = \text{Availability} * \text{Reliability} * \text{Maintainability} * \text{Capability}$$

$$= \text{Availability} * \text{Reliability} * \text{Performance}$$

$$(\text{Maintainability} * \text{Capability})$$

$$= \text{Availability} * \text{Dependability} \, (\text{Reliability} * \text{Maintainability})$$

$$* \text{Capability}.$$

The effectiveness equation is the product of the chance that the asset or system will be available to perform its duty, it will operate for a given time without failure, it will be repaired without excessive loss maintenance time, and it will perform its intended production activity according to the standard (Barringer et al., 1996).

2.3.1 Effectiveness, Benchmarks, and Trade-Off Information

The reason to quantify elements of the effectiveness equation (and their associated costs) is to find areas for improvement to increase overall effectiveness and reduce losses. For example, if availability is 98% and capability is 65%, the opportunity for improving capability is likely much greater than for improving availability.

As shown in Figure 2.7, system effectiveness equations (Effectiveness/LCC) are helpful for understanding benchmarks; past, present, and future status, and trade-off information (Barringer and Humble, 1998).

A company that finds itself in the lower right-hand corner of the trade-off square will get more "bang for its buck" (Weisz, 1996). Those in the upper left-hand corner will not be so happy. The remaining two corners raise questions about worth and value.

The system effectiveness equation is useful for trade-off studies (Brennan et al., 1985) as shown in the outcomes in Figure 2.8.

System effectiveness equations have an enormous impact on LCC because decisions made in the early periods of a project largely determine the final LCC. In fact, about two thirds of the total LCCs are fixed during project conception (Followell, 1995; Yates, 1995). Even though there will be later costs (Brennan et al., 1985), the chance to influence LCC cost reductions (Blanchard, 1991) grows smaller over time, as shown in Figure 2.9.

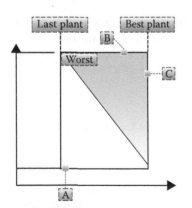

Parameter	A Last plant	B New plant	C Best plant
Availability	0.95	0.95	0.98
Reliability	0.3	0.4	0.6
Maintainability	0.7	0.7	0.7
Capability	0.7	0.6	0.8
Effectiveness	0.14	0.22	0.25
LCC	80	100	95

FIGURE 2.7
Benchmark data shown in trade-off format. (Adapted from Barringer, H.P. and Humble, T.X., Life cycle cost and good practices, in: *Presented at the NPRA Maintenance Conference*, San Antonio Convention Center, San Antonio, TX, May 19–22, 1998, http://www.barringer1.com/pdf/lcc_gp.pdf.)

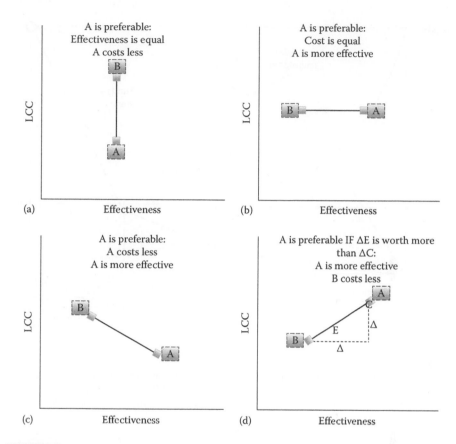

FIGURE 2.8
Possible outcomes of trade-off studies, LCC versus effectiveness. (Adapted from Barringer, H.P. and Humble, T.X., Life cycle cost and good practices, in: *Presented at the NPRA Maintenance Conference*, San Antonio Convention Center, San Antonio, TX, May 19–22, 1998, http://www.barringer1.com/pdf/lcc_gp.pdf.)

Figuratively speaking, engineering creates a cost funnel, and production/maintenance pours money into the funnel. Thus, LCC should be considered early on, when the final outcome can be influenced for better business results. Making major changes in LCC when the project is turned over to production is not possible because the die has been cast.

Breaking poverty cycles of buying cheap assets or systems and repairing them often at great expense can be accomplished in two ways: (1) using LCC techniques, or (2) extending capital projects for at least 8 years. In this way, new projects will be designed for the least long-term costs of ownership and build wealth for the company and its stockholders. Either method is effective because thoughtful value judgments are used, but both take work (Barringer and Humble, 1998).

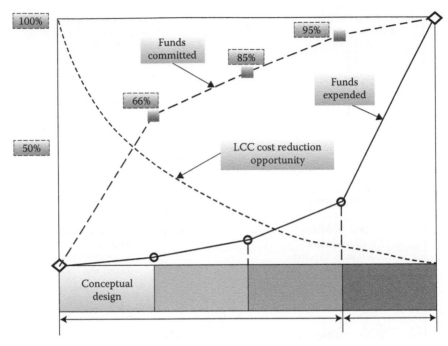

FIGURE 2.9
Funding trends by commitment and expenditure. (Adapted from Barringer, H.P. and Humble, T.X., Life cycle cost and good practices, in: *Presented at the NPRA Maintenance Conference*, San Antonio Convention Center, San Antonio, TX, May 19–22, 1998, http://www.barringer1.com/pdf/lcc_gp.pdf.)

2.4 LCC Analysis as Maintenance DSS

LCC analysis helps change perspectives on business issues with an emphasis on enhancing economic competitiveness by working toward the lowest long-term cost of ownership. Too often parochial views result in ineffective actions best characterized by short-term cost advantages and long-term cost disadvantages (Barringer et al., 1996).

Owners, users, and managers need to make decisions on the acquisition and ongoing use of many different assets, including items of equipment and the facilities that house them. The initial capital outlay cost is usually clearly defined and is often a key factor influencing the choice of the asset, given a number of alternatives from which to select.

The initial capital outlay cost is, however, only a portion of the costs over an asset's life cycle that needs to be considered to make the right choice for asset investment. As indicated earlier, the process of identifying and documenting all the costs involved over the life of an asset is called life cycle costing (LCC) (Water Research Foundation (WERF), 2011a).

A related concept, total cost of ownership (TCO), refers to all the costs associated with the use of assets including the administrative costs, training and development, maintenance, technical support, and any other associated costs. In fact, the TCO of an asset is often far greater than the initial capital outlay cost and can vary significantly between several different possible solutions that satisfy a given operational need (Voogt and Knezek, 2008).

TCO analyses serve as planning tools: until you know what you own it is hard to come up with a plan to reduce costs or to make better use of the available resources. This section outlines the purposes, reasons, and benefits of undertaking a TCO, and proposes some contextual issues that ought to be considered in the process (Moyle, 2004).

TCO analyses are undertaken for a variety of purposes:

- Identifying the components of an asset deployment
- Enabling calculations of what the total assets are worth
- Allowing the weighing of options
- Helping in the management of risk
- Enabling analysis from a system or whole school perspective

Examining TCO components and frameworks is important for the following reasons:

- The role of current assets should be outlined.
- Purchases and other associated costs (e.g. professional development) should match the roles required of the asset in the facilities, and deliver a return on investment (ROI).
- Decisions about what purchases can best be handled at different levels can be made based on documented TCO analyses.
- Different hierarchical levels must budget over time according to goals and standards.
- Completed TCO frameworks provide a basis upon which to monitor costs over time.

The following are some of the immediate benefits of performing TCO:

- Provides leaders, managers, and administrators with an oversight of expected costs
- Provides a way to measure changes and improvements in asset technology using agreed-upon benchmarks
- Enables the development of budgetary guidelines
- Supports the development of informed understandings of all costs required to support the use of an asset
- Enables insight into the longer-term costs of particular models of assets

- Enables identification of the direct and "hidden" costs of asset deployment
- Facilitates decision-making between different choices available, based on agreed-upon benchmarks
- Provides the basis for the development of business cases for asset investments
- Supports decision-making about the pros and cons of centralized deployment and management versus site-based strategies (UKESSAYS, 2015)

TCO data are context- or location-specific, and should not be undertaken in isolation to other activities occurring within the school or jurisdiction.

- The key to understanding the financial aspect of the use of assets is consideration of the range of viable options for the investments to be made.
- Business case analyses of viable solutions should include the identification of both quantifiable and intangible benefits for those respective potential solutions.

In summary, TCO is a locally focused index which has no sense in isolation inside the facility or with a lack of benchmarking capabilities. Indeed, if the end user has no chance to benchmark, the TCO value remains void.

But when this opportunity of benchmarking exists either in a harmonized or unstructured way, the asset user can consider the costs over the whole life of an asset, providing a sound basis for decision-making. With this information, it is possible to do the following:

- Assess future resource requirements by projecting itemized line item costs for relevant assets
- Assess comparative costs of potential acquisitions (investment evaluation or appraisal)
- Decide between sources of supply (source selection)
- Account for resources used now or in the past (reporting and auditing)
- Improve system design with improved understanding of input trends such as manpower and utilities over the expected life cycle
- Optimize operational and maintenance support through a more detailed understanding of input requirements over the expected life cycle
- Assess when assets reach the end of their economic life and if renewal is required by understanding changes in input requirements, such as manpower, as the asset ages

The life cycle costing process can be as simple as a table of expected annual costs or it can be a complex model that allows the creation of scenarios based on assumptions about future cost drivers. The scope and complexity of the LCCA should reflect the complexity of the assets under investigation, the ability to predict future costs, and the significance of the future costs to the decision being made by the organization.

An LCCA involves the analysis of the costs of a system or a component over its entire life span. Typical costs are acquisition, operating, and maintenance (Water Research Foundation (WERF), 2011a).

As many of the underlying possible cost drivers are connected with operations and maintenance, it is important to map the factors that influence maintenance and operations throughout the life cycle of the product before purchasing a new one.

2.4.1 Maintenance as a Value Driver

All types of assets, including manufacturing equipment, buildings, structures, and infrastructure, pass through a series of stages during their lifetime: planning, design, construction, operations and maintenance, renewal and revitalization, and disposal, as shown in Figure 2.10. The TCO is the total of all expenditures over all phases (National Research Council (NRC), 2008).

However, the amounts and distributions of expenditures are not equal. Design and construction require large capital expenditures, typically last fewer than 5 years, and account for 5%–10% of the TCO. In contrast, the operations and maintenance of assets may require annual expenditures for 30 or more years and account for as much as 80% of the TCO (National Research Council (NRC), 2008).

Assets are composed of many separate but interrelated systems, including construction, mechanical, electrical, communications, security and

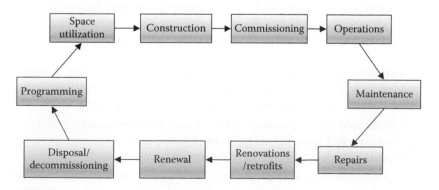

FIGURE 2.10
Asset life cycle model. (From National Research Council (NRC), *Core Competencies for Federal Facilities Asset Management Through 2020*, The National Academies Press, Washington, DC, 2008.)

safety, and control systems, as well as information technologies. Each type of system is composed of individual components (valves, switches, coils, drainage pans, materials, etc.), all of which must be kept in good working order if assets are to perform as intended. An asset's overall performance is a function of the interactions of its various systems and components, of the interactions with other assets, of the original design, and of operations and maintenance procedures.

How long asset systems and components actually perform at a satisfactory level (service life) depends on many factors, including the quality of the original design, durability of materials, incorporated technology, location and climate, use and intensity of use, and amount and timing of investment in maintenance and repair activities. The service lives of asset systems and components can be optimized or at least improved by timely and adequate maintenance and repairs.

Conversely, when maintenance and repair investments are not made, the service lives of asset systems and components are shortened.

2.4.2 Typical Outcomes of Investments in Maintenance and Repair

A typical maintenance and repair program includes several types of activities addressing different aspects of asset systems and components, with different objectives and outcomes. Maintenance is typically a continuous activity of routine work accomplished on a recurring basis and including some minor repairs. More important and often more expensive repair requirements are typically identified as separate projects. When asset managers identify specific maintenance and repair requirements in funding requests, the funding for maintenance activities is generally presented as a lump sum, with individual repair projects above some threshold identified separately.

Projects identified as required but not funded make up the bulk of the backlog of deferred maintenance and repair projects (National Research Council (NRC), 2012); see Figure 2.11.

Maintenance and repair activities include the following:

- Preventive maintenance includes planned or scheduled maintenance, periodic inspections, adjustments, cleaning, lubrication, parts replacement, and minor repair of equipment and systems.
- Programmed major maintenance includes maintenance tasks whose cycle exceeds 1 year (painting, roof maintenance, road and parking lot maintenance, utility system maintenance, etc.).
- Predictive testing and inspection activities involving the use of technologies to monitor the condition of systems and equipment and to predict their failure.

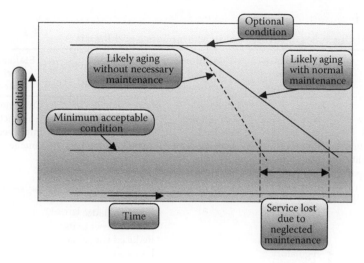

FIGURE 2.11
Impact of adequate and timely maintenance and repairs on the service life of an asset. (Adapted from National Research Council (NRC), *Predicting Outcomes of Investments in Maintenance and Repair of Federal Facilities*, The National Academies Press, Washington, DC, 2012, https://www.nap.edu/read/13280/chapter/4#32.)

- Routine repairs to restore a system or piece of equipment to its original capacity, efficiency, or capability.
- Emergency service calls or requests for system or equipment repairs that, unlike preventive maintenance work, are unscheduled and unanticipated.

All these activities and projects are intended to do one or more of the following:

- Prevent a situation, such as breakdown or failure, that could result in unplanned outages and downtime, disrupting the delivery of programs or operations undertaken in support of organizational missions or possibly resulting in the loss of life, property, or records.
- Comply with legal regulations on safety and health.
- Extend the service life of asset systems and components.
- Upgrade the condition of asset systems and components to bring them back to original operating performance.
- Avoid higher future costs through timely investment and efficient operations.
- Respond to stakeholder requests.

TABLE 2.1

Beneficial Outcomes of Investment in Maintenance and Repair

Mission-Related Outcomes	Compliance-Related Outcomes	Condition-Related Outcomes	Efficient Operations	Stakeholder-Driven Outcomes
Improved reliability Improved productivity Functionality Efficient space utilization	Fewer accidents and injuries Fewer building-related illnesses Fewer insurance claims, lawsuits, and regulatory violations	Improved condition Reduced backlog of deferred maintenance and repairs	Less reactive, unplanned maintenance and repair Lower operating costs Lower life cycle costs Cost avoidance Reduced energy use Reduced water use Reduced greenhouse gas emissions	Customer satisfaction Improved public image

Source: Adapted from National Research Council (NRC), *Predicting Outcomes of Investments in Maintenance and Repair of Federal Facilities*, The National Academies Press, Washington, DC, 2012, https://www.nap.edu/read/13280/chapter/4#32.

Because of the interrelated nature of the systems and components embedded in assets, maintenance, or repair of one system or component can result in improvements in others. For that reason, investments in maintenance and repair can result in multiple outcomes that achieve several purposes. Maintenance and repairs reducing energy and water use, for example, will also lower operating costs and provide more efficient operations.

The beneficial outcomes of maintenance and repair investments are shown in Table 2.1 and described later. Note: they are grouped by their primary purposes but an outcome can be related to more than one purpose (National Research Council (NRC), 2012).

2.4.3 Mission-Related Outcomes

2.4.3.1 Improved Reliability

Users require reliable supplies of power, heating, ventilation, air-conditioning, water, and other services to achieve their goals. For some assets, these services are required 24 × 7 to keep processes running, ensure people are safe and comfortable, or power equipment and computers.

Maintenance and repair activities are undertaken to ensure that mechanical, electrical, and other systems are reliable and can perform without substantial interruptions, so companies and their services can operate continuously both on a routine basis and during and after disasters or crises (National Research Council (NRC), 2012).

2.4.3.2 Improved Productivity

Maintenance and repair activities that support reliability also support improved productivity. Productivity for an individual or an organization has been defined as the quantity and/or the quality of the product or service delivered (Boyce et al., 2003). Productivity is most easily measured in manufacturing or similar functions where some number of units (such as cars or computer chips) with a given value can be expected to be produced per unit time. If production goals fail to be met because of equipment or mechanical downtime, it is relatively easy to assign a monetary value to the loss. For example, the number of units that are not produced because of downtime can be multiplied by the sales value or the profit margin to arrive at a monetary value of lost productivity (National Research Council (NRC), 2012).

Figure 2.12 shows the connection between profitability and maintenance expenses and/or investments (Kans, 2008).

2.4.3.3 Functionality

Functionality is an assessment of how well a facility functions in support of organizational goals. It also addresses the capacity to carry out planned activities (National Institute of Building Sciences (NIBS), 2008). Functionality loss, which is independent of condition, results from technical obsolescence, changes in user requirements, and changes in laws, regulations, and policies.

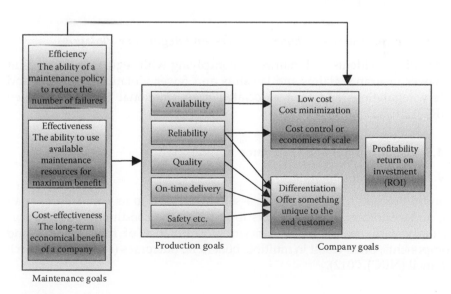

FIGURE 2.12
Connection of maintenance and profitability. (Adapted from Kans, M., On the utilization of information technology for the management of profitable maintenance, PhD dissertation (Terotechnology), Växjö University Press, Acta Wexionensia, Växjö, Sweden, 2008.)

Thus, an asset can be in good condition but inadequate for its function.

Obsolete assets that are still in use may fail to support organizational goals adequately; they may siphon off large amounts of money for maintenance and repair. In some cases, it is more cost-effective to dispose of/demolish an obsolete asset and replace it with a state-of-the-art asset than to renovate and continue to operate it (National Research Council (NRC), 1993).

2.4.4 Compliance-Related Outcomes

All organizations must comply with an array of safety and health regulations or face penalties for not doing so. These regulations are intended to protect the health, safety, and welfare of workers and the public.

Maintenance and repair activities undertaken to comply with regulatory standards include the replacement of obsolete or worn-out components to bring them up to current standards and codes, the installation or modification of equipment to support accessibility for workers, preventive maintenance, and testing of life-safety systems (National Research Council (NRC), 2012).

2.4.4.1 Fewer Accidents and Injuries

Maintenance and repair investments are made to protect the safety of asset users by eliminating hazards that can lead to accidents and injuries (National Research Council (NRC), 2012).

2.4.4.2 Fewer Insurance Claims, Lawsuits, and Regulatory Violations

Preventing accidents and injuries by complying with regulations can result in fewer insurance claims and lawsuits and fewer violations of health and safety regulations and their associated costs (National Research Council (NRC), 2012).

2.4.5 Condition-Related Outcomes

2.4.5.1 Improved Condition

Condition refers to the state of an asset with regard to its appearance, quality, and performance. Investments to improve the condition of assets, particularly with respect to the efficient performance of their systems and components, often result in multiple beneficial outcomes (National Research Council (NRC), 2012).

2.4.5.2 Reduced Backlogs of Deferred Maintenance and Repair

The total deferred maintenance and repair work necessary to bring assets back to their original designed performance capability, including updates to meet current safety codes, can be expressed as a monetary value.

Deferred maintenance "implies that the quality and/or reliability of service provided by an asset on which maintenance has been deferred is lower than it should be, and thus the asset is not or will not later be adequately serviced. In the short term, deferring maintenance will diminish the quality of services. In the long term, deferred maintenance can lead to shortened asset life and reduced asset value" (American Public Works Association (APWA), 1992). Increasing backlogs of maintenance and repair projects cause financial problems for the investors (National Research Council (NRC), 2012).

2.4.6 Outcomes Related to Efficient Operations

2.4.6.1 Less Reactive, Unplanned Maintenance and Repair

Asset management is more efficient when maintenance and repair activities are planned and scheduled to prolong the service lives of existing components and equipment and to replace equipment before a breakdown results in adverse events. Manpower is wasted when a large portion of staff time is spent reacting to unexpected breakdowns (National Research Council (NRC), 2012).

2.4.6.2 Lower Operating Costs

Operating costs include such elements as energy and water use, custodial services, security, fire suppression and detection, alarm testing, and servicing. Timely maintenance and repair investments to ensure asset systems and components are operating properly can reduce energy and water use and costs and improve performance and environmental quality (National Research Council (NRC), 2012).

2.4.6.3 Lower Life Cycle Costs

In some cases, a modest investment in maintenance and repair can result in longer-term cost savings or extended service life. For example, the replacement of a low-efficiency component or subsystem could pay for itself in a few years through savings in energy costs (National Research Council (NRC), 2012).

2.4.6.4 Cost Avoidance

Cost avoidance results from making investments in the short term to avoid the need for larger investments later, a key objective of preventive maintenance activities. Examples include lubricating equipment components to avoid replacing the entire asset system or fixing minor leaks to avoid total replacement, applying protective coatings to avoid replacing equipment because of corrosion, or realigning equipment periodically to avoid shortening of service life due to wear and tear. Timely maintenance and repair can also avoid the need to keep large inventories of spare parts on hand and prevent unplanned service calls (National Research Council (NRC), 2012).

In the case of manufacturing processes, it is reasonable to characterize the value of process, equipment, and yield changes as cost savings. However, the value of life cycle management activities (i.e., sustainment) is usually quantified as a cost avoidance: "Cost avoidance is a cost reduction that results from a spend that is lower than the spend that would have otherwise been required if the cost avoidance exercise had not been undertaken" (Ashenbuam, 2006). A simpler definition of cost avoidance is a reduction in costs that have to be paid in the future to sustain a system.

The reason cost avoidance is used rather than cost savings is that if the value of an action is characterized as cost savings, then someone wants the saved money back. In the case of sustainment activities, there is no money to give back. Unfortunately, making business cases based on a future cost avoidance argument is often more difficult than making business cases based on cost savings; therefore, there is a greater need to be able to provide detailed quantification of sustainment costs. Requesting resources to create cost avoidance is not as persuasive an argument as cost savings or return on investment arguments (Sandborn, 2013). Chapters 3 and 5 expand the idea of avoided costs and overinvestments in maintenance as intangible costs with huge positive impact.

2.4.6.5 Reductions in Energy Use and Water Use

Maintenance and repair activities, such as the replacement of malfunctioning systems, the replacement of components with more efficient ones, and the replacement of worn-out components, can result in reductions in energy and water use. These actions often help companies meet various mandates for high-performance assets, including regulatory objectives (National Research Council (NRC), 2012).

2.4.7 Stakeholder-Driven Outcomes

Stakeholders in maintenance and repair investments include not only asset managers and users, but also investors and shareholders, who are responsible for investment oversight. Each group of stakeholders has different expectations for the outcomes that should be achieved by investing in the maintenance and repair of assets (Figure 2.13).

Customer satisfaction is an outcome related to the quality of services provided to asset users. Continuous and efficient operation of systems helps to create productive, safe, and healthy environments. Conversely, system

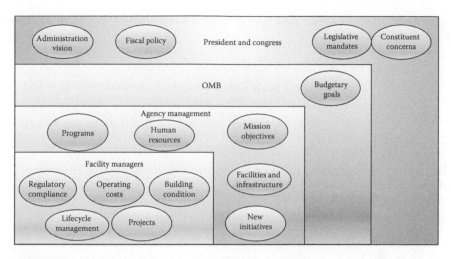

FIGURE 2.13
Stakeholders in investments in federal assets in the United States. (Adapted from National Research Council (NRC), *Investments in Federal Facilities: Asset Management Strategies for the 21st Century*, The National Academies Press, Washington, DC, 2004.)

failures cause work disruptions; inefficient operations or poor maintenance result in many adverse effects (National Research Council (NRC), 2012).

2.4.8 Risks Posed by Deteriorating Assets

The beneficial outcomes of maintenance and repair investments are related to the total resources invested and how these resources are used. Because the demands for resources for all kinds of users increase every year, priorities must be established for investments, and trade-offs must be made. Risk assessment is an important tool for decision-making in a resource-constrained operating environment.

Risk assessment processes have been used for many years and for many applications. Organizations typically use risk assessment to inform themselves and the public about hazards of their operations, including air and water quality, and the different actions or policy options available to manage the risks (National Research Council (NRC), 2009).

The essence of risk assessment as applied to asset components is captured by the three questions posed originally for the risk assessment of nuclear reactors by Kaplan and Garrick (1981):

1. What can go wrong?
2. What are the chances that something with serious consequences will go wrong?
3. What are the consequences if something does go wrong?

Equivalent questions for risk management were posed later by Haimes (1991):

1. What can be done and what options are available?
2. What are the associated trade-offs in terms of all costs, benefits, and risks?
3. What are the impacts of current management decisions on future options?

More recently, Greenberg (2009) framed the risk management questions as follows:

1. How can the consequences be prevented or reduced?
2. How can recovery be enhanced if the scenario occurs?
3. How can key local officials, expert staff, and the public be informed to reduce concern and increase trust and confidence?

Just as maintenance and repair investments can result in a plethora of beneficial outcomes, so too a lack of investment and the deferral of needed maintenance and repair projects can result in adverse events (what can go wrong).

Adverse events include more interruptions or stoppages of operations, more accidents, injuries, and illnesses, more lawsuits and insurance claims, increased operating costs, and shortened service lives of equipment and components.

Risk—a measure of the probability and severity of adverse effects—will increase for users and the assets and components will continue to deteriorate through wear and tear and lack of investment. (The likelihood of an event occurring is also related to geography, climate, and other factors). The risks associated with deteriorating assets are described in the following sections (National Research Council (NRC), 2012).

2.4.8.1 Risk to Users

The risks related to lack of reliability include unplanned interruptions and downtime of assets and components, the diversion of resources to excess, obsolete, or underutilized assets, and lowered productivity.

2.4.8.2 Risk to Safety, Health, and Security

Risks include increased injuries, illnesses, or even deaths involving company personnel or contractor personnel, as well as more lawsuits and claims; they result from asset-related hazards and failure to comply with regulations.

2.4.8.3 Risk to Efficient Operations

Risks include underperforming assets that drive up operating costs, customer dissatisfaction, cost avoidances, and other operational efficiencies.

2.4.8.4 Indexes and Models for Measuring Outcomes

Myriad indexes derived from models have been developed to measure outcomes related to asset condition, functionality, and performance.

Asset condition index (ACI): The ACI is a well-known and widely used condition index created from the ratio of two direct monetary measures: backlog of maintenance and repair (cost of deficiencies) and current replacement value (National Association of College and University Business Officers (NACUBO), 1991). Typically, when applied to an asset, the ratio ranges from 0 to 1, but it is sometimes multiplied by 100 to expand the range from 0 to 100.

Condition index (CI): The CI is defined as by the Federal Real Property Council as a general measure of a constructed asset's condition at a specific time. CI is calculated as the ratio of repair needs to plant-replacement value (PRV): CI = (1 − $ repair needs/$ PRV) × 100 (GSA, 2009). Repair needs represent the amount of money necessary to ensure the asset is restored to a condition substantially equivalent to the originally intended and designed capacity, efficiency, or capability. PRV is the cost of replacing an existing asset at today's standards (GSA, 2009). Like FCI, CI is a financial measure that is a proxy for physical condition.

Engineering research–based condition indexes: Engineering research–based indexes measure the physical condition of assets, their systems, and their components. These indexes are based on empirical engineering research and represent the driving engines for the sustainment management systems (SMS) (decision support systems and asset management systems) developed by the U.S. Army Corps of Engineers. The indexes can be applied to airfield pavements (Shahin et al., 1976; Shahin, 2005), roads and streets (Shahin and Kohn, 1979), railroad tracks (Uzarski et al., 1993), roofing (Shahin et al., 1987), and building components (Uzarski and Burley, 1997).

Each research-based condition index follows a mathematical weighted-deduct-density model in which a physical condition–related starting point of 100 points is established. Points are then deducted on the basis of the presence of various distress types (such as broken, cracked, or otherwise damaged systems or components), their severity (effect), and their density (extent).

To maximize the usefulness of these indexes, condition standards need to be established, that is, the point at which the component condition drops below a minimum desired value whereby production is adversely affected

TABLE 2.2

Asset Functionality Index Categories and Descriptions

Category	Description
Location	Suitability of asset location to mission performance
Asset size and configuration	Suitability of asset or area size and layout to the mission
Efficiency and obsolescence	Energy efficiency, water conservation
Environmental and life safety	Asbestos abatement, lead paint, air quality, fire protection, and similar issues
Missing and improper components	Availability and suitability of components necessary to support the mission
Aesthetics	Suitability of asset appearance
Maintainability	Ease of maintenance for operational equipment
Cultural resources	Historic significance and integrity issues that affect use and modernization

Source: Adapted from Grussing, M.N. et al., *ASCE J. Infrastruct. Syst.*, 371, December 2009.

or the risk of production impairment becomes unacceptable and triggers maintenance and repair requirements. The minimum value is a variable that depends on the asset, production goals, risk tolerance, redundancy, location, and other factors.

Building functionality index: Functionality is a broad term that applies to an asset, in this case a building, and its capacity to support an organization's programs and goals effectively. Functionality is related to user requirements (goals), technical obsolescence, and regulatory and code compliance, and it is independent of condition.

Asset functionality index (AFI): This index for assets follows the same form, format, and rating-scale development theory as engineering research–based physical condition indexes. However, rather than accounting for distresses, functionality issues are considered with respect to how severe (effect) and how widespread the issue is. The numerical AFI scale (0–100) is correlated to modernization needs. The model addresses 65 specific functionality issues, grouped into 14 general functionality categories, as shown in Table 2.2.

Asset performance index (API): The asset performance index combines the ACI and the AFI into a measure of the overall quality of an asset. The API is derived mathematically by taking the sum of two thirds of the lower of the ACI and AFI values and one third of the higher of the two values. The two thirds to one third split was derived through regression analysis and is intended to serve as a measure of rehabilitation needs (National Research Council (NRC), 2012).

Chapter 4 describes many different performance measurement techniques to quantify the financial dimension of maintenance costs and their impact on the overall performance.

2.5 Remaining Service Life as Gauge and Driver for Maintenance Expenses and Investments

Outcomes of maintenance and repair investments are measurable (either directly or through a model) by taking before and after measurements to gauge the effects of the investment. However, predictive models are needed to estimate the outcomes before investment (or in the absence of investment, i.e., the Do Nothing case). The prediction can be compared with the measured post-investment value to determine whether the expected outcome has been realized. Such predictions are crucial for performing a consequence analysis of maintenance and repair alternatives.

Modeling approaches have been developed to predict the remaining service lives of assets, including their systems and components, and to estimate the probability of failure; these models support risk-based decision-making related to the timing of maintenance and repair investments. Cost and budget models are also available to support the development of multiyear maintenance and repair programs (National Research Council (NRC), 2012).

2.5.1 Service Life and Remaining Service Life

Service life is the expected usable life of a component. At the end of the service life, replacement or major rehabilitation or overhaul is required. Remaining service life (RSL) is the time from today to when the service life will be expended.

Service life is based on many factors, including the manufacturer's test data, actual in-service data, and opinion based on experience. Service life is typically expressed as 5, 10, 15, 25, or 50 years. In reality, service life is variable because of operating environments, the magnitude and timing of maintenance, use and abuse of equipment, and other factors.

Knowing the service life and the RSL of a component is important to make decisions about the timing of investments and to plan maintenance and repair work on the asset in question. Service life and remaining service life models can also consider risk. If risk tolerance is high, maintenance and repair investments can be planned for the year in which the service life is expected to expire (or beyond). If risk tolerance is low, maintenance and repair may be planned to take place before the RSL expires. As risk tolerance decreases, maintenance and repair activities will be implemented sooner rather than later in relation to RSL (National Research Council (NRC), 2012).

Efforts have been made to determine the lives of assets as a whole, that is, all components comprising the asset. Such knowledge is useful for planning overall asset recapitalization and modernization, for computing commercial tax liability, and for appraising value (Whitestone Research and Jacobs Facilities Engineering, 2001).

There are two main remaining life indexes. Remaining service life is the difference between the current age and the predicted end of service life (Uzarski et al., 2007). Remaining maintenance life (RML) measures the time remaining before maintenance and repair should be performed (Uzarski et al., 2007).

2.5.2 Techniques for RSL Estimation and Maintenance Investment Outcomes

A variety of distribution models have been used to represent the probability of failure. For example, they are used in the railroad industry to predict defect formation in rails (Orringer, 1990). The presence of defects and the defect rate (defects/mile) are criteria for planning rail-defect testing and rail replacement. Because of public safety concerns, the risk tolerance for defect-caused rail breaks that result in derailments is very low (National Research Council (NRC), 2012).

2.5.2.1 Engineering Analysis

Traditional engineering analyses and models are used to predict remaining component life. Examples include fatigue analysis, wear-rate analysis, and corrosion effects on structural strength. Typically, components are analyzed only on an as-needed basis. Computations of stress or strain coupled with material properties (such as strength and dimensions) and operating conditions are often used in a model to estimate an outcome (National Research Council (NRC), 2012).

2.5.2.2 Cost and Budget Models

Predicting outcomes of maintenance and repair investments requires estimates of costs or budget needs and consequences. The traditional approach involves using cost estimates developed for individual projects. The expected outcomes of a list of projects can be married to the costs of the projects, and the outcome of maintenance and repair investment can be predicted. However, this approach has several problems. First, when the overall program and cost are being developed with planned outcomes, the projects to support the program may or may not have already been developed and their costs estimated. Second, the estimated costs may have been developed two or more years previously and may require updating. Third, developing project-level cost estimates can be expensive, so it is not typically done unless there is a high certainty of project funding. Finally, the costing approach is not particularly sensitive to what-if consequence analyses. Any change requires the cost estimator to compute iteratively the cost, incorporating the changes.

Parametric cost or budget models have been developed to overcome the problems of cost-estimating models. These models use cost correlated to particular measures to provide a reasonable estimate that is sufficiently accurate for planning purposes. They may be economic-based (involving, for example, depreciation or average service life) or engineering-based (involving, for example, actual and predicted condition or adjusted service life). Detailed project estimates are completed later in the project planning and execution process, when there is greater certainty about funding (National Research Council (NRC), 2012).

2.5.2.3 Operations Research Models

Operations research (OR) models have been applied to the management of some types of infrastructure, such as bridges (Golabi and Shepard, 1997). Although OR models are well suited to maintenance and repair investments, they are seldom applied. With OR techniques (there are many), an objective function (such as minimizing energy consumption) could be established subject to budget, labor, and other constraints. Multiple criteria can be considered, and an optimal mix of projects and a prediction of the outcomes can be identified.

Stochastic optimization models are used in situations where decision-makers are faced with uncertainty and must determine whether to act now or to wait and see. There is uncertainty in future asset deterioration, budget levels, the effects of maintenance and repair actions, and so on. Stochastic optimization models evaluate managerial recourse, and this provides an opportunity to fix problems if worst-case scenarios occur.

Another OR model, the Markov decision process (MDP), links asset condition and optimal long-term maintenance strategy. MDP models are used to manage networks of assets and components. The application of these models in asset management is limited, in that, for example, an asset's condition may result from several different deterioration processes and, thus, require different remedial actions. Deterioration processes and required remedial actions are difficult to define in a model (National Research Council (NRC), 2012).

2.5.2.4 Simulation Models

Simulation models are used to analyze the results of what-if scenarios and can be used instead of or in conjunction with OR models. Simulation models require more data than OR models, generally cannot guarantee an optimum solution (which OR solutions can), and can be very cumbersome to run (potentially long simulation times) (National Research Council (NRC), 2012). However, simulation models are generally more detailed than OR models and are used in practice for obtaining useful solutions to real problems, whereas the model simplifications required to obtain "convex" formulations in OR models result in OR models being unrealistically simplified representations of real systems in many cases.

2.5.2.5 *Proprietary Models*

The private sector has long been active in asset management and has collected large amounts of data. Some companies have used the data to develop asset management models. Generally, the models purport to predict condition and budgets and outcomes related to both. However, such models are proprietary, so little about how they work or about their assumptions, robustness, and accuracy is publicly known (National Research Council (NRC), 2012).

2.6 Uncertainty in LCC and Maintenance Cost Estimations

The work of the analyst is by no means finished when he or she has computed a series of economic measures. LCCA requires some thought as to what these measures mean and how they are going to be used. When we perform LCCA, we could be tempted to make "best-guess" estimates and use them in LCC equations as if they were certain. But investments are long-lived and necessarily involve at least some uncertainty about project life, operations and maintenance costs, and many more factors that affect project economics. If there is substantial uncertainty as to the validity of cost and time information, an LCCA may have little value for decision-making. It makes sense, then, to assess the degree of uncertainty associated with the LCC results and to take that additional information into account when making decisions (Boussabaine and Kirkham, 2004).

Even though we may be uncertain about some of the input values, especially those occurring in the future, it is still better to include them in the economic evaluation rather than basing a decision simply on the first costs. Ignoring uncertain long-term costs implies they are expected to be zero, a poor assumption to make.

As Craig points out, "A LCC analysis that does not include risk analysis is incomplete at best and can be incorrect and mis-leading at worst" (Craig, 1998). LCCA combined with risk analysis provides different decision scenarios where the consequences of the decision made are considered in depth.

Some of the available data might not be applicable to the situation because of different operational and environmental or other conditions; however, by using experts, comparison with similar systems, and parametric evaluations techniques, a basis for a decision can be established. The goal is to optimize the LCC.

Failure mechanisms, effect and criticality analysis (FMECA), fault tree analysis (FTA), event tree analysis (ETA), and hazard and operability analysis (HAZOP) are some of the tools and methods employed to reduce uncertainty and to optimize the asset life cycle with respect to maintenance and availability.

2.6.1 Approaches to Uncertainty in LCC

Numerous treatments of uncertainty and risk appear in the technical literature. Table 2.3 lists approaches used to assess uncertainty in investment decisions. When decision-makers are faced with an investment choice under uncertain conditions, they are mostly concerned about accepting a project whose actual economic outcome might be less favorable than what is acceptable. But there is also the risk of passing up a good investment. All techniques in the table provide information, albeit at different levels of detail, to account for this uncertainty. The techniques can be divided into two broad categories:

- Deterministic approaches use single-value inputs; they measure the impact of the project outcome to changing one uncertain key value or a combination of values at a time. The results show how a change in input value changes the outcome, with all other inputs held constant. The analyst determines the degree of risk on a subjective basis.

- Probabilistic approaches, by contrast, are based on the assumption that no single figure can adequately represent the full range of possible alternative outcomes of a risk investment. Rather, a large number of alternative outcomes must be considered, and each possibility must be accompanied by an associated probability. When probabilities of different conditions or occurrences affecting the outcome of an investment decision can be estimated, the expected value of a project's outcome can be determined. If the outcome is expressed in terms of a probability distribution, statistical analysis can measure the degree of risk (Fuller and Petersen, 1996).

No single technique in Table 2.3 is the "best" one in every situation. What is best depends on the relative size of the project, availability of data,

TABLE 2.3

Selected Approaches to Uncertainty Assessment in LCCA

Approaches to Uncertainty Assessment	
Deterministic	**Probabilistic**
1. Conservative benefit and cost estimating	1. Input estimates using probability distributions
2. Breakeven analysis	2. Mean-variance criterion and coefficient of variation
3. Sensitivity analysis	3. Decision analysis
4. Risk-adjusted discount rate	4. Simulation
5. Certainly equivalent technique	5 Mathematical/analytical technique
6. Input estimates using expected values	

Source: Adapted from Fuller, S.K. and Petersen, S.R., *Life-Cycle Costing Manual for the Federal Energy Management Program*, Building and Fire Research Laboratory, Office of Applied Economics, Gaithersburg, MD, 1996.

availability of resources (time, money, and expertise), computational aids, and user understanding. In this chapter, we primarily discuss sensitivity analysis and break-even analysis—deterministic approaches to uncertainty assessment. They are easy to perform and easy to understand and require no additional methods of computation beyond the ones used in LCCA. Since probabilistic methods have considerable informational requirements, they make uncertainty assessment much more complex and time-consuming, and before embarking on this course, it makes sense to test the sensitivity of the analysis results to any changes in input values.

This is not to say we should not use probabilistic methods if there is a serious question about the certainty of cost and time data, provided the size of the project or its importance warrants their use.

For the reference of readers, Marshall (1988) describes the techniques listed in Table 2.3 at greater length, discussing the advantages and disadvantages of each technique to help the decision-maker choose the appropriate one for any given problem.

As can be seen in Table 2.3, and as explained in the previous section, LCC calculation depends on many factors that are not explicit, and each cost generates an uncertainty that must be addressed and studied.

2.6.2 What Uncertain Variables Go into Life Cycle Costs?

The process of LCC is shown schematically in Figure 2.14.

LCC can also be represented as a very simple tree based on the costs for acquisition and the costs for sustaining the acquisition during its life, as shown in Figure 2.15.

Acquisition and sustaining costs are not mutually exclusive. If we acquire equipment or start a process, these always require extra sustainment costs. Acquisition and sustaining costs are determined by gathering the correct inputs, building the input database, evaluating the LCC, and conducting sensitivity analysis to identify the cost drivers.

The cost of sustaining equipment is frequently 2–20 times the acquisition cost. As an example, consider the cost for a simple American National Standards Institute (ANSI) pump. The power cost for driving the pump during its lifetime is many times larger than the acquisition cost. Are pumps bought with an emphasis on energy-efficient drivers and energy-efficient rotating parts, or is their acquisition simply based on the cheapest purchase price? The oft-cited rule of thumb is that 65% of the total LCC is set when the equipment is specified! This means we should not take the specification process lightly. The first obvious cost (pump acquisition) is usually the smallest amount of cash that will be spent during the life of the acquisition, and most sustaining expenses are not obvious.

Every example has its own unique set of costs and problems to solve if we want to minimize LCC. Remember that we want to minimize LCC because

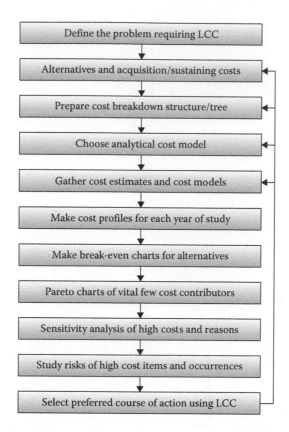

FIGURE 2.14
Life cycle costing process. (Adapted from Barringer, P.H. and Humble, T.X., A life cycle cost summary, in: *International Conference of Maintenance Societies* (ICOMS®-2003), 2003.)

this pushes up the net present value (NPV) of the asset and creates wealth for stockholders. But doing so requires finding details for both acquisition and sustaining costs—not an easy task, by any means (Barringer and Humble, 2003).

To continue the tree analogy, acquisition costs can be considered to have several different branches, as shown in Figure 2.16.

Each branch of the acquisition tree has other branches described in detail elsewhere (Fabrycky and Blanchard, 1991). Sustaining costs also have several branches, as shown in Figure 2.17.

What cost goes into each branch of the acquisition and sustaining branches? Simply stated, it all depends on the specific case and is generally driven by common sense. Building a nuclear power plant to generate electricity obviously requires special categories for each item of acquisition cost and sustaining cost. Building a pulp and paper mill or modifying coker drums at a refinery to prevent characteristic overstress that occurs

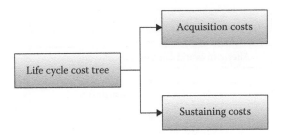

FIGURE 2.15
Top levels of LCC tree. (Adapted from Barringer, P.H. and Humble, T.X., A life cycle cost summary, in: *International Conference of Maintenance Societies* (ICOMS®-2003), 2003.)

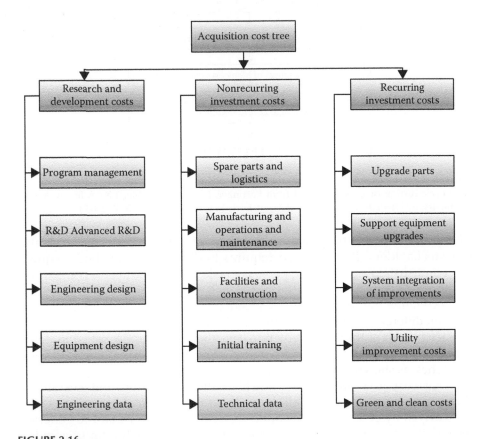

FIGURE 2.16
Acquisition cost tree. (Adapted from Barringer, P.H. and Humble, T.X., A life cycle cost summary, in: *International Conference of Maintenance Societies* (ICOMS®-2003), 2003.)

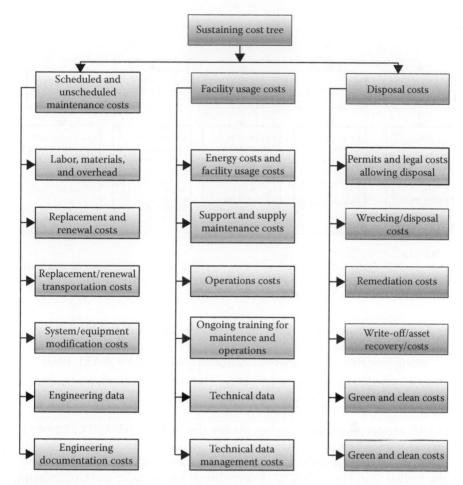

FIGURE 2.17
Sustaining cost tree. (Adapted from Barringer, P.H. and Humble, T.X., A life cycle cost summary, in: *International Conference of Maintenance Societies* (ICOMS®-2003), 2003.)

during quench cycles will have different cost structures than building a nuclear reactor. Analysts should include the appropriate cost elements and discard those elements that do not substantially influence LCC (Barringer and Humble, 2003).

Raheja (1991) lists the following alternative LCC models:

1. LCC = nonrecurring costs + recurring costs.
2. LCC = initial price + warranty costs + repair, maintenance, and operating costs to end users.

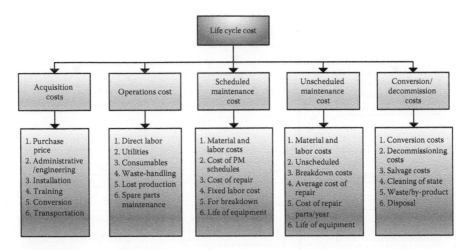

FIGURE 2.18
Model of LCC. (Adapted from Society of Automotive Engineers (SAE), *Reliability and Maintainability Guideline for Manufacturing Machinery and Equipment*, SAE, Warrendale, PA, 1993.)

3. LCC = manufacturer's cost + maintenance costs and downtime costs to end users. SAE International also has an LCC model for a manufacturing environment (Society of Automotive Engineers (SAE), 1993).

4. LCC = acquisition costs + operating costs + scheduled maintenance + unscheduled maintenance + conversion/decommission.

The SAE model mentioned earlier breaks down the various LCC costs as shown in Figure 2.18.

The LCC models described here and more complicated models described in the British Standards BS-5760 (International Electrotechnical Commission 2004) include costs to suppliers, end users, and "innocent bystanders." In short, the costs are viewed from a total systems perspective. LCC will vary with events, time, and conditions. Many cost variables are not deterministic, but are truly probabilistic. This usually requires starting with arithmetic values for cost and growing the cost numbers into the more accurate, but more complicated, probabilistic values (Barringer et al., 1996).

2.6.2.1 Application of LCC Techniques for Machine/Equipment Selection

To give an example of a specific industry's use of LCC techniques for equipment selection, consider the mining sector. The mining industry is characterized by the fact that it does not produce any machines/equipment itself, as most are bought from suppliers. The equipment is often large and expensive, and operated for long hours in a harsh and demanding environment. Many of the applications are specialized and demand special operating skills.

Research shows that 60%–70% of mining system failures can be attributed to the design and construction, 25%–30% to operating procedures, and 5%–15% to maintenance (Kumar, 1990). If we look carefully, we will find that most of the problems are due to the designer's failure to foresee the system's workload/profile and working environment. Often the operation profile of the product is considered, but the working environment is not. Maintainability issues like repair time, repair cost, accessibility, diagnostic ability, and serviceability, should be planned and designed into the system from the beginning. The same argument is valid for reliability issues like downtime, uptime, availability, dependability, and so on. Often a company has the choice of doing cost–benefit balancing between R&M issues in the design phase.

In brief, most of the problems can be tracked down to the design phase. Considering maintenance at the design stage through R&M considerations will point out the most critical properties that will affect the system failure and repair characteristics throughout the life cycle. This emphasizes the importance of mapping how the equipment is to be operated and used, the skills of the operators, the maintenance strategy, and last, but maybe most important, the work environment in which the machinery is to be operated in before purchasing it. Although we have been talking about mining, all the above points can be more generally applied throughout industry.

The life cycle cost of a product can be depicted as in Figure 2.19. The thick line shows pre-exploitation costs and accumulated exploitation costs. If R&M issues are considered in the early phases of the product life cycle, it is normally expected that the investment cost and the lead time may increase, but the accumulated exploitation costs may be reduced, as the product will have a longer life span because it is designed with maintenance in mind. Further, including R&M considerations in the design stages may not cause

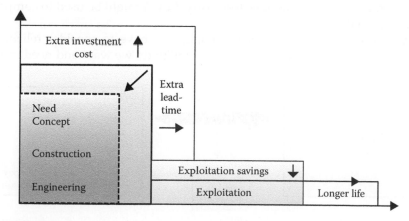

FIGURE 2.19
Including maintenance considerations in design. (Adapted from Van Baaren, R.J. and Smit, K., *Design for Reliability, Availability, Maintainability, Safety and Life Cycle Cost*, SDV Forum: Maintenance Engineering, Stavanger, Norway, February 1, 2000.)

an increase in investment cost and lead time, but actually decrease them, as shown by the dotted line in the figure (Van Baaren and Smit, 2000). Put otherwise, if the R&M issues were addressed early in the project, there would be fewer expensive and complicated design iterations.

To return to mining, if the operational and environmental characteristics are studied, degradation causes and cost drivers can be identified, studied, and evaluated; steps might even be taken to avoid the induced R&M cost by changing the design, choosing better material, removing rotating parts, choosing another design alternative, and so on (Markeset and Kumar, 2001). In the mining industry, the cost of maintenance is often very high, approximately 40%–60% (Kumar, 1990), so LCCA is essential when designing and selecting mining equipment and production systems. However, we stress that these points apply equally well elsewhere.

2.6.2.2 Application of LCC to Select Design Alternatives: To Design Out Maintenance or to Design for Maintenance

The manufacturer of a product has the choice to either design out maintenance or design for maintenance. Designing out maintenance, or creating a maintenance-free product, may be impossible because of technological limitations or cost. To design for maintenance means to balance cost and benefits with respect to R&M. The two options are rendered schematically in Figures 2.20 and 2.21. After having identified maintenance-related characteristics in the design, we may want to eliminate those characteristics that will incur maintenance costs. However, if maintenance is to be designed out, we will have to consider the cost of reliability (Markeset and Kumar, 2001).

Throughout the product's life cycle, we must also consider the state of the art of technology—lack of available technology might not allow us to eliminate maintenance, or it might be too costly. LCCA might be used to compare design alternatives.

If we decide to design out maintenance, we have to consider reliability against maintainability issues. What availability do we want and need in the

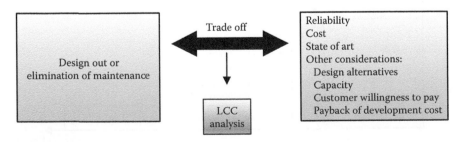

FIGURE 2.20
Design out maintenance. (Adapted from Markeset, T. and Kumar, U., R&M and risk-analysis tools in product design, to reduce life-cycle cost and improve attractiveness, *Proceedings Annual Reliability and Maintainability Symposium*, Philadelphia, PA, IEEE, 2001.)

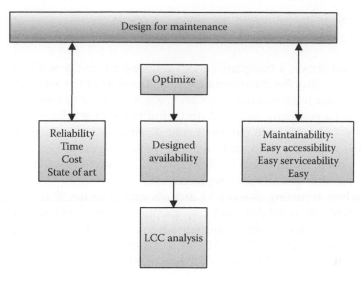

FIGURE 2.21
Design for maintenance. (Adapted from Markeset, T. and Kumar, U., R&M and risk-analysis tools in product design, to reduce life-cycle cost and improve attractiveness, *Proceedings Annual Reliability and Maintainability Symposium*, Philadelphia, PA, IEEE, 2001.)

product? What does the customer want and is willing to pay for? It is well known that increased reliability results in increased cost. If the state of the art does not allow us to design out maintenance, or if we are able to decide that the reliability is good enough throughout the product's life span, that is, acceptable availability is achieved, we may have a good design alternative. If we see that the reliability demand is too low, we have to consider maintainability issues such as accessibility to parts that need to be maintained, serviceability, and interchangeability of parts and systems (Blanchard and Fabrycky, 1997).

Many of these data may not be available to the purchaser. It is therefore important to specify to the manufacturer what data are needed and how they are to be used. An alternative would be to ask the manufacturer to analyze the LCC on the basis of both the manufacturer's and purchaser's data (Markeset and Kumar, 2000).

Sources of data for life cycle costing include the following:

- Engineering design data
- R&M data
- Logistic support data
- Production or construction data
- Consumer utilization data
- Value analysis and related data
- Accounting data

- Management and planning data
- Market analysis data

As suggested above, a company cannot increase reliability without increasing the cost or reducing the maintainability. The state of the art in technology is another factor to be evaluated. Often it is not possible to design out maintenance, and a company may end up trying to balance reliability, cost, and availability. Other ways to eliminate maintenance are to reduce capacity, to remove the weak functions, or to replace weak components by more robust ones. If a system/component is allowed to fail, a company will need to have provided for easy quick repair/replacement.

Thus, when adjusting designs to include maintenance, it is necessary to examine both the reliability and the maintainability characteristics. R&M are traded off to meet design requirements and budgetary constraints using LCCA. Any improvement in R&M will cost money, but may result in saving when the whole-life performance of the asset is taken into account (Markeset and Kumar, 2001).

2.7 LCC Data Acquisition and Tracking Systems

2.7.1 Data, Tools, and Technologies to Support Investments in Maintenance and Repair

Reliable and appropriate data and information are essential for measuring and predicting beneficial outcomes of investments in maintenance and repair and for predicting the adverse outcomes of a lack of investment. Data and information can be the basis of higher situational awareness during decision-making, of transparency during the planning and execution of maintenance and repair activities, of an understanding of the consequences of alternative investment strategies, and of increased accountability.

A 2004 National Research Council study stated that a portfolio-based asset management program requires the following elements (National Research Council (NRC), 2004):

- Accurate data for the entire portfolio of assets to enable life cycle decision-making
- Models for predicting the condition and performance of all assets
- Engineering and economic decision-support tools for analyzing trade-offs among competing investment approaches
- Performance measures to evaluate the effects of different types of actions (such as maintenance versus renewal) and to evaluate the timing of investments

This part of the chapter focuses on the data, tools, and technologies used to support portfolio-based asset management and strategic decision-making on investments in maintenance and repair. It is organized by data acquisition and tracking systems, indexes and models for measuring outcomes, and predictive models for decision support (Table 2.4).

The costs associated with data collection, analysis, and maintenance can be substantial. Costs will depend on the amount and accuracy of the data collected, how often they are collected, and the cost of the entire process, including data entry, storage, and staff time (National Research Council (NRC), 1998). Once data on an asset are created, it is necessary to update them throughout the asset's life cycle. The types of data to be maintained, their level of detail, and their currency, integrity, and attributes will depend on the outcomes to which they are related and the importance of the outcomes for strategic decision-making. The data on assets or systems that are mission-critical, for example, might need to be updated more often than data on less strategic assets.

Because of the costs, no data should be ignored. Every system and data item should be directly related to decision-making at some level. To the greatest extent possible, data should be collected in a uniform manner, and benchmarking should be used (National Research Council (NRC), 2012).

2.7.2 Technologies for Asset Data Management

Asset management data should include at least inventory data (e.g., number, locations, types, size of assets) that are relatively static once collected, and attribute data or characteristics that change (e.g., equipment and systems, condition, space utilization, maintenance history, value, age) (National Research Council (NRC), 2004). The systems described here are designed to assist asset managers gather and maintain accurate, relevant data throughout the asset life cycle. Some are traditional passive systems that rely on manual entry of data, while others collect data automatically in "real time" (National Research Council (NRC), 2012).

Systems used to collect data on assets include the following.

Computer-aided asset management (*CAAM*) *systems*: Computer-aided facility management systems have evolved over several decades and through several generations of technology. From the beginning, their focus has been on space planning and management and asset management. Applications now include energy and lease management, real estate management, maintenance and operations, and geographic information systems (GIS) integration. The latest incarnation of CAAM systems—integrated workplace management systems (IWMS)—emphasize the integration of all those applications with an organization's financial and human resources data systems. All aspects of the asset's life cycle—including planning, design, financial analysis and management, project management, operations, facilities management, and disposal—are accounted for in IWMS (National Research Council (NRC), 2012).

TABLE 2.4

Data, Tools, and Technologies to Support Strategic Decision-Making for Investment in Asset Maintenance and Repair

Data and Technologies	Primary Purpose
Data Acquisition and Tracking Systems	
Computer-aided asset management (CAAM) systems	Asset management
Computerized maintenance management systems (CMMS)	Maintenance-related work management
Asset automation systems	Monitoring and control of lighting, heating, ventilation, and other asset systems
Bar codes	Tracking of equipment, components, or other assets
Radio frequency identification (RFID) systems	Real-time asset tracking
Sensors	Monitoring of equipment and systems for vibration, strain, energy use, temperature, presence of hazardous materials, and the like
Condition assessments	Assessment of the physical condition of asset systems and components
Hand-held devices	Allowing inspectors to enter work management, condition, and other information directly into CAAM and CMMS
Automated inspections	Inspection of infrastructure, such as roads and railroads, using an array of technologies
Nondestructive testing	Monitoring of the condition of assets not visible to the human eye
Self-configuring systems	Control and other assets, such as HVAC, that can diagnose a problem and fix it with minimal human intervention
Indexes and Models for Measuring Outcomes	
Asset condition index (ACI)	A financial index based on a ratio of backlog of maintenance and repair to plan value or current replacement value
Condition index	A financial index based on a ratio of repair needs to plan value or current replacement value
Engineering research–based condition indexes	Physical condition indexes based on empirical engineering research and developed from models; indexes have been developed for buildings, some building components (such as roofs) and a variety of infrastructure, including railroad tracks, airfield pavements, and roads
Asset functionality index	An index to measure asset functionality in 14 categories; functionality requirements are independent of condition and are generally related to user requirements (mission), technological efficiency or obsolescence, and regulatory and code compliance
Asset performance index	An index based on the ratio of a physical asset condition index and an asset functionality index

(Continued)

TABLE 2.4 (*Continued*)

Data, Tools, and Technologies to Support Strategic Decision-Making for Investment in Asset Maintenance and Repair

Data and Technologies	Primary Purpose
Predictive Models for Decision Support	
Service life and remaining service life models	Models to predict the expected service life or remaining service life of systems and components; the purpose of these models is to help determine the appropriate timing of investments in maintenance and repair or replacement
Weibull models	Models that estimate the probability of failure of building or infrastructure systems or components
Engineering analysis	Analyses (such as fatigue analysis and wear-rate analysis) used to predict the remaining life of a system or component
Parametric models for cost estimating or budgeting	Economic-based (such as depreciation) or engineering-based (such as physical condition) models that can be used to develop multiyear maintenance and repair programs and cost estimates for annual budget development
Simulation models	Models used to analyze the results of "what if?" scenarios; an example is the integrated multiyear prioritization and analysis tool (IMPACT), which simulates the annual fiscal cycle of work planning and execution; it can be used to set priorities for maintenance and repair work based on different variables, including budget
Proprietary models	Asset models developed for numerous applications, including the prediction of outcomes of investments for maintenance and repair developed by private-sector organizations; relatively little information is publicly available about how they work or their assumptions, robustness, or accuracy

Source: Adapted from National Research Council (NRC), *Predicting Outcomes of Investments in Maintenance and Repair of Federal Facilities*, The National Academies Press, Washington, DC, 2012, https://www.nap.edu/read/13280/chapter/4#32.

Computerized maintenance management systems (CMMS): Computerized maintenance management systems have evolved, with maintenance-related work management as their main focus. Today's fully developed CMMS can be used for preventive maintenance scheduling, labor requirements, work-order management, material and inventory management, and vendor management. Most CMMS track asset-related materials, location, criticality, warranty information, maintenance history, cost and condition, component assembly, and safety information.

Note: One drawback of CMMS and CAAM is that data are often entered manually, and this increases the likelihood of error. In addition, the information available to decision-makers can be compromised in that manually updated data may not be entered on a timely basis. However, if the data are accurate, complete, and current, such systems provide a database that can be used for measuring outcomes, risk analysis, energy and condition assessment modeling, and investment decision support.

Real-time active asset-data acquisition systems: Several technologies bring "intelligence" to assets, systems, and components, allow automated data entry into CMMS and CAAM systems, and permit real-time monitoring of the performance of assets, systems, and components.

Asset automation systems (AAS): Asset automation systems are typically installed to monitor and control lighting, heating, ventilation, and air-conditioning (HVAC) systems, security systems, and life-safety systems, such as fire suppression. These control systems provide real-time feedback in the form of alarms based on operating characteristics (i.e., an alarm sounds when specified parameters are exceeded) and records of equipment performance. They are often the best source for early detection of equipment problems.

Bar codes: Bar codes involve an older technology that has evolved and become more complex with greater capabilities. When bar codes are placed on equipment, components, or other assets, information can be automatically scanned into a CMMS or CAAM system.

Radio frequency identification (RFID) systems: Radio frequency identification systems use two technologies: first, a radio tag consisting of a microchip that stores data on an object and an antenna that transmits data; second, a reader that creates the power for the microchip in the tag for passive RFID tags and then reads and processes the data from the tag. The radio wave data transmitted by the tag are translated into digital information, which, in turn, can be used by the software to record the status or location of an asset or component.

RFID tags can replace bar code systems for traditional inventory applications. Embedding a tab in an asset increases the durability and reusability of the tag. Using RFID tags with sensors substantially increases the number of applications for monitoring performance. Sensors are most often used to monitor motion (vibration) of and strain on assets and components or to monitor temperature. Such sensors (often in conjunction with AAS) are increasingly used for energy management systems. Other sensors can be used to detect the presence of radiation, chemicals, or other hazardous materials.

RFID systems are only now beginning to be used for asset management–related activities. Given their numerous advantages over traditional bar coding and their decreasing size and cost, it seems only a matter of time before they replace bar coding. Although some stand-alone RFID systems offer alerts, customized reports, and other features, the integration of RFID systems with real-time AAS or passive CAAM, CMMS, and IWMS makes possible a greater variety of applications, more efficient use of data, and more efficient operations.

Condition assessment: Condition is an underlying factor in the performance of most assets, systems, and components. It is also an important predictor of future performance: systems and components that are in good condition will be more reliable and perform better than those that are deteriorating. Condition assessments provide reference points for asset managers on the current condition of assets, systems, and components. Trends in condition can be used to determine whether asset systems and components are being maintained and are meeting their expected service lives or whether their performance is deteriorating faster than expected.

Information about the condition of assets, their systems, and their components can be gathered and updated using different approaches whose costs vary. The choice of approach will affect the availability, timeliness, and accuracy of data and, thus, affect the value of the data for strategic decision-making.

Depending on the information needed and how an organization uses the results, condition assessments can range from detailed assessments of individual components by engineers or technicians of various specialties to walk-through visual inspections by small teams. Condition assessments of any kind help to verify assumptions about asset conditions and to update records. The consistency and quality of condition assessments are also important in determining the usefulness of the data collected for decision support and priority setting.

Condition assessments undertaken on a multiyear cycle and conducted for an entire portfolio of assets can be inefficient and expensive and the information can quickly lose its value for decision-making (National Research Council (NRC), 2012).

Some organizations are taking a "knowledge-based" approach to condition assessment. The term knowledge-based indicates "that knowledge (quantifiable information) about an asset's system and component inventory is used to select the appropriate inspection type and schedule throughout a component's life cycle. Thus, inspections are planned and executed based on knowledge, not the calendar" (Uzarski et al., 2007).

Because different asset components have different service lives, and some may be more important than others with respect to outcomes and risks, some components are inspected more often than others and at different levels of detail. By tailoring the frequency and level of inspections, a knowledge-based approach makes better use of the available resources and provides more timely and accurate data to support investment-related decisions (Uzarski, 2006).

Automated inspections: Some technologies replace the human inspector for gathering condition-related data. One example is the International Road Roughness Method, used around the world by the highway industry (Sayers et al., 1986). For its part, the railway industry routinely uses laser optical sensors, accelerometers, displacement transducers, motion detectors, and gyroscopes to measure track quality under a moving load, looking for deviations that increase derailment risk and adversely affect operations.

Such technologies allow the collection of more data and higher-quality data at a fraction of the cost of manual data collection (Union Pacific, 2005).

Nondestructive testing: Nondestructive testing is sometimes used to collect condition-related data on components and systems not visible to the human eye. Many technologies can be used for these purposes: infrared thermography to detect excessive leaks, delamination, and defective areas and to map stress; ultrasonic testing and laser technology to detect cracks and other defects; and ground-penetrating radar to detect abnormalities in subsurface systems (National Research Council (NRC), 1998).

2.7.3 Emerging Technologies for Data Acquisition and Tracking

The technologies for various aspects of asset management are continually evolving and advancing. Three technologies that could substantially improve the acquisition and tracking of data, improve maintenance and repair activities, and provide support for decision-making are self-configuring systems, machine vision, and building information modeling (BIM) (National Research Council (NRC), 2012).

Self-configuring systems: A self-configuring (or self-healing) system is capable of responding to changing contexts in such a way that it achieves a target behavior by regulating itself (Williams and Nayak, 1996). The objective of self-configuring systems is to enable computer systems and applications to manage themselves with minimal but high-level guidance by humans (Parashar and Hariri, 2005).

Machine vision: Machine vision is an emerging technology for conducting inspections. It uses video imaging and computer software to detect component defects, such as cracks. The technology has been used for pavement inspection (Tsai et al., 2010) and railroad-car structural inspection (Schlake et al., 2010) and is under development for railroad track inspection (Resendiz et al., 2010).

Building information modeling (BIM): Building information modeling involves the generation and management of digital representations of physical and functional characteristics of places. The technology can be used to support decision-making about a place or structure (building, bridge, etc.) and/or its infrastructure (National Research Council (NRC), 2012).

2.8 Restriction of Maintenance Role in Operation Phase

In many companies, there is a permanent conflict between those responsible for production and those handling maintenance. Production workers complain that the attention received from the technicians responsible for maintenance does not correspond to the best possible practices, with an unfortunate effect on the production results. For their part, maintenance staff say production does not allow them to stop the assets to perform the necessary and often mandatory preventive maintenance activities; they are forced to make temporary interventions with poor reliability to accommodate the speed with which equipment must be redelivered to production. Further, the treatment of various assets by the production staff is far from ideal. In fact, both are often right (García, 2009).

The situation from the maintenance department's point of view:

> Production workers' treatment of assets is often out of line with the minimal care required by critical installations. No work procedures have been devised to ensure the good condition of assets; nor do production staff seem to understand that a problem in an asset is "the problem" not "someone else's problem."

The situation from production department's point of view:

> In many cases, the maintenance department does not use the best techniques, from either the corrective point of view or the preventive point of view. Workers make rapid interventions to "get by," with little concern for the ultimate solution to address the fault. They leave the provisional solution as the final solution until a new problem shows up. In addition, the maintenance department does not plan its interventions. And, in fact, in most companies where there is tension between production and maintenance, we find the maintenance is purely corrective. In such instances, maintenance generally employs a strategy of repairing faults as they emerge. The workplace quickly enters a vicious cycle: in the absence of preventive interventions, corrective interventions skyrocket, leaving even less time to think about preventive interventions. Degradation is quick, and the company becomes a battlefield: breakdowns are rampant, staff are discouraged, and production results are far from optimal.

Possible solutions to the conflict:

1. The heads of both departments should stop blaming the other department. Instead, they should start thinking about what they can do to turn the situation around. A new way of thinking, including a desire to improve the situation and taking some responsibility is fundamental to change.

2. Production operators cannot completely ignore the assets' maintenance. For example, they could do such things as lubricate equipment, monitor operating parameters, and clean the assets, as these activities do not require great technical knowledge. In fact, the most basic repairs can be done by production workers. This ensures their responsible use of assets and improves the care lavished on them. The basis of total productive maintenance (TPM) is to delegate some maintenance to the production staff.

3. Maintenance should perform interventions when they will interfere less with the production schedule. Maintenance should not be corrective, even though this maintenance strategy is useful in a very small number of cases. To preclude corrective maintenance, it is imperative to develop a series of reviews and inspections. Based on this, an intervention schedule can be designed, considering such things as pending repairs, inspections, cleaning, calibrations, and systematic replacement of worn parts. The time to perform maintenance must be restructured as well, as work should be done on nights, weekends, and so on, that is, at a time when production stops or at least slows down (García, 2009).

These solutions have been proven valid in many manufacturing environments, especially in the automotive sector, starting in Japan in the early 1970s. However, this problem reinforces the aspect of maintenance as something relevant in the O&M phase and irrelevant during the rest of the life cycle. It is still a challenge to "infect" the rest of life cycle stages with "maintenance concerns" to minimize maintenance cost and maximize productivity and availability performance.

References

American Public Works Association (APWA), 1992. *Plan. Predict. Prevent. How to Reinvest in Public Buildings.* Special Report No. 62, APWA, Chicago, IL.

Ashenbuam, B., 2006. Defining cost reduction and cost avoidance. CAPS Research.

Barringer, H. P., Weber, D. P., Humble, T. X., Mainville, O. H., 1996. Life cycle cost tutorial. In: *Fifth International Conference on Process Plant Reliability*, Houston, TX, October 2–4, 1996. Organized by Gulf Publishing Company and Hydrocarbon Processing. http://www.barringer1.com/pdf/lcctutorial.pdf (accessed April 30, 2017).

Barringer, H. P., Humble, T. X., 1998. Life cycle cost and good practices. In: *Presented at the NPRA Maintenance Conference*, San Antonio Convention Center, San Antonio, TX, May 19–22, 1998. http://www.barringer1.com/pdf/lcc_gp.pdf (accessed April 30, 2017).

Barringer, P. H., Humble, T. X., 2003. A life cycle cost summary. In: *International Conference of Maintenance Societies (ICOMS®-2003)*. http://www.barringer1.com/pdf/LifeCycleCostSummary.pdf (accessed April 30, 2017).

Blanchard, B. S., 1992. *Logistics Engineering and Management*, 4th edn. Prentice-Hall, Englewood Cliffs, NJ.

Blanchard, B. S., 1991. Design to cost, life-cycle cost. In: *1991 Tutorial Notes Annual Reliability and Maintainability Symposium*, available from Evans Associates, Durham, NC.

Blanchard, B. S., Fabrycky, W. J., 1990. *Systems Engineering and Analysis*, 2nd edn. Prentice-Hall, Englewood Cliffs, NJ.

Blanchard, B. S., Verma, D., Peterson, E. L., 1995. *Maintainability: A Key to Effective Serviceability and Maintenance Management*. Prentice-Hall, Englewood Cliffs, NJ.

Blanchard, B. S., Fabrycky, W. J., 1997. *Systems Engineering and Analysis*, 3rd edn. Prentice Hall International Series in Industrial and System Engineering. Prentice Hall, Englewood Cliffs, NJ.

Boussabaine, H. A., Kirkham, R. J., 2004. *Whole Life-cycle Costing—Risk and Risk Responses*. Blackwell Publishing Ltd., Oxford, U.K., 1-4051-0786-3.

Boyce, P. R., Hunter, C., Howlett, C., 2003. *The Benefits of Daylighting Through Windows*. Lighting Research Center, Troy, NY.

Brennan, J. R., Stracener, J. T., Huff, H. H., Burton, B. S., 1985. Reliability, life cycle costs (LCC) and warranty. Lecture notes from a General Electric in-house tutorial.

Craig, B. D., 1998. Quantifying the consequence of risk in life cycle cost analysis. In: *First International Industry Forum on Life Cycle Cost*, Stavanger, Norway, May 28–29, 1998.

Department of Energy (DOE), 1995. Posted 4/12/1995 on http://www.em.doe.gov/ffcabb/ovpstp/life.html (accessed April 30, 2017).

Fabrycky, W. J., Blanchard, B. S., 1991. *Life-Cycle Cost and Economic Analysis*. Prentice Hall, Inc., Englewood Cliffs, NJ.

Followell, D. A., 1995. Enhancing supportability through life-cycle definitions. In: *Proceedings Annual Reliability and Maintainability Symposium*, Washington, DC, available from Evans Associates, Durham, NC.

Fuller, S. K., 2010. Life-cycle cost analysis (LCCA). National Institute of Standards and Technology (NIST), Washington, DC. https://www.wbdg.org/resources/lcca. php (accessed April 30, 2017).

Fuller, S. K., Petersen, S. R., 1995. *Life-Cycle Costing Manual for the Federal Energy Management Program*, 1995 edn. NIST Handbook 135. U.S. Department of Commerce. Technology Administration National Institute of Standards and Technology, Washington, DC.

Fuller, S. K., Petersen, S. R., 1996. *Life-Cycle Costing Manual for the Federal Energy Management Program*. Building and Fire Research Laboratory, Office of Applied Economics, Gaithersburg, MD.

García, G. S., 2009. Conflict between operation and maintenance. OPEMASA general manager. http://www.mantenimientopetroquimica.com/en/maintenance-operationconflict.html (accessed April 30, 2017).

Golabi, K., Shepard, R., 1997. Pontis: A system for maintenance optimization and improvement of U.S. bridge networks. *Interfaces* 27(1), 71–88.

Grussing, M. N., Uzarski, D. R., Marrano, L. R., December 2009. Building infrastructure functional capacity measurement framework. *ASCE Journal of Infrastructure Systems* 15(4), 371–377.

Greenberg, M., 2009. Risk analysis and port security: Some contextual observations and considerations. *Annals of Operations Research* 187(1), 121–136.

GSA, 2009. Fiscal year 2008 federal real property report. GSA, Washington, DC. http://www.gsa.gov/graphics/ogp/FY_2008_Real_Property_Report.pdf (accessed April 30, 2017).

Haimes, Y. Y., 1991. Total risk management. *Risk Analysis* 11(2), 169–171.

Institute of Industrial Engineers (IEE), 1992. *Handbook of Industrial Engineering*, 2nd edn. Salvendy, G. (Ed.), John Wiley & Sons, New York.

International Electrotechnical Commission, 2004. IEC 60300-3-3: Dependability management—Part 3-3: Life cycle cost analysis—Application guide, International Electrotechnical Commission, Chicago, IL.

Kececioglu, D., 1995. *Maintainability, Availability, and Operational Readiness Engineering.* Prentice Hall PTR, Upper Saddle River, NJ.

Kans, M., 2008. On the utilization of information technology for the management of profitable maintenance. PhD dissertation (terotechnology). Acta Wexionensia, Växjö University Press, Växjö, Sweden.

Kaplan, S., Garrick, B. J., 1981. On the quantitative definition of risk. *Risk Analysis* 1(1), 11–27.

Kumar, U., 1990. Reliability analysis of load haul dump machine. PhD thesis: 88T. Luleå University, Luleå, Sweden.

Landers, R. R., 1996. *Product Assurance Dictionary.* Marlton Publishers, Marlton, NJ.

Lund, P., 1998. Life cycle costing as a decision tool. In: *First International Industry Forum on Life Cycle Cost*, Stavanger, Norway, May 28–29, 1998.

MacKay, D. J., Ronan, K., October 3, 2015. *LCCA for HVAC Systems.* KBA Building Technology, L.L.C., New York.

Markeset, T., Kumar, U., 2000. Application of LCC techniques in selection of mining equipment and technology. In: *Ninth International Symposium on Mine Planning and Equipment Selection*, Athens, Greece, November 6–9, 2000.

Markeset, T., Kumar, U., 2001. R&M and risk-analysis tools in product design, to reduce life-cycle cost and improve attractiveness. In: *Proceedings of Annual Reliability and Maintainability Symposium*, Philadelphia, PA. IEEE.

Marshall, H. E., September 1988. Techniques for treating uncertainty and risk in the economic evaluation of building investments. NIST Special Publication 757. U.S. Department of Commerce, National Institute of Standards and Technology, Gaithersburg, MD.

MIL-HDBK-259, 1993. *Military Handbook, Life Cycle Cost in Navy Acquisitions*, April 1, 1983, available from Global Engineering Documents, Washington, DC.

MIL-HDBK-276-1, February 3, 1984. *Military Handbook, Life Cycle Cost Model for Defence Material Systems*, Data Collection Workbook, Global Engineering Documents, Washington, DC.

MIL-HDBK-276-2, February 3, 1984. *Military Handbook, Life Cycle Cost Model for Defense Material Systems Operating Instructions*, Global Engineering Documents, Washington, DC.

Moyle, K., 2004. Total cost of ownership and open source software. Department of Education and Children's Services South Australia, Adelaide, South Australia, Australia.

National Association of College and University Business Officers (NACUBO), 1991. *Managing the Facilities Portfolio: A Practical Approach to Institutional Facility Renewal and Deferred Maintenance.* NACUBO, Washington, DC.

National Institute of Building Sciences (NIBS), 2008. *Assessment to the U.S. Congress and U.S. Department of Energy on High Performance Buildings.* NIBS, Washington, DC.

National Research Council (NRC), 1993. *The Fourth Dimension in Buildings: Strategies for Minimizing Obsolescence.* National Academy Press, Washington, DC.

National Research Council (NRC), 1998. *Stewardship of Federal Facilities: A Proactive Strategy for Managing the Nation's Public Assets.* National Academy Press, Washington, DC.

National Research Council (NRC), 2004. *Investments in Federal Facilities: Asset Management Strategies for the 21st Century.* The National Academies Press, Washington, DC.

National Research Council (NRC), 2008. *Core Competencies for Federal Facilities Asset Management Through 2020.* The National Academies Press, Washington, DC.

National Research Council (NRC), 2009. *Science and Decisions: Advancing Risk Assessment.* The National Academies Press, Washington, DC.

National Research Council (NRC), 2012. *Predicting Outcomes of Investments in Maintenance and Repair of Federal Facilities.* The National Academies Press, Washington, DC. https://www.nap.edu/read/13280/chapter/4#32. (accessed April 30, 2017).

Orringer, O., 1990. *Control of Rail Integrity by Self-Adaptive Scheduling of Rail Tests.* DOT/FRA/ ORD-90/05. U.S. Department of Transportation, Federal Railroad Administration, Office of Research and Development, Washington, DC.

Parashar, M., Hariri, S., 2005. Autonomic computing: An overview. *Lecture Notes in Computer Science* 3566, 257–269.

Pecht, M., 1995. *Product Reliability, Maintainability, and Supportability Handbook.* CRC Press, New York.

Raheja, D. G., 1991. *Assurance Technologies.* McGraw-Hill, Inc., New York.

RAMS, 2001. *Proceedings of Annual Reliability and Maintainability Symposium,* Philadelphia, PA. Cumulative Indexes, page cx-29 for LCC references, available from Evans Associates, Durham, NC.

Resendiz, E. Y., Molina, L. F., Hart, J. M., Edwards, J. R., Sawadisavi, S., Ahuja, N., Barkan, C. P. L., 2010. Development of a machine-vision system for inspection of railway track components. In: *Paper presented at the 12th World Conference on Transport Research,* Lisbon, Portugal. http://intranet.imet.gr/Portals/0/UsefulDocuments/documents/03355.pdf.

Sandborn, P., 2013. *Cost Analysis of Electronic Systems.* World Scientific, Singapore, 440pp.

Sayers, M. W., Gillespie, T. D., Queiroz, C. A. V., 1986. The international road roughness experiment: A basis for establishing a standard scale for road roughness measurements. *Transportation Research Record* 1084, 76–85.

Schlake, B. W., Todorovic, S., Edwards, J. R., Hart, J. M., Ahuja, N., Barkan, C. P. L., 2010. Machine vision condition monitoring of heavy-axle load railcar structural underframe components. *Proceedings of the Institution of Mechanical Engineers, Part F: Journal of Rail and Rapid Transit* 224(5), 499–511.

Shahin, M. Y., 2005. *Pavement Management for Airfields, Roads, and Parking Lots,* 2nd edn. Springer Science + Business Media, Dordrecht, the Netherlands.

Shahin, M. Y., Bailey, D. M., Brotherson, D. E., 1987. *Membrane and Flashing Condition Indexes for Built-Up Roofs,* Vol. 1: Development of the Procedure. Technical Report M-87/13/ADA190367. U.S. Army Corps of Engineers Construction Engineering Research Laboratory, Champaign, IL.

Shahin, M. Y., Darter, M. I., Kohn, S. D., 1976, *Development of a Pavement Maintenance Management System.* Vol. 1: Airfield Pavement Condition Rating. AFCEC-TR-76-27. U.S. Army Corps of Engineers Construction Engineering Research Laboratory, Champaign, IL. http://trid.trb.org/view.aspx?id=56392.

Shahin, M. Y., Kohn, S. D., 1979. *Development of a Pavement Condition Rating Procedure for Roads, Streets, and Parking Lots,* Vol. 1: Condition Rating Procedure. Technical Report M-268. U.S. Army Corps of Engineers Construction Engineering Research Laboratory. Champaign, IL, July.

Society of Automotive Engineers (SAE), 1995. *Life Cycle Cost, Reliability, Maintainability, and Supportability. Guidebook,* 3rd edn. SAE International, Warrendale, PA.

Society of Automotive Engineers (SAE), 1993. *Reliability and Maintainability Guideline for Manufcturing Machinery and Equipment.* SAE, Warrendale, PA.

Thuesen, G. J., Fabrycky, W. J., June 9, 2000. *Engineering Economy,* 9th edn. 637pp., Prentice-Hall International Series in Industrial and Systems, 978-0130281289.

Tsai, Y. C., Vivek, K., Merserau, R., 2010. Critical assessment of pavement distress segmentation methods. *ASCE Journal of Transportation Engineering* 136(1), 11–19.

UKESSAYS, 2015. Why open source for educational institute information technology essay. https://www.ukessays.com/essays/information-technology/why-open-source-for-educational-institute-information-technology-essay.php (accessed March 23, 2015).

Union Pacific, 2005. Union Pacific unveils $8.5 million state-of-the-art track inspection vehicle. http://www.uprr.com/newsinfo/releases/capital_investment/2005/1216_ec5.shtml (accessed April 30, 2017).

Uzarski, D. R., 2006. Deficiency vs. distress-based inspection and asset management approaches: A primer. *APWA Reporter* 73(6), 50–52.

Uzarski, D. R., 2004. *Knowledge-Based Condition Assessment Manual for Building Component Sections.* U.S. Army Corps of Engineers Construction Engineering Research Laboratory, Champaign, IL.

Uzarski, D. R., Burley, L. A., August 1997. Assessing building condition by the use of condition indexes. In: *Proceedings of the ASCE Specialty Conference Infrastructure Condition Assessment: Art, Science, Practice,* Reston, VA, pp. 365–374.

Uzarski, D. R., Darter, M. I., Thompson, M. R., 1993. *Development of Condition Indexes for Lowvolume Railroad Trackage.* Transportation Research Record (Transportation Research Board, National Research Council) 1381, Washington, DC, pp. 42–52.

Uzarski, D. R., Grussing, N. M., Clayton, J. B., 2007. Knowledge-based condition survey inspection concepts. *ASCE Journal of Infrastructure Systems* 13(1), 72–79.

Van Baaren, R. J., Smit, K., February 1, 2000. *Design for Reliability, Availability, Maintainability, Safety and Life Cycle Cost.* SDV Forum: Maintenance Engineering, Stavanger, Norway.

Voogt, J., Knezek, G., 2008. *International Handbook of Information Technology in Primary and Secondary Education. Part One.* Springer, New York. 978-0-387-73314-2.

Water Research Foundation (WERF), 2011a. *Overview: What is Life Cycle Costing?* GWRC, GHD Consulting, Inc., Alexandra, VA (accessed April 30, 2017).

Water Research Foundation (WERF), 2011b. Select a discount rate. GWRC, GHD Consulting, Inc., Alexandra, VA (accessed April 30, 2017). http://simple.werf.org/simple/media/LCCT/stepOneOneLinks/03.html.

Weisz, J., 1996. An integrated approach to optimizing system cost effectiveness. In: *Tutorial Notes Annual Reliability and Maintainability Symposium,* Las Vegas, NV, available from Evans Associates, Durham, NC.

Whitestone Research and Jacobs Facilities Engineering, January 2001. *Implementation of the Department of Defense Sustainment Model: Final Report.* DoD Washington, DC.

Williams, B., Nayak, P., 1996. A model-based approach to reactive self-configuring systems. *Proceedings of the National Conference on Artificial Intelligence* 2, 971–978.

Yates, W. D., 1995. Design simulation tool to improve product reliability. In: *Proceedings Annual Reliability and Maintainability Symposium,* Washington, DC, available from Evans Associates, Durham, NC.

3

Maintenance Budget versus Global Maintenance Cost

3.1 Asset Management and Annual Maintenance Budget

One of the basic responsibilities of a maintenance manager is to prepare the annual maintenance budget. Managers should approach this task with enthusiasm and confidence. However, if they come up with a forecast simply by taking the previous year's budget and adding a little here and there, they will be shirking their duty. Producing a meaningful financial forecast takes work.

Several factors should be considered in developing a thorough budget. First, a manager must know what the objectives are, what the shareholders expect, and what degree of excellence is demanded. The objectives and expectations can usually be found in the manager's job description or in the company's planning program policies. Maintenance managers should review the objectives periodically to be alerted to any changes in what the company wants. Without a definite understanding of the objectives, the manager cannot possibly prepare a sensible and functioning budget.

Second, the manager must consider labor costs. Much of the success of a maintenance program depends on the status of the working crew. It is also possible to determine whether too much time is being spent in a certain area by carrying out a time-use study.

Third, the manager must think about possible plans for change. Such plans should be classed as long-term or short-term. Long-term change is more expensive and, on occasion, requires outside consultants and contractual agreements.

Fourth, the manager should carefully evaluate the past year's actual expenses when preparing the new budget. Major variations between the previous budget and the actual expenditures should be thoroughly investigated to determine whether the problem might recur.

3.1.1 "Selling" the Maintenance Budget

The manager should provide a brief explanation for each of the budgeted items. A careful and accurate explanation is especially needed whenever large increases or decreases appear. The manager needs to explain how he or she arrived at the figures. It is also helpful when proposing a budget to individually list the past 4 or 5 years' actual expenditures in each of the budgeted areas. Remember also that a budget is much more meaningful when it is checked periodically. In other words, how much of each budgeted item is spent month by month.

The following points briefly sum up the importance of maintenance budgets:

1. First of all, the budget necessitates the establishment of a maintenance program and regulates and emphasizes performance within that program. We cannot prepare a budget without first establishing a complete program, after which changes in the program will reflect as variances in the budget.

2. Having a maintenance budget encourages participation and promotes understanding among company stakeholders. By witnessing the budgetary problems first-hand, they can come to a better understanding of the manager's position and the problems he or she faces.

3. Having a budget requires all expenditures to be specifically labelled, giving them an identity. In other words, expenses must be listed under a particular budgeted item.

4. The budget identifies immediate needs for the coming year. Budget variances point out areas of concern, such as too much labor being used in one area, or suggest the need to purchase a new piece of maintenance equipment.

5. Large budget variances reflect unusual and unexpected problems; these often occur when an asset is under stress.

6. A budget ensures continuity of operations, especially if there is a new manager.

7. A budget provides the maintenance manager with a vehicle to evaluate his or her maintenance program.

8. A good budget builds the manager's credibility with stakeholders.

9. Budgets regulate inventories. They restrict the purchase and storing of unneeded supplies.

3.1.2 Composition of Maintenance Budget

The maintenance budget refers to the portion of an operating budget that is set aside in a single fiscal year for maintenance activities on the organization's assets. The budget typically falls into the following categories:

- Routine maintenance: maintenance on an asset that is planned and performed on a regular and consistent basis.

- Time-based maintenance (TBM): maintenance based on a calendar schedule; that is, time is the trigger for maintenance. It is also called "clock-based" maintenance, "condition- independent" maintenance, or "calendar-based" maintenance.

3.1.2.1 Maintenance Budget Composition

To determine the composition of a maintenance budget, a department's maintenance funding needs should be split into the following cost components:

- Condition assessment costs
- Statutory maintenance costs
- Preventive maintenance costs
- Condition-based maintenance costs
- Unplanned maintenance costs
- Maintenance management costs

A definition for each of these budget components is presented in Table 3.1.

There should be appropriate mechanisms in place to ensure appropriate administrative and decision-making processes and systems for planning and monitoring the maintenance delivery.

Departments should structure ledgers or cost centers around the cost components listed above to get a clear indication of where maintenance funds have been spent (Maintenance Management Framework, (II), 2012).

3.1.3 Development of an Annual Maintenance Budget

Maintenance managers must take really good care when preparing the annual budget and be as accurate as possible.

A maintenance budget should be based on maintenance demand, as illustrated in Figure 3.1.

It should also support efforts to meet environmental and performance requirements. For example, sufficient funding should be allocated to replace components at the end of their useful life with modern equivalents that reduce energy and water consumption or have the potential to reduce long-term maintenance needs.

Differentiating between maintenance expenditure and capital expenditure is important because of the difference in accounting approaches (and tax liabilities) associated with these expenditures. For example, maintenance expenditure affects the cost of a department's outputs, while capital expenditure impacts the value of the department's assets and, subsequently, their depreciation and equity return. At times, the nature or intent of the work (or parts of the work) identified may extend beyond restoring an asset to its original condition, capacity, or function. In these cases, the expenditure may

TABLE 3.1

Maintenance Budget Components

Cost Component	Definition
Condition assessment cost	This is the cost of undertaking condition assessments in accordance with the MMF.
Statutory maintenance cost	This is the cost associated with undertaking maintenance to meet mandatory requirements of various regulations such as the servicing of fire protection systems.
Preventive maintenance cost	This is the cost associated with the periodic servicing of plant and equipment and preventative repairs to other building components to ensure reliable operation, comply with "duty of care" responsibilities and general good maintenance practice to preserve assets in a condition appropriate for service delivery.
Condition–based maintenance cost	Condition-based maintenance is maintenance undertaken as a result of deteriorated condition identified through condition assessments. In this regard, funding of this component is variable and less predictable.
Unplanned maintenance cost	Unplanned maintenance is reactive work undertaken as a result of breakdowns and routine failure of building components and services. Funding of this component of maintenance would fluctuate in varying degrees between agencies. However, historical data should provide guidance in terms of annual estimates of funding required.
Agency management cost	This is the cost incurred by agencies in managing maintenance and includes the costs of management personnel, maintenance management systems, financial administration and other overhead costs. Activities to be costed include: general management; administration; maintenance planning; program formulation; program management; and contract management (if maintenance is outsourced).

Source: Adapted from Maintenance Management Framework, (II), *Building Maintenance Budget*, 2nd edn., Department of Housing and Public Works, Brisbane, Queensland, Australia, 2012.

be more appropriately classified as a "capital outlay" (i.e., it increases the value of the asset on which the expenditure is incurred).

Establishing an adequate maintenance budget requires an understanding of many variables associated with maintaining assets, particularly when dealing with a portfolio that consists of a complex mix (i.e., buildings of different ages, varied geographical location/climate, intensity of use, and functional/service delivery requirements).

When formulating a maintenance budget, due consideration must be given to plans for the assets, including the following points:

- Existing assets to be maintained
- New assets requiring maintenance
- Existing assets to be upgraded or refurbished or to have components replaced (a "minimum maintenance" approach may be appropriate for these assets in the lead-up to the intended actions)

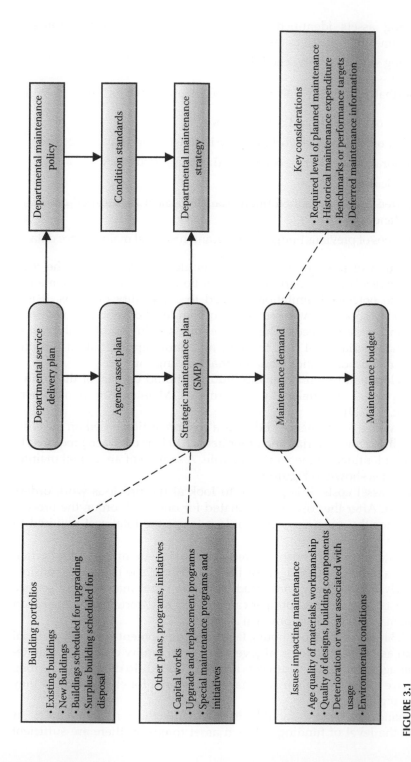

FIGURE 3.1

Development of annual maintenance budget. (Adapted from Maintenance Management Framework, (II), *Building Maintenance Budget*, 2nd edn., Department of Housing and Public Works, Brisbane, Queensland, Australia, 2012.)

- Existing assets identified for inclusion in special maintenance programs and initiatives (as applicable)
- Existing surplus assets scheduled for disposal

Key factors that impact the level of maintenance funding include the following:

- Quality of materials and components
- Quality of designs and workmanship in construction/manufacturing
- Deterioration or wear associated with usage/occupancy
- Climatic conditions
- Required level of planned maintenance to meet the desired condition standards
- Outcomes of previous budget reviews and historical maintenance data

The calculation of the funding required should be governed by the total maintenance needs of a department's portfolio (i.e., maintenance demand), not based on perceived limitations related to availability of funds. It is a department's responsibility to seek the required level of funding to address identified maintenance needs. This may require seeking additional funding from the managerial level (i.e., through an annual appropriation funding process) or reallocating funds from internal funding sources.

The development of a maintenance budget should be part of the annual budgetary process undertaken by departments (Maintenance Management Framework, (II), 2012).

Briefly stated, a maintenance budget is a cost projection based on the costs of labor, equipment, material, and other items (such as contracts) required to do all work identified in the work schedule. A sample of an annual maintenance budget is shown in Figure 3.2.

To project asset costs, it is possible to look at the previous work orders for that asset. After the costs are calculated for one work order, the process is repeated for the remaining work orders to get the total cost required to maintain the asset.

The maintenance supervisor is responsible for monitoring the actual expenditures against the budget for the year. He or she is also responsible for its yearly update using forecasted labor rates and material and service contract costs. The updated budget can be used to determine the operation and maintenance costs of all physical assets.

Figure 3.3 recaps total labor hours, labor cost, equipment cost, and material cost for an asset. At this point, all overhead costs, utility costs, and maintenance management supervision costs are also entered. To determine the total maintenance budget, a similar summary is prepared for each asset and they are added together.

Maintenance managers should discuss with their administrators or managers the level of funding for each asset to ensure there are sufficient

	Description	Crew Hours	Crew Size	Total Person Hours	LABOUR COST			TOTAL COST				
					Quantity	Rate	Cost	Labor	Equipment	Material		
					Hrs	$/hr	$	$	$	$	$	
	Daily Janitorial Activities	858,00	1.00	858,00	858,00	9.00	7722.00	7722.00	250.00	300.00		
	Weekly Janitorial Activitives	104.00	1.00	104.00	104.00	9.00	936.00	936.00	52.00	104.00		
	Monthly Janitorial Activities	36.00	1.00	36.00	36.00	9.00	324.00	324.00	24.00	100.00		
	Semi-Annual Janitorial Activities	7.00	1.00	7.00	7.00	9.00	63.00	63.00	10.00	5.00		
	Annual Janitorial Activities	4.00	1.00	4.00	4.00	9.00	36.00	36.00	0.00	0.00		
	Bldg&Equip. Repair Activities	24.00	1.00	24.00	24.00	9.85	236.00	236.00	30.00	500.00		
	Monthly Maintenance Activities	45.60	1.00	45.60	45.60	9.85	449.00	449.00	25.00	240.00		
	Quarterly Maintenance Activities	4.00	1.00	4.00	4.00	9.85	39.40	39.40	2.00	10.00		
	Semi-Ann. Maintenance Activities	12.00	1.00	12.00	12.00	9.85	118.20	118.20	10.00	50.00		
	Annual Maintenance Activities	10.10	1.00	10.10	10.10	9.85	99.49	99.49	10.00	50.00		
								Fire Extinguisher				
	Painting Activities	26.00	2.00	52.00	26.00	9.85	256.10					
					26.00	8.43	219.18	475.28	20.00	100.00		
	TOTALS			1156.70				10493.93	433.00	1459.00		

ANNUAL MAINTENANCE BUDGET-WORKSHEET

First Nation:_____   Site No.:_____   Date:_____

Asset Name: Day Care Center   Asset No.: 0050   Page:_____

FIGURE 3.2
Annual maintenance budget—worksheet. (Adapted from Maintenance Management Systems, Technical information document, RPS for INAC. TID-AM-01, October 2000.)

funds for an appropriate level of maintenance for every asset (Maintenance Management Systems, 2000).

3.1.4 Basis of a Maintenance Budget

A maintenance budget considers the amount of annual maintenance to be done, the type of maintenance, and the maintenance strategy (Maintenance Management Framework, (II), 2012).

ANNUAL MAINTENANCE BUDGET-WORKSHEET						
First Nation:_____ Site No.:_____ Date:_____						
Asset Name: Day Care Center Asset No.: 0050 Page:_____						
Description	Total Person Hour	TOTAL COST Labour $	Equipm $	Material $	Other contracts	Total $
Janitorial Activities	1009,00	9080,00	336,00	509,00	50,00	9976,00
Preventive Maintenance Activities	147,70	1417,94	97,00	950,00	175,00	2639,94
Sub-total:	1156,70	10498,94	433,00	1459,00	225,00	12615,94
Benefit Costs-Vacation Pay		419,96				419,96
Employment insurance		146,99				146,99
Worker Compensation		96,59				96,59
Utility Charges-Water					170,00	170,00
Hydro					3000,00	3000,00
Garbage Collection					150,00	150,00
Insurance Cost						700,00
Grounds Keeping Costs						950,00
Total	1156,70	12162,48	433,00	1459,00	5195,00	18249,48
Supervision of Maint. Activities & System						912,47
GRAND TOTAL						19161,95

FIGURE 3.3

Annual maintenance budget—summary. (Adapted from Maintenance Management Systems, Technical information document, RPS for INAC. TID-AM-01, October 2000.)

3.1.4.1 Maintenance Program

The projected maintenance should consist of a balance of planned and unplanned maintenance; departments should endeavor to minimize unplanned maintenance.

Planned maintenance includes preventive servicing and condition-based and statutory maintenance.

Unplanned maintenance incorporates corrective, breakdown, and incident maintenance.

If future directions for a maintenance strategy require minimal maintenance for minor and noncritical assets or assets scheduled for refurbishment, replacement, or disposal, the maintenance budget should be adjusted accordingly.

In determining a sufficient level of funding for condition-based maintenance, cost estimates provided from condition assessments need to be carefully reviewed to ensure they are up to date in relation to a department's priorities and recommended timing of maintenance. This is particularly relevant for remedial work to be undertaken in the longer term or if the economic conditions have changed (e.g., economic recession) (Maintenance Management Framework, (II), 2012).

3.1.4.2 Key Considerations in Maintenance Budget Decisions

Departments need to identify and carry out actions to reduce backlog/ deferred maintenance and regularly evaluate the risks associated with allowing maintenance to be deferred.

The reduction of backlog/deferred maintenance may be facilitated by the following:

- Seeking special funding during the annual maintenance budget process
- Reallocating funds from internal funding sources

Maintenance decisions should be coordinated with projected future major repairs or replacements over a planning period. For example, significant future equipment replacements which may impact maintenance decisions should be identified. Cost estimates for such replacements can be obtained from manufacturers, other companies using the same equipment, and/or professionals and should be factored into maintenance budgets.

Maintenance budgeting decisions should consider whole-life costs, including the initial costs of procuring the component and the long-term maintenance, operating, and disposal costs.

There are opportunities for savings when determining maintenance programs. For example, companies may consider coordinating inspections or sequencing/bundling maintenance work across departments, branch plants, etc. Consideration could also be given to opportunities for cost-effective improvements, including the adoption of innovative technologies such as energy-efficient lighting (Maintenance Management Framework, (II), 2012).

3.1.4.3 Preparing a Maintenance Budget

Managing a maintenance budget includes the following:

- Establishing priorities.
- Regular monitoring and reporting, including analysis of budget components against actual expenditure. Where maintenance services are contracted, the monitoring of budgets may include consultation with managers/maintenance service providers on scheduling and material and equipment needs.

- Establishing accountabilities and performance requirements.
- Monitoring performance against benchmarks and policy requirements.
- Managing variances and contingencies and monitoring the effects of deferred maintenance where required.

Where additional maintenance funding is allocated for new priorities (e.g., the reduction of backlog/deferred maintenance), this should be integrated into the maintenance budget (Maintenance Management Framework, (II), 2012).

3.1.4.4 Executing a Maintenance Budget

At the company level, it is necessary to monitor maintenance outputs, compare these with what was expected, identify any deviations, and redirect efforts as necessary.

"Budgetary" control is one of the key elements; the preparation of a company budget is an integral part of the company planning process. Management must plan for production volumes to meet forecasted sales demand. This, in turn, requires a sales, production, and maintenance budget.

The budget can be regarded as the end point of the company's planning process in as much as it is a statement of the company's objectives and plans in revenue and/or cost terms. It is a baseline document against which actual financial performance is measured. In control terms, budgets are based on standard costs which provide the expected (or planned) yearly expenditure profile. This is compared to the actual expenditures (cost control) and the variances—over or under budget. With this information, management can define and take corrective action.

Usually, the word budget is taken to refer to a particular financial year. However, the annual budget is often the first year of a rolling long-term budget. For example, if a company has a strategic 5-year plan, it will normally align with a 5-year rolling financial budget (Kelly and Wilmslow, 2007).

The need for a maintenance budget arises from the overall budgeting needs of upper management and involves estimating the cost of the resources (labor, spares, etc.) that will be needed in the next financial year to meet the expected maintenance workload. This is explained in Figure 3.4. The maintenance life plans and schedule are designed to achieve the maintenance objective (incorporating the production needs, e.g., operating patterns and availability). This, in turn, generates the maintenance workload (Kelly and Wilmslow, 2007).

Typical examples of maintenance workloads are shown in Figures 3.5 and 3.6; a more detailed categorization of the maintenance workload appears in Table 3.2. Essentially, maintenance budgeting is the expression of this forecasted workload in terms of the cost of internal labor, contract labor, and materials.

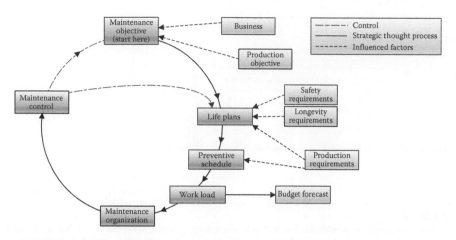

FIGURE 3.4
Relationship between maintenance strategy and budgeting. (Adapted from Kelly, T. and Wilmslow, C., Some thoughts on Maintenance Budgeting, ME CENTRAL THEME, 2007.)

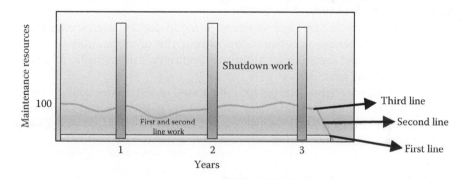

FIGURE 3.5
Typical maintenance workload for a power station. *Note:* Shutdown work can be defined as the scheduled down period for a plant for scheduled work for an extended period of time. (Adapted from Kelly, T. and Wilmslow, C., Some thoughts on Maintenance Budgeting, ME CENTRAL THEME, 2007.)

As Figure 3.5 shows, maintenance budgeting involves both the ongoing workload and the major workload (e.g., overhaul, replacement, modification).

It is important that the maintenance budget is set up to reflect the nature of the maintenance strategy and workload. In other words, there is a need for a longer-term strategic maintenance budget, as shown in Table 3.2. Much of this major long-term work involves capital expenditure that can subsequently be depreciated by upper management in the revenue budgets. In the shorter term, there is a requirement for an annual maintenance expenditure budget, as shown in Table 3.2. These costs are incorporated into the company budget operating over the financial year (Kelly and Wilmslow, 2007).

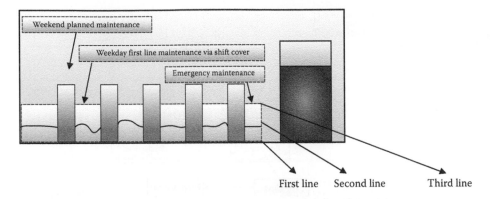

FIGURE 3.6
Typical maintenance workload for a food-processing plant. (Adapted from Kelly, T. and Wilmslow, C., Some thoughts on Maintenance Budgeting, ME CENTRAL THEME, 2007.)

Plants requiring major shutdowns to perform maintenance need specific turnaround budgets, which are an integral part of a turnaround planning procedure (Kelly and Wilmslow, 2007).

The maintenance budgeting procedure is facilitated by identifying plant cost centers and, where necessary, continuing the identification down to the unit level. A cost center in an aluminum refinery, for example, might be coded as shown in Table 3.3.

Over the designated financial period, the actual maintenance cost (of labor, spares, tools) is collected against these cost centers to enable cost monitoring and control (i.e., "cost control"). Cost control complements budgeting (Kelly and Wilmslow, 2007).

3.1.4.5 Reviewing a Maintenance Budget

The maintenance budget should be reviewed each year, with a careful assessment of the achievements of the previous year's maintenance budget against the intended outputs and impacts. The review should target the following factors:

- Adequacy
- Affordability
- Efficiency
- Effectiveness
- Competitiveness
- Compliance

First, a maintenance budget should be reviewed to identify whether previous budgets have been adequate. This includes a consideration of the need

TABLE 3.2

Categorizing Maintenance Workload by Organizational Characteristics

Main Category	Subcategory	Category Number	Comments
First line	Corrective-emergency	(1)	Occurs with random incidence and little waning and the job times also vary greatly. In some industries (e.g., power generation) failures can generate major work, these are usually infrequent but cause large work peaks.
	Corrective-deferred (minor)	(2)	Occurs in the same way as emergency corrective work but does not require urgent attention; it can be deferred until time and maintenance resources are available.
	Preventive-routines	(3)	Work repeated at short intervals, normally involving inspections and/or lubrication and/or minor replacements.
Second line	Corrective-deferred (major)	(4)	Same characteristics as (2) but of longer duration and requiring major planning and scheduling.
	Preventive-services	(5)	Involves minor off-line work carried out at short or medium length intervals. Scheduled with time tolerances for slotting and work smoothing purposes.
	Corrective-reconditioning and fabrication	(6)	Similar to deferred work but is carried out away from the plant (second line maintenance) and usually by a separate tradeforce.
Third line	Preventive-major work (overhauls, etc.)	(7)	Involves overhauls of plant or plant sections or major units.
	Modifications	(8)	Can be planned and scheduled some time ahead. The modification workload (often "capital work") tends to rise to a peak at the end of the company financial year.

Source: Adapted from Kelly, T. and Wilmslow, C., Some thoughts on Maintenance Budgeting, ME CENTRAL THEME, 2007.

for actions/corrections (such as funding replacements rather than ongoing repairs) and/or if adjustments should be made to the budget development process. Note that appropriate maintenance information systems should be in place to monitor maintenance performance (see sections on CMMS).

Second, a maintenance budget should be assessed with respect to affordability. Assets can be maintained to different standards of physical condition,

TABLE 3.3

Cost Center in an Aluminum Refinery

Cost Center	Unit	Unique Unit Number
Digestion area	Bauxite mills	
6	C	02

Source: Adapted from Kelly, T. and Wilmslow, C., Some thoughts on Maintenance Budgeting, ME CENTRAL THEME, 2007.

each of which has a different impact on the cost of maintenance. The higher the standard, the higher the budget requirement will be (all other things being equal). If the budget is not sufficient to meet the standards set, it raises the following questions:

- Are the standards realistic and can they be afforded?
- If assets fall far below the desired standards, will special funding above the normal maintenance requirements be necessary?
- Does maintenance funding need to be increased?

Third, the efficiency of maintenance planning, management, and delivery impacts maintenance budgets. A review of a maintenance budget needs to take into account the efficiency and effectiveness of maintenance demand assessment, planning, and program management. Reviewing the pattern of maintenance expenditure for existing assets and evaluating the effects of maintenance previously undertaken (including cost drivers) can highlight opportunities for improvement. To facilitate efficiency improvements, departments should monitor and review maintenance performance and encourage proactive responses from maintenance service providers.

Fourth, to assess the effectiveness of a maintenance budget, decision-makers should use relevant performance targets or benchmarks, for example, "maintenance cost," to measure actual performance against planned performance.

Fifth, the review of maintenance budgets may include comparisons with benchmarks based on technical advice or research on other similar operations elsewhere to ensure the company is remaining competitive.

Finally, maintenance budgets must support all relevant company policies and guidelines, including its financial management strategy. An appropriate budget can be achieved by integrating competing priorities and actions, addressing strategic issues, and assessing the opportunities and risks associated with the distribution of maintenance funds to areas critical to service delivery (Maintenance Management Framework, (I), 2012).

3.1.5 Control Maintenance Costs Using the Maintenance Budget

A budget developed at the asset level can be used as a tool to reduce and control costs, determine manpower requirements, identify training needs, and develop business cases (Maintenance Technology, 2004). Indeed, support systems for maintenance decisions are based on maintenance cost control in order to increase performance and reduce such costs. For that purpose, the review of the maintenance budget for a specific asset indirectly includes the review of tasks like preventive maintenance and time-based tasks together with the inclusion of cost-avoiding tasks. In this case, costs can be reduced by performing predictive maintenance tasks.

For example, vibration monitoring can be performed on a pump and the bearings can be replaced when the monitoring suggests they are starting to fail. The mean time between failures can be predicted by determining the life of the bearing, and the budget can be modified.

The actual cost of maintaining each asset should be regularly compared to the budgeted costs. Any over- or under-expenditure should be addressed on an asset-by-asset basis. If the expenditures on each asset are controlled, the overall maintenance budget will be effectively managed.

The rationalization of costs includes manpower, which must be monitored on an asset basis to treat every asset according to its criticality and expected performance as reliability-centered maintenance (RCM) states. Manpower requirements can be determined by using the individual budget for each asset. The overall manpower requirements can be developed by combining the manpower requirements required to maintain each asset.

If more resources are needed to maintain equipment, the information required to justify an increase in staffing must be available and easy to understand at the managerial level. It is much easier to convince management to support an effort if an effective business plan is developed. If a reduction in manpower is required, the maintenance manager can work with plant supervision on an asset-by-asset basis to determine which maintenance tasks will no longer be performed.

Above, we mentioned the use of vibration analysis to determine the life of pump bearings. To start a vibration analysis program at a plant, the budget for each piece of equipment with the potential to be monitored can be altered. The extra costs and savings can be identified. For example, while setting it up may incur an initial expense, there may be a decrease in manpower because the bearings are replaced before failure. Once the cost of developing the vibration analysis program is determined, the payback for implementing it can be calculated.

In summary, if a budget is created for each asset, the following can be developed:

1. A maintenance staffing plan that identifies and supports the number of technicians, by craft, required to maintain the plant.
2. A specific training plan, so that all tasks to be performed are identified, and technicians can be trained appropriately.

3. An overall maintenance budget that can be defended.

4. A justification for increasing or decreasing the maintenance budget when pieces of equipment are installed or removed.

5. A justification for increasing the maintenance budget if an asset is used more than it has been in the past. If an asset is used more frequently, the budget for the asset must be modified to reflect any increases in maintenance expenditures required to ensure it can be operated reliably.

6. A basis for a business plan that will support maintenance improvements.

In addition, budgets based on previous years and increased according to some ratios are not welcome in a rational maintenance plan, as aging and performance are not always considered within these approaches. It is highly recommended to use zero-based budgeting (ZBB) to develop life cycle costing. ZBB is a method of budgeting in which all expenses must be justified for each new period. Traditional budget methods are less effective in the maintenance arena because maintenance expenditures are made up of thousands of seemingly unrelated events. Maintenance does not seem to be volume-related (higher output equals higher maintenance). Breakdowns and other maintenance activities are hard to predict and do not necessarily reflect what happened in the previous year. To successfully budget (and therefore predict) maintenance expenditures, the whole maintenance demand must be divided into basic parts.

A ZBB breaks the overall demand for maintenance services into its constituent parts, that is, assets or areas, starting from scratch every year. Maintenance must look at each asset (or group of like assets) to determine the maintenance exposure from year to year. In essence, the budget for an asset starts from zero each year (or each business cycle).

Note that the process used to develop budgets for individual assets can be used in life cycle cost analysis (Maintenance Technology, 2004).

3.1.6 Maintenance Budgets versus Maintenance Costs

The maintenance budget is usually calculated by the person responsible for maintenance. It is not easy to make a maintenance budget as it is necessary to make future projections that may or may not materialize.

In the annual maintenance budget, part of the maintenance cost is constant or subject to very small variations, for example, the regular workforce or the cost of scheduled repairs, but other costs are variable, for example, failures. The same faults or the same severity of faults do not happen every year, so annual costs for materials and contracts can vary significantly.

But it is important to distinguish between costs related to the initial purchase of tools, purchase of stock parts, and staff training, which are not
repeated, and annual costs, which are repeated year after year, albeit with
some variations.

The maintenance budget covers the following maintenance costs:

- Implementation or mobilization of maintenance
- Maintenance personnel
- Spare parts and consumables
- Tools and technical means
- External contracts
- Stoppages and big overhauls
- Insurance, franchising, and liability
- Unexpected costs

3.1.6.1 Associated Maintenance Costs

The following are some of the maintenance costs associated with the annual
maintenance budget:

Direct costs: These are related to the performance of the company and
are lower if the maintenance of equipment is better. They are influenced by the amount of time spent on equipment and the attention
it requires. These costs are set by the number of repairs, inspections,
and other activities. They include the cost of direct labor, the cost of
materials and spare parts, costs associated directly with the execution of work, energy consumption, and equipment rental, as well as
the cost of using tools and equipment.

Indirect costs: These cannot be attributed directly to a specific operation
or to specific work. They tend to be related to the warehouse, facilities, service shop, various accessories, administration, and so on.

Overall costs: These are the costs incurred by the company to support
the areas of nonproductive functions properly. The general costs of
maintenance can also be used as a tool for analysis, and, when classified carefully, can be used to separate fixed costs from variable costs,
which, in some cases, are categorized as direct and indirect costs.

Downtime costs: These are also called the costs of lost time. They are part
of the maintenance costs but are not part of the annual maintenance
budget. Downtime costs include the following:

Production stoppages

Low effectiveness

Waste material

Poor quality

Delays in delivery

Lost sales

The maintenance department and production department must cooperate to share information about lost time or work stoppages, the need for materials, spare parts, and workforce.

3.1.7 Factors Affecting the Estimate of the Maintenance Budget

All companies use budgets to plan and manage their expenditures. In asset-intensive organizations, maintenance tends to be one of the larger cost centers, but often it is among the least well managed. It's not that they don't try; they just lack the right tools.

Asking a maintenance planner or supervisor to predict next year's spending, and deliver that information in a manner that the finance team finds useful, may be self-defeating. They may not have the time or motivation to properly capture the data, the tools to compile it, or the know-how to translate it into the proper general ledger accounts or corporate cost structures. As a result, maintenance budgets often end up as a series of very large lump-sum amounts that are highly inaccurate and nonspecific.

Those who attempt to create a maintenance budget will spend many hours crunching the numbers with homegrown tools, simple spreadsheets, or inflexible corporate budgeting systems. The finance group will dutifully load the maintenance budget into its financial accounting system and attempt to manage with it.

At the same time, maintenance personnel will spend whatever it takes to repair a production line that goes down, whether or not it is in the budget. If their budget runs dry, they may shortchange preventive maintenance or other less urgent tasks, and overlook opportunities to understand the variances or to improve their budgeting processes.

When setting out to build a budget, more than one scenario may be required and many variables need to be considered:

- Size of asset
- Type of construction, materials, and workmanship
- Quality of design
- Age of asset
- History
- How asset is used
- Location of asset

- Knowledge and dedication of staff
- Availability of spare parts and quality contractors
- Type, knowledge, expectations of users
- Laws, building codes, statutes
- Taxes
- Amount of change in the organization
- Competition
- Amount of deferred maintenance
- Hours of use per day
- Production level or speed of operation

For instance, one may represent the previous year's amount, a second may reflect a reduction in the previous year's expenditure to save money, and a third may reflect some percentage of increase over the previous year (Levitt, 2009).

3.1.7.1 Formula for Maintenance Budget Estimate

The maintenance budget can be estimated using the following formula:

$$MB = \left[(RCB * MR) + (RCSE + EMR) \right] * CR \qquad (3.1)$$

where
  RCB = replacement cost of asset.
  MR = maintenance ratio. This number is the percentage of the asset value which must be reinvested to ward off deterioration.
  $RCSE$ = replacement cost of production and mobile equipment. This includes production equipment, process boilers, mobile equipment, fleet trucks, and so on. In factories, process plants, and other equipment-intensive applications, the equipment factor far outweighs the cost of building maintenance.
  EMR = equipment maintenance ratio. This is equivalent to the MR but is used for production equipment. The EMR for all but the most unusual equipment is 7%–15% per year of replacement value. If a company has several big categories of assets, it might consider different factors for the different categories. A paper mill might have a paper machine in one group and a delivery fleet in another. An early article, "Another Way to Measure Costs" (Factory, 1972), calculates it from a 1% (labor-intensive) to 12% (capital-intensive) range.
  CR = construction ratio is the ratio of time and materials spent on renovations and new construction to the total time spent on all maintenance

activity. It must be added to the previous factor. Construction or installation and/or building of new equipment is not a maintenance function (not required to preserve the asset), but it is frequently the responsibility of the maintenance department. It includes office construction, machine building, new installations, and truck body mounting.

MB = maintenance budget with labor, materials, fringe benefits, and over-heads but without janitorial costs.

MB/Sq. ft = maintenance costs per square foot of building or facility.

MB/Rev = maintenance budget per revenue dollar.

MB/Out = maintenance costs per unit output (such as linear foot of fabric, yards of refuse, patient-days, or ton-miles) (Levitt, 2009).

3.2 Cost of Labor Force: In-House versus Outsourced, Blue versus White Collar

Outsourced contract maintenance is becoming an increasingly common method for companies to maintain their assets in industries ranging from aviation to information technology (IT) and manufacturing. There are gen-erally three approaches to maintenance management: in-house staffing, a hybrid of in-house and outsourcing, and complete outsourcing. How and to what extent these are applied is driven by each company and its unique needs (Johnston, 2014).

Not all maintenance can be performed by in-house maintenance teams. For some assets, maintenance must be outsourced because of a lack of skilled in-house staff or a lack of resources to support an employee or because the asset requires specialized equipment for proper care.

For example, it doesn't make much sense to hire a full-time elevator mechanic if a facility has only one or two elevators and these require only sporadic care. Other examples include outsourcing carpet/mat cleaning, roofing, general contractors, and maintenance service for highly specialized assets (Smith, 2012).

3.2.1 Guidelines for Choosing In-House or Outsourced Maintenance

Because it is practically impossible to have the expertise or resources neces-sary to maintain every asset with in-house maintenance staff, vendor man-agement is a crucial part of maintenance operations. Sometimes the decision is easy. At other times, maintenance management must carefully evaluate the cost versus the benefits. The following subsections offer rough guidelines for deciding to outsource maintenance (Smith, 2012).

3.2.1.1 In-House Maintenance Considerations

In general, organizations should try to keep maintenance in-house on critical assets when possible. The ability to have proactive maintenance, such as preventive maintenance or inspections, is essential to asset efficiency and effectiveness (Smith, 2012).

The following are five reasons for organizations to keep maintenance in-house:

- *Core business*: If the asset is critical to the core business operations, maintenance should be kept in-house.
- *Control*: Maintenance management will have more control of in-house maintenance and is not subject to vendor availability or reliability problems.
- *Response time*: Faster response time to problems is ensured by the use of a computerized maintenance management system (CMMS) compared to a telephone call to a contractor. All in-house work (including unplanned work) can be scheduled with a CMMS.
- *Standard operating procedures (SOP)*: This is especially effective if similar assets are spread out over multiple locations. A CMMS enables maintenance management to establish SOPs for work management.
- *Maintenance planning*: Assets maintained in-house can be set up for regular scheduled preventive maintenance, inspections, and repairs based on the historical data collected in the CMMS. Better maintenance planning results in lower capital budget requirements, lower labor costs, and lower energy bills (Smith, 2012).

3.2.1.2 Outsourced Maintenance Considerations

The following areas may benefit from outsourcing:

- *Noncore business*: If asset uptime is nonessential to an organization's core business functions, outsourcing may be a maintenance option.
- *Cleaning*: Many assets just need to be cleaned on a periodic basis; there are specialty cleaners for carpets, mats, floors, upholstered furniture, and so on.
- *Specialized assets*: These include assets requiring maintenance expertise beyond current staff abilities.
- *Predictive technology*: Infrared thermography, ultrasound analysis, vibration analysis, and laser shaft alignments for motors can either be performed by in-house staff or specialized vendors.
- *Ancillary services*: This includes landscaping, payroll, general contractors, resurfacing, security services, software vendors, cleaning services, and so on (Smith, 2012).

3.2.2 Outsourcing Maintenance Activities

There is a trend toward outsourcing maintenance work to contractors. Three main reasons for outsourcing are lack of in-house staff, lack of skilled workers, and cost-effectiveness.

Ribreau (2004) identifies the advantages and disadvantages of outsourcing maintenance. A main benefit is cost savings. The major disadvantages are deterioration of service and inefficient administration and supervision (Ribreau, 2004). Outsourced maintenance does not mean a total loss of control, however; tools for vendors and contract and document management are part of a quality CMMS (Smith, 2012).

3.2.2.1 Expected Benefits of Outsourcing

All firms seek benefits in all activities. Benefits like economic savings, flexibility, and concentration on core competencies immediately come to mind. These can be classified into strategic, management, technological, economical, and quality factors, as illustrated in Figure 3.7.

Previous literature has considered some benefits of outsourcing from the maintenance point of view. For example, Quélin and Duhamel (2003) say the main reasons for maintenance outsourcing are cost savings, focusing on value-adding activities, and access to external know-how. Overall findings include the following:

- *Focus on core activities*: The outsourcing decision allows the organization to focus on its core activities (Sislian and Satir, 2000) to ensure its competitive advantage. The decision on exactly what the core

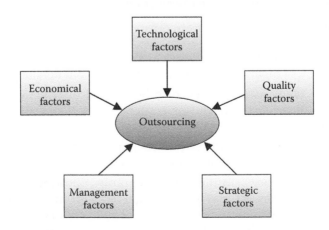

FIGURE 3.7
Expected benefits of outsourcing. (Adapted from Nili, M. et al., *Int. J. Acad. Res. Bus. Soc. Sci.*, 3(5), May 2013, ISSN: 2222–6990.)

function is will have a bearing on whether or not to outsource. This view assumes there is no competitive advantage to retaining control of all functions (Assaf et al., 2011).

- *Improve flexibility to accommodate changing market dynamics*: Flexibility can be considered in operational, resource, and demand areas (Kremic et al., 2006). The potential for improved flexibility is measured by the organization's ability to change its service range in response to market conditions. In today's rapidly changing world, an organization has to respond quickly to changing customer demands. Outsourcing helps organizations to be flexible by providing reliable workers to reduce the time needed to complete the work (Djavanshir, 2005).

- *Increase the speed of implementation*: Some services such as corrective maintenance need a rapid response to failures. A contractor should have the resources to perform a service in the agreed time (Greaver, 1999).

- *Function difficult to manage*: If a service is complex or integrated, or if there are no qualified management staff, the organization may get appropriate help from the service provider (Kremic et al., 2006).

- *Safety management*: The use of skilled external management may reduce the exposure to legal liability for accidents in work situations. It must be noted that an organization is not able to contract out its duty of care to occupiers and users of the facility and must maintain adequate risk management procedures to ensure contracted services are performed (Assaf et al., 2011).

- *Need for specialized expertise*: Specialist contractors can afford to invest in new technologies and innovative practices, because they perform only one service and have all the means to perform it. They can focus on identifying areas of improvement and acquire the knowledge needed to act (Alexander and Young, 1996). They are, however, susceptible to a short-term focus commensurate with the term of the contract (Nili et al., 2013).

- *Acquire new skills or technical knowledge*: Outsourcing may help an organization gain new skills and knowledge so that it can develop its own expertise to maintain high-level technology. When some services are outsourced, an organization may gain new skills or new technical knowledge from the outside supplier (McDonagh and Hayward, 2000).

- *Save the overall cost*: The key driver for many outsourcing decisions is the reduction in the cost of labor, materials, and parts (Nili et al., 2013). A function is outsourced when the in-house costs are higher than the anticipated costs for outsourcing it. Therefore, the higher the internal cost to perform the function relative to the anticipated cost of outsourcing, the greater is the probability of outsourcing (Kremic et al., 2006).

- *Reduce the labor and operating cost*: Costs can be reduced either by saving on labor costs or by using new technology. Djavanshir (2005) says the major benefits of outsourcing are the reduction of labor and operating costs and the acquisition of a competitive advantage. The decrease in labor and operating costs is based on a contractor's ability to provide a service more efficiently and effectively (Assaf et al., 2011).

- *Make fixed costs into variable costs*: Outsourcing helps an organization move fixed costs (such as payroll or labor productivity and materials) so that they become variable costs (Anderson, 1997). Costs for operating resources and fixed infrastructures can be reduced step by step after the services have been outsourced. The payment to the contractor will convert the fixed costs into variable costs (Blumberg, 1998).

- *Improve cash flow*: An organization's cash flow improves when it has fewer employees; it requires less infrastructure and fewer support systems, resulting in greater efficiency by reducing variable and managed costs (Fontes, 2000). Some organizations outsource to achieve better cost control (Anderson, 1997). Outsourcing can be long-term if contractors can offer quality services more cost-effectively than in-house (Yik and Lai, 2005).

- *Infuse cash into the company*: All tools, equipment, vehicles, and facilities used in the current operation have value if they improve cash infusion by being transferred to contractors (Corbett, 1999).

- *Make capital funds more available for core activities*: Outsourcing can reduce the need to invest capital funds in noncore functions and make them available in core areas. Accordingly, organizations may consider outsourcing to increase flexibility in finance and to free up capital funds for core activities (Djavanshir, 2005).

- *Increase economic efficiency*: Organizations specializing in particular services have a relatively large volume of business; this allows them to take advantage of economies of scale and, thus, to operate and maintain services more cost-effectively (Nili et al., 2013).

- *Improve service quality*: If the organization's maintenance service quality is held in high regard, outsourcing the service may represent an improvement (Anderson, 1997). The quality factors influencing the decision to outsource services are the ability to reach a higher service level, improve service quality, meet special requirements, and achieve a competitive advantage (Assaf et al., 2011). When an organization is recognized for its quality, decision-makers may fear outsourcing will affect the quality of its services (Kremic et al., 2006), but if it needs to react rapidly to user requirements, outsourcing is a good solution. In this way, the quality of services is improved at a lower cost (Campbell, 1995).

- *Maintain required quality*: Maintenance is required to bring facilities and equipment to a condition that meets acceptable standards. When maintenance services are outsourced, the quality of services rendered should be measured against the standards (Campbell, 1995). Maintenance requirements continually change because of wear and tear, technological developments, and evolving operational requirements. Quality requirements involve statutory and regulatory compliance with minimum standards of material and implementation (Campbell, 1995).

- *Ensure higher reliability and competency*: The reliability of processes and services may be improved by engaging a contractor (Al-Najjar, 1996). This, in turn, generates satisfaction among users (Kremic et al., 2006).

3.2.2.2 Potential Risks of Outsourcing

Despite its potential benefits for asset management, outsourcing can be a risky decision. Because outsourcing is a tool used only recently by maintenance managers, the complete maintenance costs are not yet known. The lack of a working methodology may cause some outsourcing failures (Bounfour, 1999; Lonsdale, 1999). Lonsdale suggests that outsourcing failures do not reflect an inherent problem with outsourcing but, rather, the lack of a guiding methodology among managers (Lonsdale, 1999).

Baitheiemy (2003) says outsourcing is a powerful tool to cut costs, improve performance, and refocus on the core business, but outsourcing initiatives often fall short of expectations. In a survey of nearly 100 outsourcing efforts in Europe and the United States, Baitheiemy (2003) finds one or more of seven possible "deadly sins" underlies most failed outsourcing efforts. The seven sins are:

1. Outsourcing activities that should not be outsourced.
2. Selecting the wrong vendor.
3. Writing a poor contract.
4. Overlooking personnel issues.
5. Losing control of the outsourced activity.
6. Overlooking the hidden costs of outsourcing.
7. Failing to plan an exit strategy (i.e., changing vendors or reintegrating an outsourced activity). Outsourcing failures are rarely reported because firms are reluctant to publicize them.

3.2.3 Cost Determination Methodology

A study by Martin (1993) seeks to determine the true cost of using in-house and outsourced services. For in-house services, direct costs are defined as fully dedicated costs for a target service; indirect costs are those benefiting from more than one target service. The indirect costs for personnel must be proportionally allocated to target services. The total cost for in-house services is the sum of the direct costs and a proportional share of the indirect costs.

According to Martin (1993), three types of costs are associated with contracts: contract administration, one-time conversion, and new revenue. A "contract administration cost" refers to all expenditures from the contract start to the contract end. A "one-time conversion cost" occurs when a target service is converted from an in-house to a contract service and must be amortized over a certain period. For example, the salary of workers is a "one-time conversion cost" because the existing workers cannot be removed immediately (i.e., they have contracts). A "new revenue cost" occurs when services are contracted out, and the contracting company does not need to use some of the resources or equipment; the owner can then sell these resources or equipment. The total cost incurred in a private contract is the sum of the "contract administration cost" and the "one-time conversion cost" minus the "new revenue cost" (Martin, 1993).

Total cost in outsourced environments, especially the maintenance of physical assets, consists of the line activity cost, program support cost, and enterprise support cost. Line activity costs are direct costs. Program support costs are those costs that do not deliver any specific work product of construction or maintenance, but do support one or more line activity, such as maintenance staff, office, or utilities. Enterprise support costs include head office administration, information technology, planning and research, and legal advice.

Five steps are involved in determining the share of a particular support cost in the direct costs:

1. Collect and separate the various maintenance program costs.
2. Determine the share of support program costs in the costs of the line activities.
3. Collect and separate the various support costs.
4. Determine the share of enterprise support costs in the costs of the line activities.
5. Add the costs of the line activities, the share of support costs, and the share of enterprise support costs to determine the full cost.

A percentage share of the costs for both the support program activity and the enterprise support activity to a line activity is calculated based on the ratio of the various line activity costs to the total line activity cost.

3.2.4 Performance-Based Contracts

Studies conducted on using performance-based contracts (PBC) for maintenance projects focus on four aspects:

1. PBC contracting process
2. Advantages and disadvantages of PBC
3. Development of performance measures for PBC
4. Lessons learned using PBC

Details are summarized in the following sections.

3.2.4.1 Performance-Based Contracting Process

PBC is a form of contracting with three characteristics:

- Clear definition of a series of objectives and indicators by which to measure contractor performance.
- Collection of data on the performance indicators to assess the extent to which the contractors are successfully implementing the defined services.
- Consequences for the contractor, such as provision of rewards or imposition of sanctions. Rewards can include continuation of the contract, provision of performance bonuses, or public recognition. Sanctions can include termination of the contract, financial penalties, public criticism, and debarment from future contracts (World Health Organization [WHO]).

To give an example of how a PBC works, the World Bank (2002) has prepared a sample of PBC for road maintenance projects. It includes samples of performance specifications, notes the criteria for service quality, explains the inspection methods for the levels of service quality, and discusses timeliness, payment reductions, and liquidated damages. These specifications cover both paved and unpaved roads. The quality inspections of paved and unpaved roads are similar; for each, inspections are carried out as directed by the project manager (Shrestha et al., 2015).

Several studies recommend pay reductions in the event of noncompliance with the quality specified in PBC contracts (Zietsman, 2004; Stankevich et al., 2009; Gharaibeh et al., 2011), including a percentage of the monthly lump-sum amount of the contractor's pay. To continue the example of road maintenance, for an unpaved road, if the contractor cannot meet the "road usability" criterion or if the road is closed for traffic, 1% of the monthly lump-sum amount for the entire project or for the affected road section will be reduced (World Bank, 2002). Pay reductions for all criteria are set as a percentage of the contractor's monthly payment.

Stankevich et al. (2009) differentiate PBCs from traditional contracts. When using a PBC, the contractor is paid based on work performance, and the company will not specify either the methods or the materials to be used by the contractor. During the contract selection process, the "best value" method is normally used for PBC, whereas traditional contracts rely on low bids. Stankevich et al. (2009) identify two types of PBC: pure PBCs and hybrid PBCs, i.e., a combination of pure PBCs and maintenance-based contracts (MBCs). According to the authors, pure PBCs are based entirely on the outcome of the projects; in hybrid PBCs, some activities are paid based on the PBC and the remainder are paid based on the MBC.

3.2.4.2 Advantages and Disadvantages of Performance-Based Maintenance Contracts

The main advantage of a performance-based maintenance contract (PMBC) is reduced maintenance costs (Shrestha et al., 2015). The following list gives the advantages and disadvantages of PBMCs (Hyman, 2009).

Advantages:

- Potential reduction in costs
- Improved level of service (but could cost more)
- Transfer of risk to the contractor
- More innovation
- More integrated services
- Enhanced asset management
- Ability to reap the benefits of partnering
- Ability to build a new industry
- Ability to achieve economies of scale

Disadvantages:

- Costlier procurement process
- Longer procurement process
- Reduction in competition
- Uncertainty associated with long-term contracting relationships
- Challenges in mobilizing
- Loss of control and flexibility, for example, in the reallocation of funds when there are large long-term commitments

3.2.4.3 Development of Performance Indicators for Performance-Based Maintenance Contracting

Developing performance indexes for PBMC requires a clear understanding of the fundamental types of measures (Hatry et al., 1990; Government Performance and Results Act, 2003; Hyman, 2004). The basic categories of measures are as follows:

1. *Inputs*: These usually consist of labor, equipment, materials, and the associated financial expenditures. In some instances, they can include facilities or land.

2. *Outputs*: These represent how much work gets done. Traditional maintenance management systems record accomplishments (outputs) and resources used (inputs) upon completion of work. Some PBCs specify the outputs to be achieved.

3. *Outcomes*: These are the results of or changes caused by maintenance. To an increasing degree, PBMC is concerned with outcomes important to customers. Another class of outcomes is expressed in economic terms. These economic impacts are important outcomes, but they are difficult to measure and incorporate into a PBC. Chapter 5 deals with intangible costs like these when calculation is challenging but quantification is a real issue and must be assessed by the means of indirect variables.

4. *Explanatory variables*: The company and the contractor should keep track of variables that can help explain resource utilization, outputs, and outcomes. Many explanatory variables are outside the control of the contractor and the company; they include emergencies and contingencies. Accounting for explanatory variables outside the contractor's control provides a basis for adjusting incentives and disincentives and more fairly allocating risk.

PBCs may include more than one type of index indicators. For example, there may be a combination of outcome and output measures along with method specifications (possibly including equipment requirements). However, the trend is toward reducing or eliminating method specifications and increasing the customer-oriented outcomes as opposed to outputs. In sum, PBMC is evolving into a contracting procedure that provides both disincentives and incentives for achieving measurable targets or standards based on outcome-oriented performance specifications (Hyman, 2009).

Many experts on performance measurement advocate using a few good or vital measures, sometimes called key performance indicators (KPIs). However, PBCs can become excessively complex because of the large number of maintenance activities they address. There is a tension between having just a few good measures because of their simplicity and manageability and having many measures to be complete and thorough. Contracts focused on a single maintenance activity may use only a few measures, whereas contracts involving virtually all types of maintenance and operations are likely to have many measures (Hyman, 2009).

3.2.4.4 Lessons Learned Using Performance-Based Contracting in the Maintenance Function

- PBMC involves politically and socially sensitive decisions.
- A significant cultural shift of both the owner–contracting company and the contractor is usually required for PBMC to be successful.
- Adequate contractor capacity is necessary to ensure meaningful competition and to be confident that a contractor and its subcontractors can achieve the performance standards.
- The more a maintenance department uses PBMC, the more its role shifts from managing and performing maintenance work to planning, contract administration, and contractor oversight. Workers' skills must shift accordingly.
- The request for a proposal (RFP) from contractors and the final contract both require a clear expression of the scope of work.
- Companies starting to use PBMC could begin with projects with a limited scope, such as one maintenance activity or relatively few activities.
- Performance measures must be clearly defined, the measurement process repeatable, and targets realistic and in line with the company's goals. In short, performance specifications must be clearly defined.
- The contract must have the proper incentives and disincentives.
- A firm funding commitment is required for multiyear PBCs.
- Cost savings are highly desirable but difficult to document. Cost savings are often claimed based on the difference between the company's estimated cost and the amount of the contract award.
- Quality and/or the level of service (LOS) sometimes suffer during the first year on long-term total asset management contracts. Quality is likely to improve the first year on PBCs if serious maintenance has been deferred or LOS has been low.
- If an asset is severely deteriorated, it needs to be reconstructed or rehabilitated before standard performance-based maintenance procedures begin. In numerous cases around the world, the contractor has been responsible first for a rehabilitation phase and then for a maintenance phase.

- Partnering and trust are imperative between the maintenance organization and the contractor.

- If the contract is poorly written, or either party misreads significant portions of the contract to serve its own interests or point of view, the result may be failure.

- Failure is likely to occur if the company's staff believes strongly that contractors are taking their jobs. If the company's workers are responsible for monitoring the contractor's performance, they may be overzealous in holding contractors to timeliness requirements and other performance standards.

- Failure is likely to occur occasionally because success cannot happen 100% of the time. Certain events and conditions may prevent PBC from working. A company whose contractors perform poorly or whose outcome is disappointing might attempt to learn from its experience and try again.

- Warranties and performance bonds can help mitigate failures. Bids will be higher if contractors have to ensure against failure.

- Innovation will be the most significant under long-term, performance-based, lump-sum agreements with selection based predominantly on best qualifications.

- The client will experience a perceived loss of control and flexibility.

- There are three ways to monitor contractor performance; each has different implications. First, the contractor can monitor itself. This approach requires a strong bond of trust with the contracting company and is generally the least expensive. Second, the contracting company can monitor the contractor's performance. Companies using this approach believe they can observe contractor performance effectively and have a good deal of control. Third, an independent third party can monitor the contractor. This method provides the most objectivity but is the most expensive.

- A program plan that lists upcoming performance-based procurements is valuable for alerting qualified contractors to bidding opportunities. This could be part of a long-term procurement strategy that sends a message to industry more generally.

- Criteria for the prequalification of contractors should be developed and used to identify a suitable number of competitors.

- Risks must be shared in an equitable manner; the contractor should not bear all the risks.

- Even though a mandate to downsize does not trigger the decision to use PBMC in the first place, pressures to downsize may build up as more maintenance work is outsourced (Hyman, 2009).

3.3 Spare Parts Policies for Cost Savings

Spare parts are stored in the inventory to replace or support an existing part for a production asset. Spare parts include such things as bearings, hydraulic cylinders, and electric motors. They may also refer to infrastructure parts, such as ducting, expansion joints, and conveyor components. Finally, spares include disposable goods, such as nuts and bolts, lubrication, welding rods, air filters, and safety equipment.

A bill of material (BOM) is a list of items or ingredients needed for a product and is always used for spare parts. A company's CMMS will use a BOM to maintain an inventory for an asset. Workers can simply key in the equipment they are using and a list of parts will be displayed. Some systems even allow the storekeeper to build an order of all the parts he or she has. This order can be pulled out by other people and delivered electronically through the web. If equipment is due to retire or be removed from service, the BOM permits its quick identification.

A maintenance department must maintain equipment to extend its life at minimal cost. Machine reliability is difficult to predict without proper monitoring of equipment and software to ensure they are in good condition. Many factors could lead to machine failure, including environmental issues, process problems, poor operator procedures, or even accidents. It is imperative to store spare parts. Storing spare parts on site could decrease the time required to collect a new part and replace the failed one. Many specialized parts need a long time to access, however, and for equipment that has been used for many years, it can be difficult to find the replacement part. Unavailability and long delivery times are two additional problems. All these factors suggest the need for a large spare parts department.

That said, it is costly and not feasible to have a spare for each piece of equipment. A company should determine the criticality of equipment listed in the BOM and stock its spares accordingly (Shangguan, 2013).

A good maintenance strategy is based on good maintenance principles. The following are eight basic operating principles:

1. Reduce and minimize production downtime.
2. Reduce costs of production.
3. Control spares and inventory.
4. Control repair and maintenance costs.
5. Eliminate emergency repair actuations.
6. Reduce spoilage and enhance quality of plant and machinery performance output.

7. Make fewer large-scale repairs and fewer repetitive repairs.
8. Increase employee morale and enhance worker safety (Srinivasan and Srinivasan, 1986).

While the maintenance department has a great impact on what is included in the spares inventory, the engineering department also has a say. Engineering should ensure that the proper equipment is used for the maintenance process and is cost-effective. In addition, the department considers the timely and efficient acquisition of machinery and spare parts.

Acquiring spare parts always takes time and money. To reduce the spare parts purchasing cost, the following principles should be taken seriously:

- The schedule for delivery should be agreed on and written into the contract.
- Suppliers should provide stock at agreed rates.
- Slow-moving "safety stock" items should be held at one location and drawn from stores as required (Shangguan, 2013).

3.3.1 Spare Parts Management

Dependability is a "collective term used to describe the availability performance and its influencing factors: reliability performance, maintainability performance and maintenance support performance" (IEV, 2007).

The goal of dependability management is to decrease the risk of health, safety, and environmental problems throughout the life cycle. Production performance is related to customers, markets, regulations, demands, and requirements. Dependability directly affects availability and indirectly affects production. The relationship between dependability and other factors is illustrated in Figure 3.8.

Spare parts management is an aspect of spare parts logistics (product support logistics). It is a broad field that includes several different topics, such as maintenance, reliability, supply chain management, inventory control, and other strategic aspects. It is essential to have an adequate supply of spare parts, especially critical parts. Lack of spare parts could lead to lower availability and increased operational risk. Indeed, spare parts management plays a key role in maintenance support performance and therefore in the achieved availability performance or dependability (Figure 3.8).

Spare parts management is an important part of achieving the desired plant availability at an optimum cost. Downtime costs for plants and machinery are extremely expensive. In many industries, the nonavailability of spare parts contributes to 50% of the total maintenance cost (Shangguan, 2013).

A company always wants to achieve a sufficient service level while keeping the inventory investment and administrative cost at a minimum

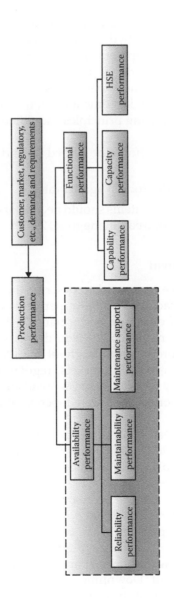

FIGURE 3.8
Dependability and production performance. (Adapted from Markeset, T., Design for performance: Review of current research in Norway, in: *Proceedings of Condition Monitoring and Diagnostic Engineering Management*, Nara, Japan, 2010.)

(Huiskonen, 2001). Each spare part being held and not used represents money spent in the form of insurance and maintenance costs. The part needs storage space, which takes money as well. Because of poor planning and inadequate management, some spare parts may not be used for many years.

Modern corporations emphasize good spare parts management. Procurement cost savings always outweigh reverse logistic costs. IBM (Kennedy and Patterson, 2002) is one of the pioneering companies to recognize the importance of a closed-loop supply chain and spare parts management.

The goal of spare parts management is to ensure the availability of spares for maintenance and repairs at an optimum cost. It may be difficult to find a balance between the non-availability of spare parts to meet requirements and soaring capital costs of acquiring and storing spare parts.

Effective spares management will include the following:

- Identifying spare parts
- Forecasting spare parts requirements
- Carrying out inventory analyses
- Formulating selective control policies for various categories
- Developing inventory control systems
- Creating stocking policies for routable spares or subassemblies
- Creating replacement policies for spares
- Inspecting spare parts
- Reconditioning spare parts
- Establishing a spare parts bank
- Creating and using computer applications for spare parts management (Shangguan, 2013)

3.3.2 Spare Parts Evaluation and Optimization

The spare parts evaluation process is a significant part of spare parts management. Seen from a broader perspective, spare parts management is the link between maintenance planning and maintenance execution. It also connects the suppliers, manufacturers, and contractors. The risk of equipment failure is considered during maintenance planning. It allows companies to provide sufficient stock for future requirements. Factors such as lead time, unit price, delivery condition, and purchasing costs should be taken into consideration as well to ensure a timely supply. These are shown diagrammatically in Figure 3.9.

If a spare part is being kept in inventory, it is a good idea to figure out the quantity needed. When the inventory drops below a certain level, the stock must be replenished to the desired number of parts to avoid the risk of the machine stopping (due to waiting for spare parts). The replenishment should

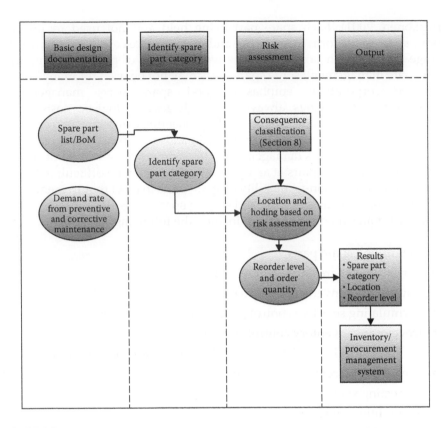

FIGURE 3.9
Spare parts evaluation process. (Adapted from Norsok, Z., Risk based maintenance and consequence classification, 2011.)

be determined on the basis of expected usage rate of the part and economic risk of its unavailability. The annual usage rate can be calculated from the part failure rate, the usage rate per component, and the number of similar components in the system. This will indicate the total usage rate and suggest the need for replenishment. Figure 3.10 shows the relationship between these factors.

3.3.3 Inventory Analysis

An inventory analysis considers annual consumption value, criticality, lead time, unit cost, and frequency of use. Optimum replacement policies for selective items for which the costs of downtime and replacement are high should be given special attention. In addition, some spare parts come from a considerable distance and may even be imported from foreign countries.

For these spare parts, it is essential to extend the life cycle by appropriate reconditioning or repair techniques. For individual industries, it would

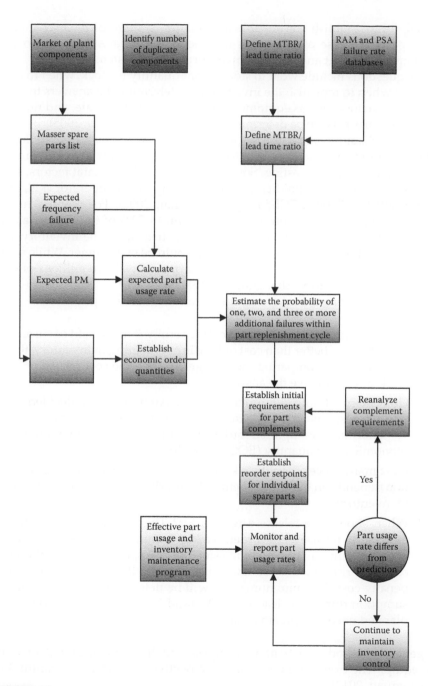

FIGURE 3.10
Management process to spare parts optimization: determination of complement. (From IAEA, Reliability Assurance Programme Guidebook for Advanced Light Water Reactors, IAEA-TECDOC-1264, IAEA, Vienna, Austria, 2001.)

be helpful to establish one suitable information system for the exchange of spare parts. The use of computers to process spare parts information will facilitate an efficient and effective inventory control system.

Key questions include what to stock, what quantity to stock, when to reorder, and when to replenish the inventory. To determine the answers to these questions, factors such as equipment failure rate, demand rate, and number of similar parts need to be determined (Ghodrati and Kumar, 2005).

System reliability characteristics and factors such as mean time to failure (MTTF) and mean time to repair (MTTF) are required for reliability analysis and spare parts forecasting. Some operating environmental factors, such as dust, temperature, humidity, pollution, vibration, and operator skill, may greatly affect reliability (Ghodrati and Kumar, 2005). For manufacturing organizations, inventory could account for up to 50% of the current assets of the business. In other words, up to 50% is tied up in the inventory. For wholesale and retail businesses, this figure could be even bigger. While some companies do not reach this level, many corporations have millions of dollars tied up in inventory (Shangguan, 2013).

Normally, maintenance departments face the following questions:

1. Do we keep a spare in stock or not? Generally, if the benefit of current availability is better than cost of holding something in inventory, the answer is yes. Comparing the storage cost and the cost related with stock-out could give the answer.

2. How many should we order at once? After we have decided to buy a spare, it is time to figure out how many to order at one time. To determine an optimal order quantity, a well-known model, economic order quantity (EOQ), is useful.

3. How many pieces should we keep in stock? The answer to this question depends on the annual demand, ordering cost, and holding cost of inventory.

4. When should we release a new order? The re-order point, or the moment to release a new order, is a key parameter in inventory control. If a company has too much stock of one spare part, the holding cost may rise. If it has too few items in stock, there could be a high penalty cost. The minimum stock will be determined based on consumption during the lead time. Demand is calculated based on the planned need and previous data.

A general method of inventory analysis and the replenishment of critical equipment from an engineering perspective are shown in Figure 3.11 (Shangguan, 2013).

One proper criterion or standard is needed to calculate or quantify the demand, but maintenance personnel often have various theories about what this criterion ought to be.

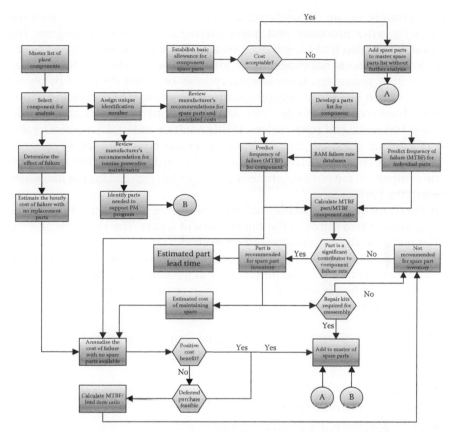

FIGURE 3.11
Spare parts optimization process: determination of inventory. (Adapted from IAEA, Reliability Assurance Programme Guidebook for Advanced Light Water Reactors, IAEA-TECDOC-1264, IAEA, Vienna, Austria, 2001.)

3.3.4 Determining Optimal Parameters for Expediting Policies

One study of expediting policies (Özsen and Thonemann, 2015) was motivated by an inventory optimization problem encountered by a major equipment manufacturer. The manufacturer wanted to analyze the inventory performance of the spare parts network. The study began with an analysis of the inventory performance at the central warehouse; researchers collected data on demands, lead times, and lead time variability. They used standard inventory optimization approaches, which assume lead time variability is caused by external factors, such as delays in the manufacturing processes of suppliers. In these models, lead times are independent of the inventory levels at the central warehouse.

The results of optimization were surprising. To achieve the service level desired by the company, the models suggested higher inventory levels than those currently in use. To understand this result, researchers reviewed the company's order processes and interviewed the inventory managers. The interviews revealed that a key model assumption was violated. More specifically, lead times were not exogenously given, but were actively controlled by the inventory managers. If the inventory level of a part dropped below an "alert-level," an inventory manager received a message from the information system. The inventory manager then analyzed the part's historical demands, demand forecasts, and open orders. If this analysis indicated a considerable risk of a future stock-out, the inventory manager contacted the supplier to negotiate an expedited delivery of outstanding orders. Thus, the lead times at the company were not exogenously given, but depended on the inventory levels of the parts.

At the company, the "alert-levels" were set using rules-of-thumb, such as setting them at two thirds of the safety stock of a part. The riskiness and the cost of potential stock-outs were not quantified.

The optimization problem inspired Özsen and Thonemann (2015) to build a model corresponding to the policy used at the company: "If the inventory level in the warehouse drops below a certain level, and if the costs of order expediting are less than the expected costs of a stock out, then open orders are expedited."

The model considers various costs that can be associated with order expediting:

- Fixed cost per expediting
- Fixed cost per order that is expedited
- Fixed cost per batch of expedited orders
- Cost per unit and period of expediting

At the company, the fixed cost of expediting was one of the main cost drivers, because expediting required contacting a supplier, negotiating with the supplier, and potentially paying for express shipments. The model takes such costs into account and balances them against the expected cost of future backorders to determine optimal base-stock levels and optimal expediting levels.

Özsen and Thonemann (2015) that the model efficiently determines the optimal parameters of an expediting policy. Finding the optimal parameters takes only a few milliseconds, and the policy is easy to implement.

To quantify the benefits of using the expediting policy as opposed to a standard inventory policy that does not expedite open orders, they performed extensive numerical experiments using data from the equipment manufacturer. The numerical experiments indicated that the expediting policy offers substantial cost savings. The cost savings are in the double-digit percentage domain and stay in this dimension for a wide range of cost parameter values, demand rates, and lead times.

Order expediting is widely implemented in industry, and most inventory managers expedite orders to avoid backorders. Some use information system support for order expediting. Unfortunately, they seldom use mathematical approaches to optimize expediting even though these tools provide a decision support model that can enable companies to realize a substantial part of the cost savings offered by order expending (Özsen and Thonemann, 2015).

3.4 Overinvestments in Maintenance and Avoided Costs

Costs and expenses are two terms that are often used incorrectly or interchangeably, both in the literature and among practitioners, partly because of language simplifications and partly because of ignorance (Emblemsvåg, 2003).

Cost is a measure of resource consumption related to the demand for jobs to be done, whereas expense is a measure of spending that relates to the capacity to do a job (Cooper, 1990). For example, a stamping machine with daily operational costs of $100 can stamp 10,000 coins per day. One day, only 5000 coins are stamped. The expense is $100 because that is the capacity provided, which shows up in the books, but the cost of stamping the coins is only $50. Hence, on this day there was a surplus capacity of $50.

It is the resource consumption perspective that counts, because management must match capacity to demand and not the other way around. In our example, management must consider if the excess capacity should be kept or removed. They should not stamp more coins than demanded because that only drives costs up, and the risks of producing obsolete coins increases. However, to most companies, producing to capacity rather than to demand often appears to reduce costs. In fact, it leads to overinvestment and surplus capacity for companies with significant free cash flow, and this further erodes profits.

While a life cycle cost (LCC) model should represent the cost/resource consumption perspective, in the literature most LCC models are actually cash flow models or expense models at best.

Cash flow models are needed in situations where revenues and their related costs occur in different time periods. Cash flow models are important to ensure sufficient liquidity, but they cannot replace cost models. It is also important to take the time value of money into account using a discounting factor, because, as the saying goes, "it is better to earn one dollar today than one dollar tomorrow." Although the time value of money is not confined to cash flow models, such models are always used in investment analyses, which, in some ways, are like LCC analyses (Emblemsvåg, 2003).

In the wider sense, an investment is a sacrifice of something now for the prospect of something later. Furthermore, as Park points out, "we see two different factors involved, time and risk. The sacrifice takes place in the present and is certain. The reward comes later, if at all, and the magnitude may be uncertain" (Park and Sharp-Bette, 1990). Indeed, many costs are turned into rewards or higher costs and the confusion increases, especially from an accounting point of view and LCC calculations.

The principles of cost avoidance and the metric of avoided costs are central features of least-cost or integrated resource planning (IRP). The benefits of avoiding unnecessary costs seem obvious, sensible, and straightforward. In the realm of utility planning, however, the concept is relatively new, and with no clear consensus about its relevance or use.

Comparing resource options in the context of IRP requires a methodology for measuring costs (or savings). When conservation or demand-management strategies enter the mix of options as least-cost planning dictates, the concept of avoided cost becomes relevant. Demand-management strategies range from efficiency-oriented pricing to customer education programs to rebates and retrofits (Beecher, 2011).

The applicability of the avoided-cost principle extends well beyond designing and justifying demand management. Avoided costs can be used to evaluate the benefits of resource alternatives on the supply side, including repair programs. Avoided costs can be used to evaluate complex management issues, such as the potential benefits of interconnection, partnerships, and mergers with other companies (Beecher, 1996). Such evaluations are well within the spirit of comprehensive and integrated resource planning (Beecher, 2011).

3.4.1 Cost Savings, Avoided Costs, and Opportunity Costs

Cost savings, avoided costs, and opportunity costs can play an important role in maintenance planning, budgeting, and decision support. Whereas most business people readily accept cost savings as a legitimate concept, avoided costs and opportunity costs can be more problematic—often either unknown or unusual in the maintenance domain. This is unfortunate because they all provide benefits and decision support when understood and used properly. One reason for the confusion is that all three are relative terms. They are real but can be measured only when one maintenance scenario is compared to another scenario.

The three terms are briefly defined as follows:

- *Cost savings* refers to cost (expense) already incurred or being paid. If a driver trades a currently owned vehicle for a more fuel-efficient one, while maintaining the same driving habits, the driver can expect a cost saving in fuel costs.

- *Avoided cost* is also cost saving, but it refers to cost (expense) not yet incurred. Preventive maintenance for an asset (e.g., regular oil changes) avoids the future cost of parts replacement, which is certainly imminent if preventive maintenance is omitted.

- *Opportunity cost* refers to a foregone gain that follows from choosing an outcome. Suppose a collector of classic automobiles offers a very large sum to purchase a driver's car. The driver must choose between two outcomes:
  1. Continuing to own and drive the car
  2. Selling the car to the collector

The driver may have many reasons to choose option 1 and turn down the offer, but option 1 also brings a very large and real opportunity cost.

Cost savings, avoided costs, and opportunity costs are all based on similar reasoning. They are relative terms that have meaning only when one outcome is compared to another. When any of these terms appears in business planning or decision support, two key questions arise:

1. Which courses of action are really possible?
2. What are the outcomes under each option?

3.4.2 Meaning of Cost Savings

Most people readily accept cost savings as a legitimate benefit in business when a proposed action will clearly reduce costs. If, for instance, we plan to lower the electric bill for office lighting by switching to energy-saving fluorescent bulbs, no one rejects the legitimacy of the cost savings benefit.

Of course, the analyst has to estimate kilowatt hour consumption as it is and as it will be, make assumptions about light usage under the new plan, and consider all the costs of switching. Any of those points might be debated or challenged, but the idea of cost savings itself is acceptable and legitimate. No one doubts that the savings are real and measurable, and next year's operating budget may be adjusted downward based on this belief.

Many people are less comfortable, however, when avoided costs and opportunity costs enter the picture. The rationale for these costs is similar to the reasoning for cost savings, but they cannot be granted legitimacy until a few additional assumptions are made (Schmidt, 2004).

3.4.3 Concept of Avoided Cost

An avoided-cost analysis compares the incremental savings associated with not producing an additional unit of output through a specific method to the incremental cost of supplying the equivalent unit through an alternative method. For example, conservation is justified when the unit cost of freeing up existing supply capacity through demand management is lower than the unit cost of adding new supply capacity.

To give an example, many companies can avoid costs by avoiding additions to supply capacity through conservation or load management strategies, including efficiency-oriented pricing. Companies experiencing rapid demand growth and companies with a history of underpricing services may have the most to gain through demand management. Smaller increments of demand-side resources, compared with large-scale supply-side resources, can help some companies respond to change with more flexibility and lower risk.

Theoretically, avoided costs can be divided into three types of savings:

1. Direct costs (capital and operating costs)
2. Indirect costs (corollaries and externalities)
3. Opportunity costs

Most avoided-cost studies emphasize direct costs associated with resource alternatives, which are easiest to measure, analyze, and compare. Direct costs accrue to a company and include the cost of cancelled or deferred capital investments (including financing costs), as well as operating costs. Only costs directly associated with the specified avoided capacity are included. Significant operating costs for water utilities, for example, will include energy costs (for pumping), chemical costs (for water treatment), and labor costs.

Another cost concept related to cost avoidance is opportunity cost. When companies invest in any project, the required resources cannot be used elsewhere. Resource expenditures, in other words, also constitute opportunity costs because investing in one option closes opportunities to invest in another. With pressure to make improvements and satisfy demand, today's companies face difficult investment choices. Costs avoided are savings achieved. Opportunity savings may accrue to customers, the company, or society. Regardless, these savings constitute resources that can be invested in other pursuits.

Analysts bear considerable responsibility in informing decision-makers about the nature of costs included in their studies. It may be advisable to report a range of results, beginning with direct costs, and expand them to reflect other types of costs. The analyses may point to different solutions and make decision-making more challenging, but the reasons for the differences should be clear (Beecher, 2011).

3.4.4 Maintenance: Investment or Expense?

Maintenance expenditure can range from a few dollars to fix a broken fitting to many thousands of dollars to replace a significate part of a large asset. In most cases, work undertaken is readily identified as maintenance and treated as an expense. However, at times, the nature or intent of the work (or parts of the work) extends beyond restoring the asset to its original condition, capacity, or function. In these cases, managers must decide whether the expenditure is most appropriately classified as "maintenance" or as a "capital outlay" that increases the value of the asset on which the expenditure is incurred.

Treating expenditure as maintenance (i.e., as an expense) affects the cost of a department's outputs. Capital expenditure, meanwhile, has an impact on the value of the department's assets and, subsequently, on depreciation and equity return. Accounting for expenditure on assets in an appropriate and consistent manner will provide a more accurate indication of a department's output costs and the value of its assets (Maintenance Management Framework, (I) 2012).

3.4.4.1 What Are Maintenance Expenses?

Maintenance expenses refer to the costs incurred to keep an item in good condition and/or good working order. When purchasing an item that requires upkeep, consumers should consider not just the initial price tag, but also the item's ongoing maintenance expenses. Maintenance expenses are a major reason why an owned home can be more expensive than a rented one, for example. However, sometimes items that are merely leased, not owned, such as a leased car, will require the owner to pay maintenance expenses.*

3.4.4.2 Accounting of Maintenance Expenses

Maintenance can be defined as work on existing assets undertaken with the intention of reinstating their physical condition to a specified standard, preventing further deterioration or failure, restoring correct operation within specified parameters, replacing components at the end of their useful/economic life with modern engineering equivalents, making temporary repairs for immediate health, safety, and security reasons (e.g., after a major failure), and assessing maintenance requirements (e.g., to obtain accurate and objective knowledge of physical and operating conditions, including risk and financial impact, for the purpose of maintenance).

The Accounting Standards define "expenses" as "decreases in economic benefits during the accounting period in the form of outflows or depletion of assets or incurrences of liabilities that result in decreases in equity, other than those relating to distributions to equity participants." The first part of this definition (i.e., "decreases in economic benefits… in the form of outflows or depletion of assets") is relevant here.

In the context of these two definitions, maintenance is a reflection of the consumption (through usage) of the asset. As this consumption results in a reduction in the value of the asset, it meets the definition of an expense. However, work undertaken in the course of maintenance may include activities that result in the expenditure being classified as a capital one (Maintenance Management Framework, (I) 2012).

---

* http://www.investopedia.com/terms/m/maintenance-expenses.asp.

3.4.5 When Is Maintenance Work Classified as a Capital Expenditure?

Expenditure on assets is a capital expense (i.e., added to the carrying amount of the asset) when it improves the condition of the asset beyond its originally assessed standard of performance or capacity. In general, work that includes upgrades, enhancements, and additions to an asset falls into the category of capital expenditure when it results in any of the following:

- An increase in the asset's useful function or service capacity
- An extension of its useful life
- An improvement to the quality of the service(s) delivered through use of the asset
- A reduction in future operating costs
- The upgrade or enhancement becoming an integral part of the asset

Conversely, work falls into the category of maintenance expenditure when it does not result in an improvement to the asset (i.e., it simply preserves the asset's original serviceability). Consequently, expenditures on this kind of work can be capitalized when the performed tasks have increased the useful function of the asset (Maintenance Management Framework, (I), 2012).

3.4.5.1 Extension of Useful Life

The remaining useful life (RUL) of an asset can be defined as "the period over which an asset is expected to be available for use by an entity or the number of production or similar units expected to be obtained from the asset by an entity."

An asset's useful life may be estimated on the basis of its expected running hours and its expected workload.

RUL is not homogeneous, however; it depends on the end user. The useful life of an asset to one end user may well differ from its useful life to another, or even differ between business units within the same entity (Maintenance Management Framework, (I), 2012).

Significant components within large assets (e.g., electronic systems versus mechanical ones) are identified, recognized, and depreciated separately. The useful life of each of these components is generally different from that of the asset from each other.

The following factors influence the useful life of assets and their components:

- Environmental conditions
- Technical obsolescence
- Commercial obsolescence

- Legal compliance issues
- Other limitations on the continued safe and legal use of the asset

In the context of these factors, an extension of the useful life of an asset may result from incorporating the following:

- A more robust material than that used in the original structure
- A component that benefits from an improved design

As such, expenditure in these instances should be carefully reviewed with respect to its categorization as either capital expenditure or expense or a combination of both (Maintenance Management Framework, (I), 2012).

3.4.5.2 Extension of Useful Life of an Asset

Most commonly, the useful life of an asset is assessed and expressed on a time basis. To determine the useful life, the following factors must be considered:

1. In the case of physical assets, useful life is the potential physical life of the asset, that is, the period of time over which the asset can be expected to last physically, at a projected average rate of usage and assuming adequate maintenance.
2. In all cases, the useful life is the potential technical life of the asset, that is, the period of time over which the asset can be expected to remain efficient and not become technically obsolescent.
3. In all cases, the useful life is the expected commercial life of the asset, corresponding to the commercial life of its product or output (the possibility of an alternative use for the asset needs to be kept in mind).
4. In the case of certain rights and entitlements, the useful life is the legal life of the asset, that is, the period of time during which the right or entitlement exists.

The basis for allocating depreciation ought to be appropriate to the nature of the asset and its expected use. It must reflect the underlying physical, technical, commercial, and, where appropriate, legal facts.

The useful life of an asset is normally the shortest of the applicable alternatives. These alternatives relate to an asset's expected useful life in one company, which may be different from its useful life in another company. An addition or modification to an existing asset may, on occasions, extend its useful life. Opportunities for the renewal or extension of a right or entitlement are another factor to consider.

Where the useful life is estimated on a time basis, several methods are available for allocating the depreciation, according to whether the asset's service potential will remain constant from financial year to financial year or will increase or decrease over time. The straight line method is a means of determining systematic allocations which are constant from financial year to financial year; it is popular because of its simplicity. The reducing-balance method is one of several methods used when the asset's service potential decreases from financial year to financial year. Decreasing allocations are justified when an asset is expected to yield more service in the earlier financial years than in the later ones; it could be argued that the earlier financial years ought to bear a larger allocation of the depreciable amount of the asset.

An alternative basis for determining the useful life of an asset is the overall output or service the asset is expected to yield, for example, estimated production units, operating hours, or distance travelled. This is also a means of determining systematic allocations of depreciation. It is appropriate in situations when the service potential of an asset can be expected to diminish in direct proportion to its use, and before the asset becomes technically or commercially obsolete. When the estimated useful life is assessed in terms of expected output or service, depreciation is based on the actual output or service quantities in each financial year and, hence, is likely to vary from financial year to financial year. A variant suitable for certain specialized plants or equipment is to base depreciation on measurement of the specific wear that has occurred in each financial year through use (Depreciation, 1997).

The bases for allocating depreciation need to be selected with proper regard for the underlying facts. Whichever basis is selected, it is essential that it be applied consistently, from financial year to financial year, irrespective of the following (Depreciation, 1997):

- The profit or loss for any one financial year
- The amount that may be claimed as an allowable deduction when computing taxable income for any one financial year
- Whether any of the assets have been revalued

3.4.6 Reduction in Future Operating Costs: Effect of Overinvestments

Reductions in the future operating costs of building assets may occur as a result of repairs that incorporate new materials, more efficient components, or new technology. For example, the integration of more durable or weather-resistant materials into an existing building may reduce maintenance costs, while the installation of a more modern air-conditioning plant may lead to a decrease in energy costs.

Even if these types of replacements are precipitated by maintenance requirements and fail to result in an increased output capacity or improvement in service quality, the expenditure should still be reviewed in terms of its capital content. In such cases, the intent of the work may be relevant. If the primary intent is to reduce future operating costs, the expenditure should be classified as capital. However, the replacement of an asset component purely for maintenance reasons—even if the replacement is a modern engineering equivalent with the potential to reduce future operating costs—should be categorized accordingly unless there is a material change or enhancement in the physical characteristics of the asset (Maintenance Management Framework, (I), 2012).

Asset and maintenance managers need to consider issues of capital versus expense when assessing both the short- and the long-term maintenance requirements of their assets.

In summary, the following are key considerations for asset and maintenance managers making decisions on maintenance:

- Financial management and accounting policies and guidelines
- Value-for-money principles (when ascertaining whether it is more economical to upgrade, replace, or overhaul rather than to make ongoing repairs)
- Value of the asset
- Intent of the work
- Scope of the work
- Outcome of the work
- Impact of the work on asset value, depreciation, and equity return
- Consistency in decision-making

3.4.7 Optimizing Maintenance as a Cost Control Measure

The goal is to come in at or below budget on expense items and at or above budget on revenue items. Incentives like bonuses and commissions are paid for outperforming the budget, with constraints such as meeting production and quality goals applied to ensure ineffective corner cutting isn't used to undercut a budget-line item. Front-line managers are evaluated on their adherence to budgets; after all, cost overruns directly impact a company's income. As a result, cost management on the shop floor often involves attempts to control activities that are beyond local control.

In addition, the things a maintenance manager can do to keep labor costs minimized are often at odds with his or her primary function. When a failure occurs, he or she can assign a mechanic who has the lowest hourly wage, or can deny overtime and let production wait until morning before sending someone to make repairs. Of course, keeping the equipment from failing will allow the maintenance manager to avoid ever having to knowingly "break the budget."

Should we simply ignore cost control measures? The answer obviously is no. Manufacturing companies must create a culture of thrift and continuous improvement, reinforced by a long-term, organization-wide reward system,

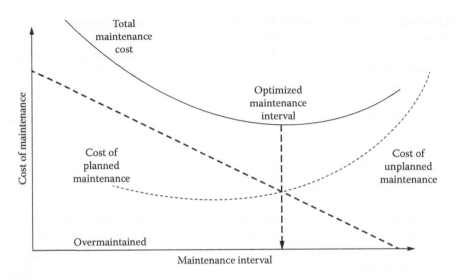

FIGURE 3.12
Cost minimization. (From Smith, R. and Hawkins, B., *Lean Maintenance: Reduce Costs, Improve Quality, and Increase Market Share*, Elsevier Butterworth–Heinemann, Burlington, MA, 2004.)

recognizing which work adds value and identifying and eliminating non-value-adding work. The emphasis should be on managing value up rather than managing costs down.

Cost minimization in maintenance is a matter of not performing unnecessary maintenance (increased labor costs, more off-line production time, etc.) and not missing required maintenance (reduced equipment reliability, equipment failures, production downtime). This is a simple concept, but achieving this balance requires sound reliability engineering applied to a total productive maintenance (TPM) system employing predictive maintenance (PdM) techniques and condition monitoring (CM). Utilizing elements of reliability-centered maintenance (RCM) is perhaps the best method to arrive at this balance. Balancing all of this to make sense out of optimum maintenance intervals is the secret to controlling maintenance costs (see Figure 3.12) (Smith and Hawkins, 2004).

3.4.8 Role of Avoided and Overinvestment Costs in Global Maintenance Cost

The global cost of maintenance, C_g, is the addition of four components:

- Costs of intervention (C_i)
- Costs of failure (C_f)
- Costs of storage (C_a)
- Cost of overinvestment (C_{si})

and can be stated as:

$$C_g = C_i + C_f + C_a + C_{si} \quad \text{(AFNOR, 1994)}.$$

This global cost can be calculated for a specific machine, group of machines, or whole plants.

3.4.8.1 Costs of Intervention

The intervention cost (C_i) includes the expenses related to preventive and corrective maintenance. It does not include costs of investment or those related directly to production: adjustments of production parameters, cleaning, and so on.

The intervention cost can be decomposed into the following components:

- Internal or external manpower
- Stock spare parts or bought for an intervention
- Required expendable equipment for an intervention

It is important to give a realistic value to the costs of intervention by unit of time and man-hour because they directly influence the global cost of maintenance.

3.4.8.2 Costs of Failures

These costs correspond to the losses of margin of operation due to a maintenance problem that has caused a reduction in the production rate of products in good condition. The loss of this margin can include an increase of operation costs or a loss of business.

Maintenance problems are caused by the following:

- Preventive maintenance badly defined
- Preventive maintenance badly executed
- Corrective maintenance badly executed, that is, conducted in very long terms, realized with bad spare parts or of low quality

It is important to stress that the cost of failure of the equipment corresponds to the losses of operation margin caused by a defect that incurs losses of production of acceptable quality.

The failure cost can be calculated with the following formula:

$$C_f = \text{Income not perceived} + \text{extra expenses of production}$$
$$- \text{not used raw material}$$

This cost comprises the following components:

- *Income not perceived*: This factor will depend on the possibility of recovering the production over diverse schedules, weekends, and so on. In case of continuous production, there is evidently no capacity to recover the losses; hence, the production of that time slot and the associated incomes should be imputed into this item.
- *Extra expenses of production*: The use of temporary slots to recover production will have additional costs, including the following:
  - Necessary energy for the production
  - Raw materials
  - Fungibles
  - Expenses of services such as quality, purchases, maintenance, and so on
- *Not used raw material*: This will become a factor if it is not possible to recover production. This cost will be subtracted from the failure cost, because the consumption of the raw material has not taken place. The raw material might, however, be used if the productive plan is recovered, possibly with some cost overrun of storage, transport, or degradation (unless it is a perishable product that must be rejected if not processed).

Chapter 5 explains some common models used to calculate failure cost when assets fail or do not deliver the expected function.

3.4.8.3 Cost of Storage

Inventories represent nearly a third of the assets of a typical company (Diaz and Fu, 1997). In fact, in maintenance, 70% of the budget is manpower and 30% is spare parts. The storage cost represents the costs incurred in financing and regulating the necessary inventory of spare parts for maintenance.

3.4.8.4 Cost of Overinvestments

The overall goal of most companies is to diminish the global maintenance cost of an asset during its entire service life. The initial investments in equipment may be bigger than they should be simply because the costs of associated intervention and storage are considered cheaper, creating overinvestment in equipment. To include overinvestment, in cost calculations, the difference is amortized over the life of the equipment.

Similarly, the indiscriminate application of expensive technologies and methodologies like monitoring unnecessary condition parameters can be considered overinvestments, creating a burden on the maintenance budget. Finally, all investments on assets which await return belong in this category.

3.4.8.5 Avoided Costs

One of the most frequent financial problems in modeling maintenance systems is that the original costs can be modified in applications of methodologies or technologies that reduce the global cost using the concept of avoided costs.

In this type of policy, three of the four parameters that constitute the global cost are affected:

- *Costs of interventions* (C_i): Normally, if these are reduced in frequency and in volume, there has been a reduction of corrective and an increase of preventive maintenance.
- *Costs of failures* (C_f): These are reduced in predictive maintenance policies where complete overhaul is replaced by small inspections performed without shutting down the production process.
- *Cost of oversized investments* (C_{si}): Useless expensive equipment and plans of inspection are perhaps the most noticeable items in this cost because the budget is increased, but the items are rarely used and, in consequence, there is no added value to the process itself.

The following equation shows the impact of the costs avoided. The cost of intervention and failure is reduced by a certain percentage with intervention; however, the cost of overinvestment will increase if the technique implemented does not work for the company and does not result in a return of investment.

$$C_g = C_i + C_f + C_a + C_{si} - C_{av}$$
$$C_g = \left(C_i - C_{av\_i}\right) + \left(C_f - C_{av\_f}\right) + C_a + \left(C_{si} + C_{av\_si}\right)$$

3.5 CMMS as a Maintenance Cost Control Tool

A CMMS database is the lifeblood of maintenance and engineering departments. Its data indicate the general condition of facilities and systems, reveal small problems that could become major headaches, and form the basis of every decision a manager must make, from staffing and training needs to inventory levels and big-ticket purchases.

3.5.1 Uses of Computerized Maintenance Management System

CMMS is a software that enables companies to operate and maintain their buildings and technical assets in an efficient, cost-effective, and compliant way. It supports forecasting, planning, and evaluating any type of maintenance work, either contracted or delivered by internal maintenance staff.*

In addition, CMMS is a tool to help manage and track maintenance activities, such as scheduled maintenance, work orders, parts and inventory, purchasing, and projects. It gives full visibility and control of maintenance operations, so everyone can see what has been done and what needs to be done. It helps identify tasks that need to be done or prioritized, ensuring nothing is overlooked. One of the biggest benefits of CMMS is increased labor productivity. The system can help plan and track work so technicians can complete their tasks without interruption. With proper planning and tracking, the maintenance team is more organized and less stressed.

CMMS can help an organization become more safety-compliant in a number of ways. Safety procedures can be included on all job plans, ensuring technicians are aware of the risks. Safety checks, such as fire equipment inspections, can be scheduled and tracked in CMMS, ensuring the organization is compliant and ready for those audits.

In the past, CMMS was complex and difficult to use but modern CMMS applications like Maintenance Assistant are simple to use, intuitive, and accessible on any device such as mobiles, tablets, or PCs. CMMS helps maintenance managers become more organized as it reduces the dependence on paper and memory by automating many mundane daily activities. Rather than trolling through receipts and dockets at the end of a year, the manager can simply run a costing report in CMMS to see where the budget was spent. Over time, using CMMS can help drive down the cost of maintenance, increase asset life, improve productivity, reduce downtime, and lower the total cost of ownership of assets.

3.5.1.1 CMMS Needs Assessment

In determining the need for CMMS, managers should assess their current mode of operation. The following are key questions that should be asked:

- Do you have an effective way to generate and track work orders? How do you verify the work was done efficiently and correctly? What is the notification function upon completion?
- Are you able to access historical information on the last time a system was serviced, by whom, and for what condition?
- How are your spare parts inventories managed and controlled? Do you have excess inventories or are you consistently waiting for parts to arrive?

---

* http://www.investopedia.com/terms/m/maintenance-expenses.asp.

- Do you have an organized system to store documents (electronically) related to O&M procedures, equipment manuals, and warranty information?
- When service staff are in the field, what assurances do you have that they are compliant with health and safety issues and are using the right tools/equipment?
- How are your assets, that is, equipment and systems, tracked for reporting and planning?

If the answers to these questions are not well defined or lacking, it may be worth investigating the benefits of a well-implemented CMMS (Sullivan et al., 2010).

3.5.1.2 CMMS Capabilities

CMMS automates most of the logistical functions performed by maintenance staff and management. It has many options and many advantages over manual maintenance tracking systems. Depending on the complexity of the system, typical CMMS functions include the following:

- Work order generation, prioritization, and tracking by equipment/component
- Historical tracking of all work orders generated, sorted by equipment, date, person responding, and so on
- Tracking of scheduled and unscheduled maintenance activities
- Storing of maintenance procedures, as well as all warranty information by component
- Storing of all technical documentation or procedures by component
- Real-time reports of ongoing work activity
- Calendar- or run time–based preventive maintenance work order generation
- Capital and labor cost tracking by component, as well as shortest, medium, and longest times to close a work order by component
- Complete parts and materials inventory control with automated reorder capability
- Digital interfaces to streamline input and work order generation
- Outside service call/dispatch capabilities

Many CMMS programs can now interface with existing energy management and control systems (EMCS) as for interaction from managerial to shop-floor level in maintenance decisions. Coupling these capabilities allows condition-based monitoring and component use profiles (Sullivan et al., 2010).

While CMMS can go a long way toward automating and improving the efficiency of most maintenance programs, there are some common pitfalls (Sullivan et al., 2010). These include the following:

- *Improper selection of CMMS*: This is a site-specific decision. Time should be taken to evaluate initial needs and look for the proper match.
- *Inadequate training of staff on proper use of CMMS*: Staff need dedicated training on input, function, and maintenance of CMMS.
- *Lack of commitment to properly implement CMMS*: A commitment needs to be in place for the startup/implementation of CMMS.
- *Lack of commitment to persist in CMMS use and integration*: While CMMS provides significant advantages, it needs to be maintained.

It is important to remember that CMMS is an information repository whose outcomes strongly depend on the quality of the data stored in the system. If managers do not have the right kind of data or enough of it, they must make decisions without all the facts. The wrong decision could be costly.

To maximize CMMS use, managers will need to revisit the system's features and functions, review and upgrade training for users, and reinforce among users the need for a steady stream of reliable maintenance and repair data (Hounsell, 2008).

3.5.1.3 CMMS Benefits

One of the greatest benefits of CMMS is the elimination of paperwork and manual tracking activities, thus enabling maintenance staff to become more productive, especially tracking the maintenance cost and using the tool as a real cost control. It should be noted that the functionality of CMMS lies in its ability to collect and store information in an easily retrievable format besides the connectivity with other deployed systems. CMMS does not make decisions; rather, it provides the manager with the best information to make maintenance decisions (Sullivan et al., 2010).

Benefits to implementing CMMS include the following:

- Detecting impending problems before a failure occurs, resulting in fewer failures and customer complaints.
- Achieving a higher level of planned maintenance activities to enable a more efficient use of staff resources.
- Improving inventory control by enabling better forecasting of spare parts to eliminate shortages and minimize existing inventory.
- Maintaining optimal equipment performance to reduce downtime and increase equipment life

- Keeping a detailed record of maintenance costs and investments, tracking the consequences and fulfilling the annual budget. Some CMMS applications may perform LCC calculations for asset renovations, overhauls, and replacements.

3.5.1.4 Use CMMS to Save Costs by Avoiding Maintenance

Reactive maintenance work, mostly caused by asset failures or maintenance backlogs, is expensive because it is not planned or budgeted but needs immediate action. Time-based maintenance allows maintenance experts to plan and execute maintenance activities based on warranties, qualitative requirements, or compliancy and legislation. This type of maintenance reduces reactive work drastically, including the corresponding costs.

However, more than 40% of this planned maintenance is still not executed on time, and is either too early or too late. If it is too late, it causes expensive reactive work; if too early, it is a waste of investment and time. CMMS ensures a timely execution of maintenance work by providing objective asset condition assessments. However, such cost assessment comprises many intangible items that are really challenging to quantify.

Top-notch CMMS applications are able to assess the actual condition of an asset and automatically calculate and schedule the needed maintenance. Extensive maintenance libraries of activities, deterioration curves, and cost catalogues feed the CMMS software with data and intelligence to ensure just-in-time maintenance execution.

The ultimate just-in-time maintenance is supported with real-time tracking of assets' actual statuses and data. Intelligent buildings and smart assets are connected with the IT network and send data about production hours, failure statuses, production numbers, and energy consumption in real time to CMMS. Based on preset thresholds or historical information, CMMS can accurately forecast relevant maintenance jobs, their costs, and timing.*

3.5.1.5 Line Maintenance Costs

Line maintenance is a special type of maintenance, typically used in the aviation industry but found elsewhere as well, for example, in transportation (i.e., highways). It refers to maintenance done during the time of one work shift, and in aviation, it would include a check of an aircraft between flights. It is not to be confused with regularly scheduled maintenance. In aviation, the bulk of maintenance costs listed in CMMS reports are typically line

---

* http://planonsoftware.com/us/whats-new/knowledge-center/glossary/cmms/.

costs. These comprise costs related to labor, equipment, material, and others, including payments to contractors (Markow and Prairie, 2011).

- *Labor costs*: The sum of payments to employees for performing line maintenance jobs.
- *Equipment costs*: The total equipment charges incurred in performing line maintenance jobs, whether for company-owned equipment or equipment owned by contractors.
- *Material costs*: The total charges for materials and supplies in performing maintenance jobs. These charges can be for consumption of materials maintained in inventories or stockpiles, the use of materials fabricated in company shops, or materials purchased specifically from outside vendors to complete a particular maintenance job. The costs applied should be the costs of materials actually used in maintenance jobs, not the overall cost of bulk purchases. Additional costs such as overheads for inventory/stockpile operation and management should be treated as support costs, not line costs.
- *Other costs*: Total charges for other items associated with maintenance jobs that do not fit into the labor, equipment, or material categories, for example, utility charges, private equipment rental, and the sum of payments made to contractors to complete maintenance jobs in cases where maintenance activities are delivered through a combination of company and contractor resources. A challenge that many companies may face in treating contract costs within a cost determination process is to identify the maintenance units of accomplishment that relate to contract expenditures for each maintenance activity. Ideally, total units of accomplishment will be calculated before the contract is awarded rather than leaving the calculation as an afterthought.

Information on maintenance line costs is typically available from a company's CMMS and, potentially, its financial management system. Steps in preparing this information for the cost determination process are as follows:

1. Identify which activities are line cost items. Other activities (e.g., training, maintenance management, building, and yard maintenance) should be included with program support costs.
2. Develop a good understanding of the relationship between costs reported in the MMS and maintenance costs reported in the financial accounting system.

Companies with a well-integrated system architecture and a financial system that tracks specific maintenance activities should consider using the financial system data for the cost determination of line maintenance.

An important prerequisite for this decision is that financial system totals for the maintenance program must be close to, and preferably match exactly, the corresponding CMMS totals. If this is not the case, a reconciliation and adjustment review should be done to bring the respective totals closer together. Typical reasons for such differences include variations between the two systems in how individual costs are recorded, inclusion of projects for which judgments differ on whether they are part of the company's routine maintenance program, and errors by company personnel in reporting work (and whether the MMS and the financial system have isolated and corrected these errors). Once adjustments have been identified and agreed to, the financial system data can be used in further calculations.

If the financial system does not break down maintenance expenditures by activity, a hybrid approach should be investigated in which the financial system provides estimates of overall line costs, and CMMS is used to disaggregate these costs by line activity. A prerequisite is to reconcile differences between financial and CMMS data before proceeding. The objective is to identify the full, complete set of line maintenance costs as closely as possible with a reasonable level of effort. Estimates are appropriate when needed; it is more important to account for all likely sources of line costs in a realistic way than to spend excessive time determining every cost to the penny.

If the financial system does not address line maintenance costs directly (e.g., it includes only a line item for "maintenance labor" that encompasses both line and support costs), CMMS should be used as the primary source of line cost data by activity. It would be advisable—once maintenance program costs are also computed—to compare total estimated maintenance line labor plus maintenance program support labor costs from CMMS and other sources to the total "maintenance labor" item in the financial system to determine the required adjustments (Markow and Prairie, 2011).

References

Alexander, M., Young, D., 1996. Strategic outsourcing. *Long Range Planning* 29(1), 116–119.

Al-Najjar, B., 1996. Total quality maintenance: An approach for continuous reduction in costs of quality products. *Journal of Quality in Maintenance Engineering* 2(3), 4–20.

Anderson, M. C., 1997. A primer in measuring outsourcing results. *National Productivity Review* 17(1), 33–41.

Assaf, S., Hassanain, M. A., Al-Hammad, A.-M., Al-Nehmi, A., 2011. Factors affecting outsourcing decisions of maintenance services in Saudi Arabian universities. Architectural Engineering Department, King Fahd University of Petroleum and Minerals, Dhahran, Saudi Arabia.

Baitheiemy, J., 2003. The seven deadly sins of outsourcing. *Academy of Management Executive* 17(2), 87–98.

Beecher, J. A., 1996. *Regionalization of Water Utilities: Perspectives and Annotated Bibliography*. The National Regulatory Research Institute, Columbus, OH.

Beecher, J. A., 2011. Avoided cost: An essential concept for integrated resource planning. Senior Research Scientist, Center for Urban Policy and the Environment, Indiana University-Purdue University, Indianapolis, IN.

Blumberg, D. F., 1998. Strategic assessment of outsourcing and downsizing in the service market. *Managing Service Quality* 8(1), 5–18.

Bounfour, A., 1999. Is outsourcing of intangibles a real source of competitive advantage? *International Journal of Applied Quality Management* 2(2), 127–151.

Campbell, J., 1995. *Uptime: Strategies for Excellence in Maintenance*. Productivity Press, Portland, OR.

Cooper, R., November 1990. Explicating the logic of ABC. *Management Accounting* 68(9), 58–60.

Corbett, M. F., 1999. Multiple factors spur outsourcing growth. pp. 1–6. www.OutsourcingJournal.com/issues/jan. Accessed April 30, 2003.

Depreciation, 1997. Accounting Standard AASB 1021. Issued by the Australian Accounting Standards Board. Caulfield Victoria, Australia, August 1997.

Diaz, A., Fu, M. C., 1997. Models for multi-echelon repairable item inventory systems with limited repair capacity. *European Journal of Operational Research* 97(3), 480–492.

Djavanshir, G. R., 2005. Surveying the risks and benefits of IT outsourcing. *IT Professional* 7(6), 32–37.

Emblemsvåg, J., 2003. *Life-Cycle Costing: Using Activity-Based Costing and Monte Carlo Methods to Manage Future Costs and Risks*. John Wiley & Sons, Inc., Hoboken, NJ.

Fontes, R., 2000. The outsource option. *Folio: The Magazine for Magazine Management* 2000, 112–113.

Gharaibeh, N. G., Shelton, D., Ahmed, J., Chowdhury, A., Krugler, P. E., 2011. Development of performance-based evaluation methods and specifications for roadside maintenance. Report no. FHWA/TX-11/0-6387-1. Texas Transportation Institute, The Texas A&M University System, College Station, TX.

Ghodrati, B., Kumar, U., 2005. Reliability and operating environment-based spare parts estimation approach. *Journal of Quality in Maintenance Engineering* 11(2), 169–184.

Government Performance and Results Act, 2003. http://www.whitehouse.gov/omb/mgmt-gpra/gplaw2m.html.

Greaver, M., 1999. *Strategic Outsourcing: A Structured Approach to Outsourcing Decisions and Initiatives*. American Management Association, New York.

Hatry, H. P., Fountain, J., Sullivan Jr., J., Kremer, L., 1990. Service efforts and accomplishments reporting, its time has come: An overview, Governmental Accounting Standards Board, Norwalk, CT.

Hounsell, D. (Ed.), September 2008. Tapping the power of a CMMS. http://www.facilitiesnet.com/software/article/Justify-Budgets-with-a-CMMS-Facility-Management-Software-Feature--9600. Accessed April 30, 2017.

Huiskonen, J., 2001. Maintenance spare parts logistics: Special characteristics and strategic choices. *International Journal of Production Economics* 71(1–3), 125–133.

Hyman, W. A., 2004. *A Guide for Customer-Driven Benchmarking of Maintenance Activities*. Transportation Research Board, National Research Council, Washington, DC.

Hyman, W. A., 2009. *Performance-Based Contracting for Maintenance. NCHRP SYNTHESIS 389. A Synthesis of Highway Practice.* Transportation Research Board of the National Academies, Washington, DC.

IAEA, 2001. *Reliability Assurance Programme Guidebook for Advanced Light Water Reactors.* IAEA-TECDOC-1264. IAEA, Vienna, Austria.

IEV, 2007. International Electrotechnical Vocabulary (IEV) Online, Chapter 191: Dependability and quality of service, March 2012.

Johnston, M., January 5, 2014. The pros & cons of outsourced vs. in-house maintenance. http://www.chem.info/article/2014/05/pros-cons-outsourced-vs-house-maintenance. Accessed April 30, 2017.

Kelly, T., Wilmslow, C., 2007. Some thoughts on Maintenance Budgeting. ME CENTRAL THEME.

Kennedy, W. J., Patterson, W. J., 2002. An overview of recent literature on spare parts inventories. *International Journal of Production Economics* 76(2), 201–215.

Kremic, T., Tukel, O. I., Rom, W. O., 2006. Outsourcing decision support: A survey of benefits, risks, and decision factors. *Supply Chain Management: An International Journal* 11(6), 467–482.

Levitt, J. D., 2009. Maintenance budget based on replacement asset value. Springfield Resources, Maintenance Management Consultation and Training, Lafayette Hill, PA.

Lonsdale, C., 1999. Effectively managing vertical supply relationships: A risk management model for outsourcing. *Supply Chain Management: An International Journal* 4(4), 176–183.

Maintenance Management Framework, (I), 2012. Capital or expense? A guide for asset and maintenance managers. Department of Housing and Public Works, Brisbane, Queensland, Australia.

Maintenance Management Framework, (II), 2012. *Building Maintenance Budget,* 2nd edn. Department of Housing and Public Works, Brisbane, Queensland, Australia.

Maintenance Management Framework (MMF policy document), October 2012. Policy for the maintenance of Queensland Government buildings. Department of Housing and Public Works, Brisbane, Queensland, Australia. www.hpw.qld.gov.au. Accessed April 30, 2017.

Maintenance Management Systems, October 2000. Technical information document. RPS for INAC. TID-AM-01.

Maintenance Technology, November 19, 2004. Instituting a zero-based maintenance budget based on equipment requirements. The source for reliability solutions. http://www.maintenancetechnology.com/2004/11/instituting-a-zero-based-maintenance-budget-based-on-equipment-requirements/.

Markeset, T., 2010. Design for performance: Review of current research in Norway. In: *Proceedings of Condition Monitoring and Diagnostic Engineering Management,* Nara, Japan.

Markow, M. J., Prairie, E., 2011. *Determining Highway Maintenance Costs.* NCHRP REPORT 688. Cambridge Systematics, Inc., Cambridge, MA.

Martin, L., 1993. How to compare costs between in-house and contracted services. http://www.ipspr.sc.edu/publication/FINAL%20On%20Cost%20Analysis%20Comparisons.pdf. Accessed April 30, 2017.

McDonagh, J., Hayward, T., 2000. Outsourcing corporate real estate asset management in New Zealand. *Journal of Corporate Real Estate* 2(4), 351–371.

Nili, M., Shekarchizadeh, A., Shojaey, R., Dehbanpur, M., May 2013. Outsourcing maintenance activities or increasing risks? Case study in oil industry of Iran. *International Journal of Academic Research in Business and Social Sciences* 3(5), ISSN 2222-6990.

Norsok, Z., 2011. Risk based maintenance and consequence classification.

Özsen, R., Thonemann, U., 2015. *Determining Optimal Parameters for Expediting Policies*. Manufacturing & Service Operations Management. https://www.informs.org/Blogs/M-SOM-Blogs/M-SOM-Review/Determining-Optimal-Parameters-for-Expediting-Policies. Accessed April 30, 2017.

Park, C. S., Sharp-Bette, G. P., 1990. *Advanced Engineering Economics*. John Wiley & Sons, New York, p. 740.

Pettit's, J., 2000. *EVA & Strategy*. Stern Stewart & Co., New York, p. 17.

Quélin, B., Duhamel, F., 2003. Bringing together strategic outsourcing and corporate strategy: Outsourcing motives and risks. *European Management Journal* 21(5), 647–661.

Ribreau, N., 2004. Highway maintenance outsourcing experience—Synopsis of Washington State Department of Transportation's review. Maintenance Management and Services, Transportation Research Board National Research Council, Washington, DC, pp. 3–9.

Schmidt, M., 2004. *Avoided Cost, Cost Savings, and Opportunity Cost Explained*. Solution Matrix Limited, Boston, MA. https://www.business-case-analysis.com/avoided-cost.html. Accessed April 30, 2017.

Shangguan, Z., April 2013. Spare parts management in Bohai bay. Master's thesis in offshore technology. University of Stavanger, Stavanger, Norway.

Shrestha, P., Said A., Shrestha, K., June 2015. *Investigation of an Innovative Maintenance Contracting Strategy: The Performance-Based Maintenance Contract (PBMC)*. Nevada Department of Transportation, Carson City, NV.

Sislian, E., Satir, A., 2000. Strategic sourcing: A framework and a case study. *The Journal of Supply Chain Management* Summer, 4–11.

Smith, R., Hawkins, B., 2004. *Lean Maintenance: Reduce Costs, Improve Quality, and Increase Market Share*. Elsevier Butterworth–Heinemann, Burlington, MA.

Smith, S., February 3, 2012. *Choosing In-House or Outsourced Maintenance*. Mintek Mobile Data Solutions, Inc. http://www.mintek.com/blog/eam-cmms/choosing-in-house-outsourced-maintenance/. Accessed April 30, 2017

Srinivasan, M. S., Srinivasan, S., 1986. *Maintenance Standardization for Capital Assets: A Cost-Productivity Approach*. Praeger Publishers, New York.

Stankevich, N., Qureshi, N., Queiroz, C., 2009. *Performance-Based Contracting for Preservation and Improvement of Road Assets*. (Transport note. TN-27). The World Bank, Washington, DC. http://www.esd.worldbank.org/pbc_resource_guide/Docslatest%20edition/PBC/trn_27_PBC_Eng_final_2005.pdf. Accessed April 30, 2017.

Sullivan, G. P., Pugh, R., Melendez, A. P., Hunt, W. D., August 2010. Computerized Maintenance Management System. In: *Operations & Maintenance Best Practices. A Guide to Achieving Operational Efficiency, Release 3.0*. Pacific Northwest National Laboratory for the Federal Energy Management Program U.S. Department of Energy August 2010. http://energy.gov/sites/prod/files/2013/10/f3/OM_4.pdf.

World Bank, 2002. Sample bidding document: Procurement of performance-based management and maintenance of roads, Washington, DC.

World Health Organization (WHO), 2012. Guidance on performance based contracting.http://www.who.int/management/resources/finances/ Section2-3.pdf. Accessed April 30, 2017.

Yik, F. W., Lai, J. H., 2005. The trend of outsourcing for building services operation and maintenance in Hong Kong. *Facilities* 23(1/2), 63–72.

Zietsman, J., 2004. Performance measures for performance based maintenance contracts. Texas Transportation Institute, Houston, TX. http://www-esd. worldbank.org/pbc_resource_guide/Docs-latest%20edition/cases-and-pdfs/ ZietsmanTexas.pdf.

4

Maintenance Performance Measurement: Efficiency

4.1 Key Performance Indicators for Maintenance

> It is not possible to manage what you cannot control and you cannot control what you cannot measure! (Peter Drucker)

Performance measurement is a fundamental principle of management. It identifies gaps between current and desired performance and indicates progress toward closing the gaps. Carefully selected key performance indicators (KPIs) identify precisely where to take action to improve performance (Weber and Thomas, 2005).

4.1.1 Physical Asset Management

The purpose of most equipment in manufacturing is to support the production of products destined for downstream customers. Ultimately, the focus is on meeting customer needs. This is illustrated in Figure 4.1. Customer expectations are normally defined in terms of product quality, on-time delivery, and competitive pricing. By reviewing the composite requirements of all current customers and potential customers in the markets we wish to penetrate, we can define the performance requirements of our physical assets. Manufacturing performance requirements can be associated with quality, availability, customer service, operating costs, safety, and environmental integrity (Weber and Thomas, 2005).

To achieve the required performance, three inputs must be managed. The first is *design practices*. Design practices ensure the equipment is able to meet the manufacturing performance requirements "by design" (inherent capability).

The second input is *operating practices*, which make use of the inherent capability of process equipment. The documentation of standard operating practices assures the consistent and correct operation of equipment to maximize performance.

FIGURE 4.1
Managing manufacturing performance requirements to meet customer needs. (Adapted from Weber, A. and Thomas, R., Key performance indicators. Measuring and managing the maintenance function, Ivara Corporation, Burlington, Ontario, Canada, November 2005.)

The third input is *maintenance practices*. These maintain the inherent capability of the equipment. Deterioration begins to take place as soon as equipment is commissioned, but in addition to normal wear and deterioration, other failures may occur, especially when equipment is pushed beyond the limitations of its design or there are operational errors. Degradation in equipment condition results in reduced equipment capability. Equipment downtime, quality problems, and the potential for accidents and/or environmental disasters are the visible outcomes. All can negatively impact the operating cost.

Asset capability, operating practices, and the maintenance of asset condition contribute to the ability to meet performance requirements. Some typical KPIs used to measure these are operating cost, asset availability, environmental incidents, and asset utilization (Weber and Thomas, 2005).

As depicted in Figure 4.2, asset utilization is a function of many variables. For example, it is impacted by both maintenance- and non-maintenance-related downtime. Non-maintenance-related downtime may be attributed to lack of demand, an interruption in raw material supply, or production scheduling delays beyond the control of the maintenance function. Asset utilization is also a function of operating rate, quality, and yield losses, among other factors. In each of these areas, maintenance may be a factor but it is not the only contributor. In order to maintain and improve performance, each function in the organization must focus on the portion of the indicators it influences.

Similarly, asset capability, operating practices and the maintenance of asset condition all contribute to the ability to meet performance requirements but are not necessarily controlled by maintenance.

In all cases, if a manufacturing-level indicator is used to measure maintenance performance, improved maintenance may not result in a proportional

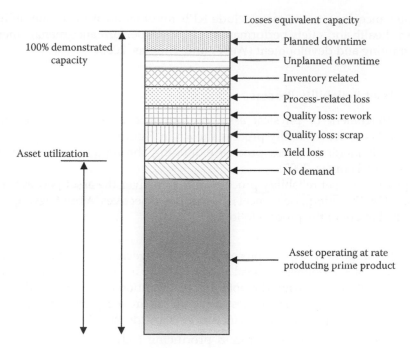

FIGURE 4.2
Asset utilization. (Adapted from Weber, A. and Thomas, R., Key performance indicators. Measuring and managing the maintenance function, Ivara Corporation, Burlington, Ontario, Canada, November 2005.)

improvement in the manufacturing metric. For instance, in the asset utilization example cited earlier, the maintenance contributors may all be positive, but the resulting asset utilization may not improve.

That said, maintenance performance contributes to manufacturing performance. The KPIs for maintenance are related to the KPIs for manufacturing.

For accurate measurements, however, there must be a direct correlation between the maintenance activity and the KPI measuring it. When defining a KPI for maintenance, a good test of the metric validity is to seek an affirmative response to the question "If the maintenance function does 'everything right', will the suggested metric always reflect a result proportional to the change; or are there other factors, external to maintenance, that could mask the improvement?"

This section focuses on defining KPIs for the maintenance function, not the maintenance department. The maintenance function can involve other departments beyond the maintenance department. Similarly, the maintenance department has added responsibilities beyond the maintenance function and, as such, will have additional KPIs to report. The KPIs for the maintenance department may include KPIs for other areas of accountability, such as health and safety performance, employee performance management, and training and development (Weber and Thomas, 2005).

4.1.2 Asset Reliability Process

The management of physical asset performance is integral to business success. The asset reliability process shown in Figure 4.3 is an integral part of a much larger business process extending throughout the enterprise (Weber and Thomas, 2005).

A proactive asset reliability process aims to deliver the asset performance required by the enterprise to meet its corporate objectives. A brief description of each element of the process follows:

- The *business focus* is represented by the green box on the left in Figure 4.3. It defines physical asset reliability in terms of the business goals of the company. The potential contribution of the asset to these goals is evaluated. The largest contributors are recognized as critical assets, and specific performance targets are identified.
- *Work identification* is a process producing technically based asset reliability programs to identify and control the failure modes impacting the equipment's ability to perform the intended function at the required performance level. Activities are evaluated as to whether they are worth conducting based on the consequences of failure.
- *Planning* develops procedures and work orders for the identified work activities. The procedures have specific resource requirements, safety precautions, and special work instructions.
- *Scheduling* evaluates the availability of all resources required for work "due" in a specified time frame. Often this work requires the equipment to be shut down, so a review of production schedules is required. Resources are attached to a specific work schedule; this allows the use of resources to be balanced out.
- In the *execution* process, trained, competent personnel carry out the required work.
- The *follow-up* process responds to information collected in the execution process. Work order completion comments outline what was done and what was found. Actual time and manpower to complete the job are documented, job status is updated as complete or incomplete, corrective work requests resulting from the analysis of inspection data are created, and requests are made for changes to procedures.

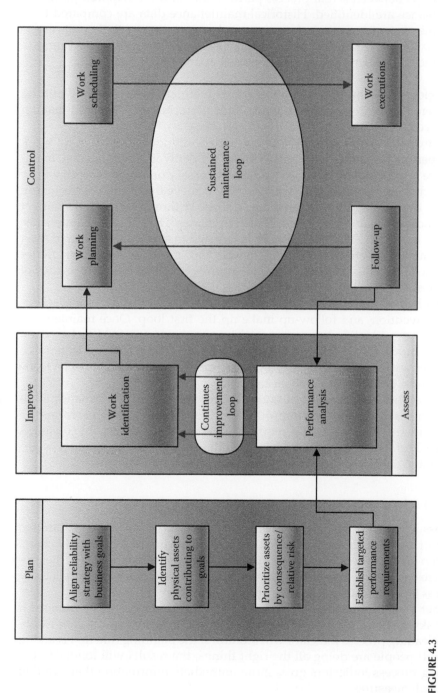

FIGURE 4.3

Asset reliability process. (Adapted from Weber, A. and Thomas, R., Key performance indicators. Measuring and managing the maintenance function, Ivara Corporation, Burlington, Ontario, Canada, November 2005.)

- *Performance analysis* evaluates maintenance program effectiveness. Gaps between actual process performance and the required performance are identified. Historical maintenance data are compared to the current process performance, and maintenance activity costs are reviewed. Significant performance gaps are addressed by revisiting the work identification function (Weber and Thomas, 2005).

Each element is important in an effective maintenance strategy. Omitting any element will result in poor equipment performance, increased maintenance costs, or both.

For example, work identification systematically identifies the right work to be performed at the right time. Without proper work identification, maintenance resources may be wasted, and unnecessary or incorrect work will be planned. Once executed, this work may not achieve the desired performance results, despite significant maintenance costs. Without planning, the correct and efficient execution of the work is left to chance. The planned maintenance process is a cycle. Maintenance work is targeted to achieve the required asset performance. Its effectiveness is reviewed and improvement opportunities identified. This guarantees continuous improvement in the performance of processes impacted by maintenance.

The planned maintenance process has two internal loops. Planning, scheduling, execution, and follow-up make up the first loop. Once maintenance activities are identified, an asset maintenance program (AMP), based on current knowledge and requirements, is initiated. The selected maintenance activities will be enacted at the designed frequency and maintenance tolerance limits. The process is self-sustaining.

The second loop consists of work identification and performance analysis. This is a continuous improvement loop. Actual asset performance is monitored relative to the required performance (driven by business needs). Performance gaps are identified; the "cause" of these gaps is established and corrective action recommended (Weber and Thomas, 2005).

4.1.3 Performance Metrics for the Maintenance Function

The asset reliability process represents the collection of "all" tasks required to support the maintenance function. The process is much like a supply chain. If a step in the process is skipped or performed at a substandard level, the process creates defects known as failures. But the output of a healthy reliability process is optimal asset reliability at optimal cost.

Asset reliability process measures are leading indicators. They monitor if the tasks are being performed that will lead to the desired results. For example, a leading process indicator will monitor if the planning function is taking place. If people are doing all the right things, the results will follow. Using leading process indicators gives more immediate information than waiting for result measures.

Result measures monitor the products of the asset reliability process. These measures include maintenance cost (a contributor to total operating cost), asset downtime due to planned and unplanned maintenance (a contributor to availability), and number of failures on assets (this can be translated into mean time between failures). Results are lagging measures. Failure is a good example of this. Typically, the same piece of equipment doesn't fail day after day. Take a pump for example. Say the pump fails, on average, once every 8 months. If we improve its reliability by 50%, it will now fail every 12 months. We have to wait at least 12 months to see the improvement.

KPIs for the maintenance function need to include both leading (maintenance process) measures and lagging (result) measures. Collectively, these measurements are the KPIs for the maintenance function (Weber and Thomas, 2005).

4.1.3.1 Reliability Process Key Performance Indicators: Leading Measures

The maintenance process is made up of elements required to complete a supply chain. KPIs of the maintenance process are process assurance measures. They answer the question "How do I know this maintenance process element is being performed well?" As discussed in the preceding section, the day-to-day execution of maintenance is addressed through the seven elements of the reliability process: business focus, work identification, work planning, work scheduling, work execution, follow-up, and performance analysis. Each element should have its own KPIs (Weber and Thomas, 2005).

It should be noted that variations of these metrics may be defined or additional performance metrics may be used (Weber and Thomas, 2005). However, the metrics presented here show whether the requirements of each element are being satisfied and, if not, what action should be taken. The following subsections go through each of the seven elements in more detail.

4.1.3.1.1 Work Identification

The function of work identification is to identify the right work at the right time. Initiating a work request is one method of identifying work. Once a work request is submitted, it must be reviewed, validated, and approved before it becomes an actual work order. If the work request process is performing well, its validation and approval/rejection should occur promptly.

A suggested measure of the work request process is the percentage of work requests remaining in "Request" status for less than 5 days, over a specified time period (e.g., the last 30 days). The world-class maintenance expectation is that most work (>80%) requests will be reviewed and validated within a maximum of 5 days. Work requests rely on the random identification of problems or potential problems wherein they are brought to the attention of

The scheduling of properly planned work is also important to maximize maintenance efficiency.

A high percentage of the available maintenance man-hours should be committed to a schedule. A scheduling KPI measures the following:

- The percentage of scheduled available man-hours to total available man-hours over the specified time period.

A world-class target of >80% of man-hours should be applied to scheduled work.

With a high quality of work identification, planning, and scheduling, maintenance should be done according to the plan and schedule. That said, it is not desirable to schedule 100% of available man-hours because additional work will arise after the schedule has been set up. This includes both emergency work and add-ins that must be accommodated (Smith and Mobley, 2011).

4.1.3.1.4 Work Execution

Work execution begins with the assignment of work to the people responsible for executing it and ends when the individuals charged with responsibility for execution provide feedback on the completed work. Work execution quality is measured by the following indicators:

- The percentage of rework. World-class levels of maintenance rework are less than 3%.

The idea that the job is not done until the work order is completed and returned is a significant challenge to many organizations. For this reason, it is important to have a KPI on work order completion. This metric should look at the following:

- A returned work order should indicate the status of the job (complete, incomplete), the actual labor and material consumed, an indication of what was done and/or what was found, and recommendations for additional work. In addition, information about process and equipment downtime and an indication of whether the maintenance responded to a failure should be provided.

A KPI of execution is schedule compliance. This refers to the following:

- The percentage of work orders completed during the scheduled period before the late finish or required by date. World-class maintenance should achieve >90% schedule compliance during execution.

As this section and the previous ones suggest, the ability to successfully monitor and manage the maintenance process and measure the results is highly dependent on gathering correct information. The vehicle for collecting this information is the work order. Work orders should account for all work executed on assets (Smith and Mobley, 2011).

4.1.3.1.5 Follow-Up

In the follow-up element of the maintenance process, actions are initiated to address the information identified during execution. Some key follow-up tasks include reviewing work order comments and closing completed work orders, initiating corrective work, and initiating part and procedural updates as required. Timely follow-up and closure of completed work orders is essential to maintenance success (Smith and Mobley, 2011).

These are the KPIs for follow-up:

- The percentage of work orders turned in with all the data fields completed. World-class maintenance organizations archive 95% compliance.
- The percentage of work orders closed within a maximum of 3 days over the specified time period. The expectation is that >95% of all completed work orders should be reviewed and closed within 3 days (Smith and Mobley, 2011).

4.1.3.2 Performance Analysis

The performance analysis element of the maintenance process evaluates maintenance effectiveness by focusing on KPIs of maintenance results. At this point, any gaps between the actual and required performance of the maintained asset are identified. Significant performance gaps are addressed by initiating work identification improvement actions to close the performance gap (Smith and Mobley, 2011).

From a maintenance process perspective it is important to have the results driving the subsequent actions. Therefore, a KPI for performance analysis is a measure of the quality of the performance analysis itself (Smith and Mobley, 2011). One indication that performance analysis is taking place is the existence of maintenance result metrics (described in the next section), that is, KPIs of maintenance effectiveness (result measures) (Weber and Thomas, 2005).

Measures of the effectiveness of performance analysis include the following:

- The number of reliability improvement actions initiated through performance analysis during the specified period. No absolute number is correct but no number suggests inaction.

- The number of asset reliability actions resolved over the last month; in other words, a measure of how successful the organization is in performance gap closure (Weber and Thomas, 2005).

4.1.3.3 Key Performance Indicators of Maintenance Effectiveness

The product of maintenance is reliability. A reliable asset is an asset able to function at the level of performance that satisfies the needs of the user. Paradoxically, reliability is assessed by measuring failure (Weber and Thomas, 2005).

4.1.3.3.1 Failures

The primary function of maintenance is to reduce or eliminate the consequences of physical asset failure, with failure considered to occur any time asset performance falls below its required performance. Therefore, a KPI for maintenance effectiveness is some measurement of failure of the asset.

If the maintenance function is effective, failures of critical assets and, thus, their consequences will be reduced or even eliminated. This is important, as failure impacts manufacturing-level KPIs as well. Failure classification, by consequence, identifies the contribution of the maintenance function to manufacturing level performance (Weber and Thomas, 2005).

Failure consequences are classified into the following categories:

- *Hidden consequence*: there is no direct consequence of a single-point failure other than exposure to the increased risk of a multiple failure (a second failure has to occur for a consequence to occur).
- *Safety consequence*: a single-point failure results in a loss of function or other damage which could injure or kill someone.
- *Environmental consequence*: a single-point failure results in a loss of function or other damage which breaches any known environmental standard or regulation.
- *Operational consequence*: a single-point failure has a direct adverse effect on operational capability (output, product quality, customer service, or operating costs in addition to the direct cost of repair).
- *Nonoperational consequence*: a single-point failure involves only the cost of repair.

Therefore, it is important to track the following:

- *Number and frequency of asset failures by area of consequence*: there is no universal standard for this metric because of the diversity of industries and even plants within industry segments. It is reasonable, however, to expect a downward trend and to set reduction targets based on current performance levels and business needs (Weber and Thomas, 2005).

4.2 Maintenance Costs

Maintenance costs are another direct measure of maintenance performance. Maintenance costs are impacted by both maintenance effectiveness and the efficiency with which maintenance is performed. There are several useful maintenance cost–related measures:

- Increasing the efficiency of maintenance through improved planning and scheduling of the right work at the right time.
- Performing proactive maintenance or intervening before the failure event occurs. The impact of proactive maintenance is not only to minimize the safety, environmental, and operational consequences of failure but also to reduce the cost of maintenance by reducing secondary damage.

For example, if the potential failure of a pump bearing is detected proactively, the catastrophic failure of the bearing can be prevented. Catastrophic failure would likely result in damage to the casing, wear rings, impeller, mechanical seals, and so on. Corrective repair of the damaged pump will require an extensive rebuild. Using a proactive task such as vibration monitoring to detect the bearing deterioration would permit the scheduled replacement of the bearing before the occurrence of secondary damage. Less secondary damage means that it takes less time to repair (labor savings) and consumes fewer parts (material savings). The overall effect is decreased repair costs (Weber and Thomas, 2005).

The target maintenance cost depends on the asset and its operating context (how the asset is applied and used). It is measured in the following ways:

- *Maintenance cost/replacement asset value*: This metric is a useful benchmark at a plant and corporate level. The world-class benchmark is between 2% and 3%.
- *Total maintenance cost/total manufacturing cost*: This metric is a useful benchmark at a plant and corporate level. The world-class benchmark is <10% to 15%.
- *Total maintenance cost/total sales*: This metric is a useful benchmark at a plant and corporate level. The world-class benchmark is between 6% and 8% (Smith and Mobley, 2011).

4.2.1 Maintenance-Related Downtime

The maintenance function has an impact on asset availability by minimizing downtime attributed to maintenance. This includes both scheduled and unscheduled maintenance-related downtime.

A key objective of proactive maintenance is to identify potential failures with sufficient lead time to plan and schedule the corrective work before actual failure occurs. If the maintenance function is successful, unscheduled maintenance-related downtime will be reduced.

The work identification element of the maintenance process strives to eliminate unnecessary scheduled maintenance by focusing on performing the right work at the right time. Through more formal work identification and enhanced planning and scheduling, shutdown overruns should be minimized (Weber and Thomas, 2005).

Therefore, these are useful KPIs associated with asset downtime attributable to maintenance:

- Unscheduled downtime (hours)
- Scheduled downtime (hours)
- Shutdown overrun (hours)

Note: It is useful to distinguish between equipment downtime where a specific piece of equipment is unavailable and process downtime where production has stopped (Weber and Thomas, 2005). Chapter 5 explains the idea of downtime cost and all consequential effects of a machine shutdown in terms of efficiency loss.

4.2.2 Summary of KPIs for Maintenance

Although many writers propose lists of KPIs, they often fail to offer a methodological approach to select or derive them. As a result, users are left to decide the relevant KPIs for their own situation (Galar and Kumar, 2016).

As the previous discussion has made clear, maintenance, reliability, engineering, and operations departments need to work together to define and measure the indicators for the asset reliability process (the seven elements required to support the maintenance function mentioned earlier), as these apply to their particular company (Galar and Kumar, 2016). They must cooperate because any gaps in the execution of the maintenance process will eventually lead to asset failure, downtime, and economic losses. If they do work together, however, the result will be optimal asset reliability at optimal cost (Weber and Thomas, 2005).

The commonly used maintenance performance indicators can be divided into two major categories, as mentioned previously. The maintenance process or effort indicators are defined as the leading indicators, and the maintenance result indicators are lagging indicators (Weber and Thomas, 2005). This division is shown in Figure 4.4, along with suggested performance indicators for each category.

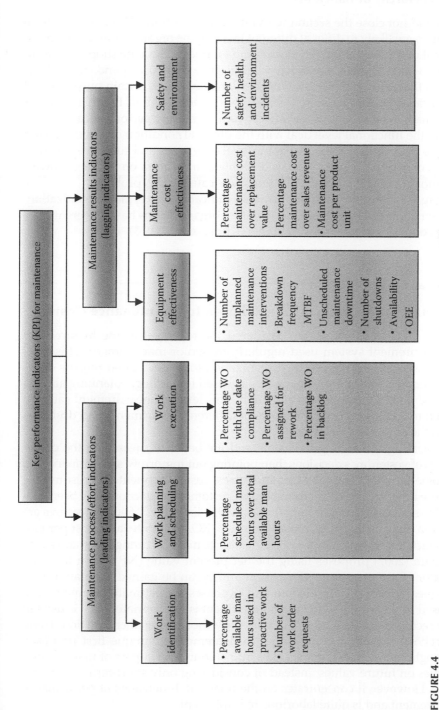

FIGURE 4.4

Key performance indicators for maintenance. (Adapted from Galar, D. and Kumar, U., *Maintenance Audits Handbook: A Performance Measurement Framework*, CRC Press, Taylor & Francis Group, Boca Raton, FL, April 6, 2016, 609pp.)

4.2.3 Hierarchy of Indicators

We should not close the section without a brief mention of how maintenance indicators are likely to differ at different levels of the organization. Upper management may have different concerns and emphases than the shop floor itself. In other words, there is a hierarchy of indicators that corresponds to the hierarchy of the company. Each level of the hierarchy serves certain purposes and is applicable to certain users. Users at the highest level of the management traditionally refer to aspects that affect overall company performance, whereas those at the functional level deal with the physical condition of assets. Having multiple performance measures at the level of systems and subsystems is useful. If a corporate indicator shows a problem, indicators on the next (lower) level should define and clarify the weakness causing this problem.

Figure 4.5 shows levels of performance indicators in a typical organization, as shown in Figure 4.5. Of course, different organizations have different hierarchies of performance measurements (Kumar et al., 2013).

4.3 Financial KPIs and Their Relation to Maintenance Costs

Financial measures are often considered the top level in the hierarchy of the measurement system used regularly by senior management. This level of KPIs shows the ability of the organization to achieve good returns on its assets and to create value. The metrics are used for strategic planning and are the backbone of the entire organization. This level of measurement can also be used to compare the performance of different departments and divisions within the parent organization.

Financial figures are lag indicators and are better at measuring the consequences of yesterday's decisions than pointing out tomorrow's performance. To overcome the shortcomings of lag indicators, customer-oriented measures like response time, service commitments, and customer satisfaction have been proposed to serve as lead indicators (Eccles, 1995). Examples of such measures are return on investment (ROI), return on assets (ROA), maintenance cost per unit of product, total maintenance costs in relation to manufacturing costs, and so on.

Vergara (2007) proposes using the net present value (NPV) as a financial indicator of maintenance. NPV is how much can be gained from an investment (i.e., an asset) if all income and expenses are estimated immediately. It is used to determine whether an investment is appropriate. NPV is used in many sectors and areas but is rarely used for the maintenance function. Tsang et al. (1999) present a performance measurement technique first proposed by Dwight (1994). The method takes into account the impact of maintenance activities on future values, instead of considering only short-term or present values. However, it concentrates on the financial dimensions of performance measurement and is quite laborious to implement.

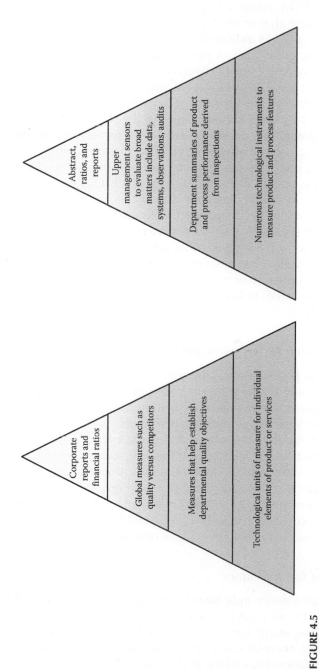

FIGURE 4.5
Pyramid used in all levels of the company. (Adapted from Kumar, U. et al., *J. Qual. Mainten. Eng.*, 19(3), 233, 2013.)

Hansson (2003) proposes using a battery of financial indicators to study the results of the maintenance department. Hansson suggests benchmarking the maintenance function, considering such measures as the percentage change in sales, ROA, return on sales, the percentage change in total assets, and the percentage change in the number of employees. These indicators are generally accepted as indicators of financial results but can be applied to maintenance. The correlation of such indicators with more tactical maintenance indicators reveals trends between maintenance and corporate strategy.

Coelo and Brito (2007) also suggest incorporating the use of financial indicators for maintenance management. They discuss the need for integration and the creation of a harmonious balance of the financial performance indicators of the organization; they take this a step further to consider how the company's strategic vision affects the efficiency of the maintenance sector.

Cáceres (2004) analyzes the financial outlook of maintenance KPIs. All planning systems should show the history of strategy and corporate positioning indicated by the organization's financial goals, linking them to the sequence of actions to be undertaken with customers, internal processes, and, finally, with the employees themselves. This perspective is focused on ROI, added value to the organization, and reduced unit costs. When it is applied to maintenance, the costs of each activity and the incidence rate of maintenance costs per unit of production and maintenance costs on the value of assets are monitored, indicating their global position in the organization (Kumar et al., 2013).

4.3.1 Maintenance Performance Indicators

The objectives of performance indicators in maintenance include providing relevant information on the status of the equipment or machine, the level of preventive or corrective maintenance within the plant, incurred costs, and the performance level of maintenance personnel.

Once defined, the indicators help detect and correct problems to prevent poor-quality products or shutdowns, thereby reducing the waste of time, cost, and effort. Some authors divide the indicators by level, depending on what is important to measure and the expected results.

Not all indicators must be known and managed by all workers, however. This calls for custom scorecards, depending on the relevant aspects of each user and their impact on the overall objectives of the company (Galar and Kumar, 2016).

4.3.1.1 World-Class Indicators

World-class maintenance indicators (or world-class indexes) are the same in all countries. Of the six world-class indexes, four are related to the analysis of equipment management Reliability Availability Maintainability and Safety (RAMS), and two are related to cost management (cost model), according to the following relationships (Galar and Kumar, 2016).

4.3.1.1.1 Mean Time between Failures

This is the relationship between the operating time of items over a certain period and the total number of failures detected in those items in the observed period. This index should be used for items that are repaired after failure (Galar and Kumar, 2016).

4.3.1.1.2 Mean Time to Repair

This is the relationship between the total corrective intervention time in a set of items with failure, and the total number of failures detected in those items in an observed period. This index should be used for items in which the repair time is significant in relation to the operating time (Galar and Kumar, 2016).

4.3.1.1.3 Mean Time to Fail

This is the relationship between the total operating time of a set of non-repairable items and the total number of failures detected in those items in the observed period. This index should be used for items that are replaced after a failure.

It is important to note the conceptual difference between the mean time to fail (MTTF) and the mean time between failures (MTBF). MTTF is calculated for items that are not repairable or repaired after the occurrence of a failure; that is, when they fail, they are replaced by new ones and, consequently, the repair time is zero. MTBF is calculated for items repaired after the occurrence of the fault. Therefore, the indexes are mutually exclusive; that is, the calculation of one excludes the calculation of the other.

The calculation of MTBF should be associated with the calculation of the mean time to repair (MTTR). Because these indexes have an average result, accuracy is associated with the number of items observed and the observation period (Galar and Kumar, 2016).

4.3.1.1.4 Equipment Availability

This is the difference between the total number of hours (calendar hours) and the number of hours for maintenance (preventive maintenance, corrective maintenance, and other services) for each observed item.

The availability of an item represents the percentage of time it is available to carry out its activity. The availability index is called "equipment performance" and can be calculated as the ratio of the total operation time of each item and the sum of this time with the total maintenance time in the considered period (Galar and Kumar, 2016).

4.3.1.1.5 Billing Maintenance Cost

This is the relationship between total maintenance cost and billing for maintenance during a certain period. This index is easy to calculate because the values of both the numerator and the denominator are normally processed by the accounting department (Galar and Kumar, 2016).

4.3.1.1.6 Maintenance Cost for Replacement

This is the relationship between total cumulative cost of maintaining a certain piece of equipment and the purchase price of the same equipment (replacement value). This index should be calculated for the most important items of the company (i.e., those affecting billing, product quality or services, safety, and environment) (Galar and Kumar, 2016).

Besides the world-class indexes for maintenance cost, there are a number of traditional financial indicators. Listed here are the most classic ones. They are referred to as FINI*n*.

- FINI1 Maintenance costs per unit processed, produced, or manufactured
- FINI2 Maintenance costs totaling process, production, or manufacturing costs
- FINI3 Sales maintenance costs
- FINI4 Maintenance costs per square meter
- FINI5 Maintenance cost by estimated replacement value of the plant or installation assets
- FINI6 Storage investment by estimated replacement value
- FINI7 Value of the asset upon which maintenance is performed by a maintenance worker
- FINI8 Subcontracting costs by total maintenance costs

Some of these indicators have been incorporated into recently published regulations. For example, maintenance costs per unit produced or maintenance costs for replacement value are incorporated into all panels, standards, and recommendations published to date (Galar and Kumar, 2016).

4.3.2 Economic Indicators in EN 15341

The European Standard EN 15341 provides maintenance performance indicators to support management in achieving excellence in maintenance and the deployment of technical assets in a competitive manner. Most of these indicators apply to all buildings, spaces, industrial services, and support (buildings, infrastructure, transport, distribution, networking, etc.). These indicators should be used to conduct the following functions:

- Measure condition.
- Make comparisons (internal and external references).
- Perform diagnostics (analysis of strengths and weaknesses).
- Identify objectives and define goals to be achieved.
- Plan actions for improvement.
- Measure changes continuously over time.

The standard describes a system for managing indicators to measure the maintenance performance within the context of the factors that influence it, such as economic, technical, and organizational aspects, in order to evaluate and improve efficiency and effectiveness and achieve excellence in asset maintenance (Galar and Kumar, 2016).

European Standard EN 15341 says maintenance performance depends on both external and internal factors, such as location, culture, transformation process and service, size, rate of use, and age. It is achieved through the implementation of corrective and preventive maintenance (PM) activities, and improvement by applying labor, information, materials, organizational methodologies, tools, and operational techniques. When a factor is defined using the terms "internal" or "external," the corresponding indicator should be used for "internal" or "external" influences respectively.

Figure 4.6 illustrates the external and internal factors that influence maintenance performance. External factors are variable conditions beyond the control of the company management. Internal factors are specific to the group, company, factory, and facilities that are beyond the control of maintenance management, but within the control of company management.

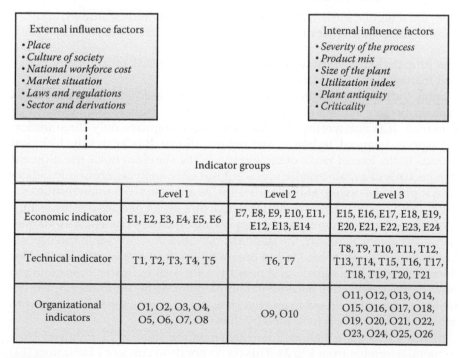

External influence factors	Internal influence factors
• *Place*	• *Severity of the process*
• *Culture of society*	• *Product mix*
• *National workforce cost*	• *Size of the plant*
• *Market situation*	• *Utilization index*
• *Laws and regulations*	• *Plant antiquity*
• *Sector and derivations*	• *Criticality*

Indicator groups			
	Level 1	Level 2	Level 3
Economic indicator	E1, E2, E3, E4, E5, E6	E7, E8, E9, E10, E11, E12, E13, E14	E15, E16, E17, E18, E19, E20, E21, E22, E23, E24
Technical indicator	T1, T2, T3, T4, T5	T6, T7	T8, T9, T10, T11, T12, T13, T14, T15, T16, T17, T18, T19, T20, T21
Organizational indicators	O1, O2, O3, O4, O5, O6, O7, O8	O9, O10	O11, O12, O13, O14, O15, O16, O17, O18, O19, O20, O21, O22, O23, O24, O25, O26

FIGURE 4.6
Economic indicators for maintenance and factors influencing maintenance and key performance indicators as noted in EN 15341. (Adapted from Galar, D. and Kumar, U., *Maintenance Audits Handbook: A Performance Measurement Framework*, CRC Press, Taylor & Francis Group, Boca Raton, FL, April 6, 2016, 609pp.)

When maintenance KPIs are used, it is important to consider these influencing factors as prerequisites to avoid misinterpretations.

Traditionally, only internal factors were considered in the maintenance field and the indicators corresponded with cost, labor hours, and little else. Dean (1986) highlights the focus on the balance of effectiveness and efficiency proposed by EN 15341; it also suggests new indicators, such as the success of reliability-centered maintenance (RCM) programs or maintenance planning capability.

To cover this mixed aspect of maintenance efficiency and effectiveness, a system of KPIs is structured into three groups: economic, technical, and organizational. These proposed indicators can be evaluated as a relationship between factors (numerator and denominator), measurement activities, resources, or events according to a given formula. The rates proposed in this standard are useful for measuring quantitative aspects or features and to make homogeneous comparisons.

These indexes espoused by EN 15341 can be used, for example:

- On a periodic basis to prepare and monitor a budget or during performance evaluation
- On a regular basis within the frame of specific audits, studies, and/ or comparisons for improvement

The time period to consider for measurement depends on company policy and management approach.

Some indicators presented in the standard are used in other approaches such as Society for Maintenance and Reliability Professionals (SMRP) or Wireman. It is noteworthy that the standard recognizes only three indicator aspects: financial, technical, and organizational. It relegates the technical aspects to the lowest ranks of the pyramid. The standard notes the different perspectives of a maintenance scorecard but only analyzes economic indicators; it proposes organizational indicators as an aspect of human resources.

EN 15341 leaves many aspects untreated, however. For example, the technical financial and economic aspects appear insufficient to function as maintenance indicators. Similarly, the standard classifies indicators into three levels but does not specify what kind of organization must assume the levels. Nor does it mention benchmarks and objectives for each indicator, thus failing to provide the perspective necessary to implement them. Indicator implementation is inexorably linked to an objective; the success of the strategy used to achieve the objective is measured by these indexes.

In the following equations, the nomenclature used is EI followed by a sequential number from 1 to 24. This corresponds to efficiency indicators (EI) used in cost models. Traditionally, these indicators are considered superior to operational indicators and are intended for measurement of financial or corporate management; they are rarely used by maintenance management (Galar and Kumar, 2016).

$$EI1 = \frac{\text{Maintenance total costs}}{\text{Assets replacement value}} \tag{4.1}$$

$$EI2 = \frac{\text{Maintenance total costs}}{\text{Added value plus external maintenance costs}} \tag{4.2}$$

$$EI3 = \frac{\text{Maintenance total costs}}{\text{Quantity produced}} \tag{4.3}$$

$$EI4 = \frac{\text{Maintenance total costs}}{\text{Transformation cost to production}} \tag{4.4}$$

$$EI5 = \frac{\text{Maintenance total costs + Unavailability costs linked to maintenance}}{\text{Quantity produced}} \tag{4.5}$$

$$EI6 = \frac{\text{Availability linked to maintenance}}{\text{Quantity produced}} \tag{4.6}$$

$$EI7 = \frac{\text{Average inventory value of maintenance articles}}{\text{Assets replacement value}} \tag{4.7}$$

$$EI8 = \frac{\text{Total cost of internal staff used in maintenance}}{\text{Maintenance total costs}} \tag{4.8}$$

$$EI9 = \frac{\text{Total cost of external staff used in maintenance}}{\text{Maintenance total costs}} \tag{4.9}$$

$$EI10 = \frac{\text{Total contracting cost}}{\text{Maintenance total costs}} \tag{4.10}$$

$$EI11 = \frac{\text{Total cost of maintenance articles}}{\text{Maintenance total costs}} \tag{4.11}$$

$$EI12 = \frac{\text{Total cost of maintenance articles}}{\text{Average inventory value of maintenance articles}} \tag{4.12}$$

$$EI13 = \frac{\text{Cost by indirect maintenance staff}}{\text{Maintenance total costs}} \tag{4.13}$$

$$EI14 = \frac{\text{Maintenance total costs}}{\text{Total energy used}} \tag{4.14}$$

$$EI15 = \frac{\text{Corrective maintenance cost}}{\text{Maintenance total costs}} \tag{4.15}$$

$$EI16 = \frac{\text{Preventive maintenance cost}}{\text{Maintenance total costs}} \tag{4.16}$$

$$EI17 = \frac{\text{CBM maintenance cost}}{\text{Maintenance total costs}} \tag{4.17}$$

$$EI18 = \frac{\text{Systematic maintenance cost}}{\text{Maintenance total costs}} \tag{4.18}$$

$$EI19 = \frac{\text{Improvement maintenance cost}}{\text{Maintenance total costs}} \tag{4.19}$$

$$EI20 = \frac{\text{Cost of programmed maintenance stops}}{\text{Maintenance total costs}} \tag{4.20}$$

$$IE21 = \frac{\text{Maintenance staff training cost}}{\text{Effective cost of maintenance staff}} \tag{4.21}$$

$$EI22 = \frac{\text{Total contracting cost of mechanical maintenance}}{\text{Maintenance total costs}} \tag{4.22}$$

$$EI23 = \frac{\text{Total contracting cost of electrical maintenance}}{\text{Maintenance total costs}} \tag{4.23}$$

$$EI24 = \frac{\text{Total contracting cost of instrumentation maintenance}}{\text{Maintenance total costs}} \tag{4.24}$$

4.3.3 Financial Perspective

Financial indicators have been used for a long time. There are economic indexes on profitability, solvency, and liquidity, and these can be applied to all types of companies. Two fundamental aspects can be improved, however. The first concerns the matching of indicators to the business unit, and the second is placing them within the life cycle of the company and its productive assets.

Occasionally, the same type of financial metric is applied to different business units, especially if an organization is looking for a certain level of return on invested capital and/or is assuming they yield the same percentage of added value, without considering that they may have different strategies. In addition, the life cycle of a product will include a succession of phases, each of which may require a unique strategy.

In the introduction phase of a business, the strategy involves expending higher resources compared to results/sales. Even though this may be a period of intense growth, the production cost of each unit is high, and sometimes the performance is negative. Investment in projects is high and maintenance of equipment produces large expenses.

When a business is in the development stage, large investments are still required, not so much in development but in logistics and promotion, but sales are higher. At this point, it is necessary to determine whether spending more resources is intended to reduce the price or increase product promotion. The goal is to produce sustained maintenance expenditure and technologies and to employ methodologies that maximize system availability.

In the maturity phase, a product has conquered a market and costs continue to decline. This, therefore, is the phase of greater profitability. Sales stabilize, and after a period determined by the nature of the competition and the possibility of specialization, it is time to start thinking about renewal. At this stage, the ability to maintain optimum levels with present assets is important, so appropriate replacement or repair policies are instituted. Changing market circumstances and possible saturation may cause a decline in sales. Sales may decrease but the benefits are maintained, and no investment is required (Figure 4.7) (Galar and Kumar, 2016).

Clearly, for each of these phases, the economic objectives are different not only in what constitutes the ROI but also in cash flow, working capital, and the relationship between promotion resources and increasing sales (Galar and Kumar, 2016). Equally obvious is the fact that maintenance requirements change in each phase.

Visualization of the different stages determines the indicators that should define company financial objectives and suggests the correct decisions for each situation. In particular, a company can estimate the stages when it should make major investments until machinery maintenance is eliminated and machines are abandoned.

The financial perspective of maintenance lies in the definition of its function: safety for staff and facilities, respect for the environment, and availability at the lowest possible cost. Briefly stated, the financial goal of maintenance is to achieve maximum plant availability at the lowest possible cost (Galar and Kumar, 2016).

Regardless of the size, age, and industry, each and every company needs to be conscious of its financial performance. Although accountants deal with all the expenses, income, and budgets, the company's leadership needs to be informed about important financial measures.

The fastest and most efficient way to keep track of an organization's business performance is to set up a KPI dashboard that displays financial KPIs and another metrics. The perfect financial KPI report presents real-time

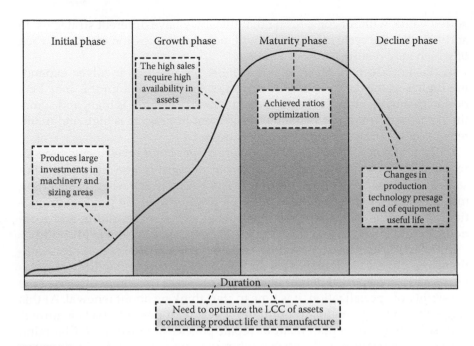

FIGURE 4.7
Product cycles. (Adapted from Galar, D. and Kumar, U., *Maintenance Audits Handbook: A Performance Measurement Framework*, CRC Press, Taylor & Francis Group, Boca Raton, FL, April 6, 2016, 609pp.)

updates on a company's important financial figures, such as the operating cash flow, the current ratio, burn rate, and so on. The next section will walk us through each of these in turn (Kessler, 2016).

4.4 Benchmarking Financial KPIs

4.4.1 Benchmarking Fundamentals

Benchmarking, best practices, and competitive analysis are terms used in business today. But are they just buzzwords, or do the words have real meaning? Are they useful tools? Can they be used to improve business practices (Wireman, 2004)?

4.4.1.1 Language of Benchmarking

Xerox Corporation defines benchmarking as follows:

> The search for industry best practices which lead to superior performance.

To understand this definition, we must first be clear what we mean by best practices. Best practices are those practices that enable a company to become a leader in its respective marketplace. However, best practices are not the same for all companies. For example, if a company is in a declining market, and the pressures are to maximize profits with a fixed sales volume, one set of best practices might allow market leadership. But the company is in a growth mode with profits dictated by gaining a rapid market share, a different set of best practices would be appropriate. Therefore, best practice is determined by business conditions; it is not fixed.

The second key term in the Xerox definition is superior performance. Many companies use benchmarking so that they can be as good as their competitors. But a company can gain very little if its goal for benchmarking is merely to achieve the status quo. Benchmarking is a continuous improvement tool used by companies striving to achieve superior performance in their respective marketplaces.

Here is an alternative definition of benchmarking:

> An ongoing process of measuring and improving business practices against the companies that can be identified as the best worldwide.

This definition emphasizes the importance of improving rather than maintaining the status quo. It addresses searching worldwide for the best companies and their best practices. It would be naive to think that best practices are limited to one country or one geographical location. Information that allows companies to improve their competitive positions must be gathered from best companies, no matter where they are located.

Companies striving to improve must not accept past constraints, especially the "not invented here" paradigm. Those failing to develop a global perspective will soon be replaced by competitors with the ability to do so. To ensure rapid continuous improvement, companies must be able to think outside the box. The more innovative the ideas, the greater the potential rewards.

The following is a third perspective on benchmarking:

> Benchmarking sources "Best Practices" to feed continuous improvement.

This statement adds another dimension to benchmarking—an external perspective. Research shows major innovations can come from outside a company's sector and be adapted to improve its practices. In today's competitive business environment, the need to develop an external perspective is critical to survival.

Still another perspective defines benchmarking as follows:

> The process of continuously comparing and measuring an organization with business leaders anywhere in the world to gain information that will help the organization take action to improve its performance.

The common thread of studying other companies to gain information to become more competitive is clear. Unless a company clearly understands the processes and procedures that allow it to become the best, little value is derived from benchmarking (Wireman, 2004).

4.4.1.2 Competitive Analysis

The terms benchmarking and competitive analysis are often confused. Benchmarking researches external business sectors for information, whereas competitive analysis shows how companies compare with their competitors. A competitive analysis produces a ranking with direct competitors; it does not show how to improve business processes.

Benchmarking provides a deep understanding of the processes and skills creating superior performance. Without this understanding, little benefit is achieved from benchmarking. Competitive analysis is less likely to lead to significant breakthroughs that will change long-entrenched paradigms of a particular market segment. Business paradigms tend to be similar for comparable businesses in similar markets. While competitive analysis often leads to incremental improvements for a business, breakthrough strategies are derived from taking an external perspective.

During the past 20 years, competitive analyses have helped companies improve their respective market position, but benchmarking takes over where this opportunity for improvement ends. Benchmarking enables companies to move from a parity business position to a superiority position. Observing best practices can help any company.

Another difference between benchmarking and competitive analysis is the type of data gathering required. Competitive analysis often focuses on meeting some specific industry standard. All that may be required is meeting a published number. By comparison, benchmarking focuses not on a number, but on the process that allows such a standard to be not only achieved, but surpassed. Process enablers and critical success factors must be clearly understood for any permanent improvement. This understanding will require extensive data collection, both internally and from the benchmarking partners (Wireman, 2004).

4.4.1.3 Enablers

Enablers are a broad set of activities or conditions that enhance the implementation of a best business practice. An essential part of a true benchmarking approach is analyzing the management skills and attitudes that combine to allow a company to achieve best business practices. This often hidden narrative is as important during the benchmarking exercise as are the visible statistical factors and hard processes.

Enablers, then, are behind-the-scene or hidden factors. They allow the development and/or continuation of best practices. Examples include leadership, motivated workforces, management vision, and organizational focus. Although these factors are rarely mentioned by specific statistics, they have a direct impact on the company's quantified financial performance. When properly used, they lead to a company's exceptional performance. Note that enablers are relative, not absolute. In other words, they are not perfect; they too can be improved.

Enablers, or critical success factors, can be found anywhere. They are not constricted by industrial, political, or geographical boundaries.

How does one company compare itself to another by enablers? Quite simply, it starts with an internal analysis. For any company to be successful, it must have a thorough knowledge and understanding of its own internal processes. Otherwise, it will be impossible to recognize its differences from its benchmarking partners. Consequently, it will not recognize and integrate the innovations of best practice companies (Wireman, 2004).

4.4.2 Defining Core Competencies

As a continuous improvement tool, benchmarking is used to improve core competencies, the basic business processes that allow a company to differentiate itself from its competitors. A core business process may have an impact by lowering costs, increasing profits, providing improved service to a customer, improving product quality, or improving regulatory compliance.

Several authors have defined core competencies for businesses. In his 1997 text *Operations Management*, Richard Schonberger defines a core competency as a key business output or process through which an organization distinguishes itself positively. He specifically mentions expert maintenance, low operating costs, and cross-trained labor.

Gregory Hines, in his text *The Benchmarking Workbook*, defines a core competency as a business process that represents core functional efforts and is usually characterized by transactions that directly or indirectly influence the customer's perception of the company. Such processes include the following functions:

- Procuring and supporting capital equipment
- Managing and supporting facilities

The maintenance function directly fits into his definition of a core business process.

In the American Productivity and Quality Center's text *The Bench-Marking Management Guide*, core competencies are identified as business processes that should impact the following business measures:

- Return on net assets
- Customer satisfaction
- Revenue per employee

- Quality
- Asset utilization
- Capacity

Again, the maintenance function in any plant or facility fits this definition.

Other authors point to a core competency as any aspect of the business operation that results in a strategic market advantage. The maintenance process in any company provides this advantage in many ways. These include enhancing any quality initiative, increasing capacity, reducing costs, and eliminating waste.

The investment a company makes in its assets is often measured against the profits the company generates. This measure is called return on fixed assets (ROFA). It is often used in strategic planning when a company picks what facility to occupy or the plant in which to produce a product.

Asset management focuses on achieving the lowest total life cycle cost (LCC) to produce a product or provide a service. The goal is to have a higher ROFA than the competitor in order to be a low-cost producer of a product or service. A company in this position attracts customers and garners a greater market share. A higher ROFA also attracts investors, ensuring a sound financial base on which to build further business.

All departments or functions within a company have the responsibility of measuring and controlling their costs, since they will ultimately impact the ROFA calculation. Only when everyone works together can the maximum ROFA be achieved (Wireman, 2004).

How does maintenance management impact the ROFA calculation? Two indicators, in particular, show the impact:

- Maintenance costs as a percentage of total process, production, or manufacturing costs: Maintenance costs are an accurate measure of manufacturing costs. They should be used as a total calculation, not a per-production-unit calculation. Maintenance will be a percentage of the cost to produce, but is generally fixed. This stability makes the indicator more accurate for the financial measure of maintenance, because it makes trending maintenance costs easier. If the maintenance cost percentage fluctuates, the efficiency and effectiveness of maintenance should be examined to find the cause of the change.
- Maintenance cost per square foot maintained: This indicator compares the maintenance costs to the total amount of floor space in a facility. It is an accurate measure for facilities because the cost is also usually stable. This indicator is easy to use to trend any increases over time. If the percentage of maintenance costs fluctuates, the efficiency and effectiveness of maintenance should be examined to find the cause of the change.

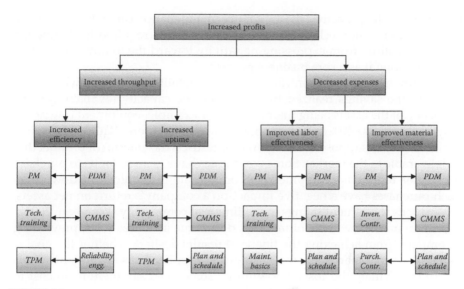

FIGURE 4.8
Total impact of maintenance on ROFA. (Adapted from Wireman, T., *Benchmarking Best Practices in Maintenance Management*, Industrial Press, Inc., New York, 2004, 256pp.)

These two indicators show that traditional maintenance labor and material costs will have an impact on ROFA. However, ensuring assets are available and operating efficiently also has an impact. The total impact of the maintenance function on ROFA is illustrated in Figure 4.8.

Overall, the goal for any company is to increase profitability. This is true whether the company is public with shareholders or is privately owned. The maintenance or asset management function can increase profits in two main ways: decreasing expenses and increasing capacity.

Estimates suggest that one-third of all maintenance expenditures are wasted through inefficient and ineffective use of the maintenance resources. Maintenance costs consist of two main divisions: labor and materials. If a maintenance labor budget for a company is $3M annually, and one-third is wasted, $1M could be saved by reducing wastage. This saving would not necessarily be achieved by eliminating workers. It may involve reducing overtime, reducing the use of outside contractors, or performing deferred maintenance without additional expenditures.

If the maintenance labor budget is $3M annually, studies show that the materials budget will be a similar amount. If the materials budget can also be reduced by one-third, the total savings for improving maintenance efficiency and effectiveness could approach $2M per year. *Note*: this saving is actually expense dollars that will not be required. Expense dollars do not translate into profit dollars.

Also note that reactive maintenance is being improved, and savings are not immediate. Time is needed to realize total savings. Changing reactive maintenance to proactive, best practice maintenance can take 3–5 years. The transition is not technically difficult, but time is required to change the corporate culture from expressing negativity toward the maintenance function to treating it as a core business process.

The pure maintenance contribution to profitability is dwarfed when compared to the savings realized by increasing the capacity (availability) and efficiency of the assets being maintained. For example, equipment downtime may average 10%–20% in some companies, or even more. Equipment that is down when it is supposed to be operating restricts the amount of product that is deliverable to the market. Some companies have gone so far as to purchase backup or redundant equipment to compensate for equipment downtime. Such purchases have a negative impact on a company's return on net assets, lowering its investment ratings in the financial community (Wireman, 2004).

Even in markets with a volume cap, downtime increases costs, preventing a company from achieving the financial results desired, whether the goal is to increase profit margins or to be a low-cost supplier. Yet some organizations refuse to calculate the cost of downtime; some even say there is no cost to downtime. They fail to consider the following factors:

- Utility costs
- Cost of idle production/operations personnel
- Cost of late delivery
- Overtime costs to make up lost production to meet schedules

The true cost of downtime is the lost sales for products not made on time. A company needs to have a clear understanding of this cost to make good decisions on its assets and how they are operated.

Suppose that a company discovers a considerable amount of unplanned downtime for the previous year, only part of which can be corrected by improving maintenance. Some other causes for equipment downtime could be related to raw materials, production scheduling, quality control, and operator error. However, if the maintenance portion of the downtime is valued at $38M and a 50% reduction could be achieved by improving maintenance, the savings could be $19M. Even if only 10% of this amount is spent improving maintenance, the total savings will still be more than $17M.

Another cost is the cost of lost efficiency. When one company examined the efficiency of its gas compressors on an off-shore operation, it found that due to age and internal wear, its compressors were operating at only 61% efficiency. This cost them approximately $5.4M annually. An overhaul would cost $450K, including labor, materials, and downtime production losses. They decided to overhaul the compressors serially to avoid total shutdown. This proved to be a good decision. The compressor overhaul was paid back only 28.1 days after restart, and the company realized $5.4M in increased production in the next 12 months.

Many Japanese studies related to total productive maintenance (TPM) have shown that efficiency losses are always greater than pure downtime losses. This finding becomes even more alarming when we consider that most efficiency losses are not measured or reported. In addition, many chronic problems are never solved until a breakdown occurs. Some chronic problems with a dramatic impact on equipment efficiencies are never even discovered. Only when accurate maintenance records are kept can these problems be revealed. By combining the maintenance data with the financial data, a company can find and solve the root cause of the efficiency problem.

If asset management is a focus for an organization, the maintenance function can contribute to overall profitability. Although the cooperation of all departments and functions within an organization is required, if this happens, the maintenance department can have a dramatic, positive impact on ROFA.

Because maintenance is typically viewed as an expense, any maintenance savings can be viewed as directly contributing to profits. By achieving maximum availability and efficiency from assets, a plant or facilities manager ensures a company does not need to invest in excess assets to produce its products or provide its services. Eliminating investment in unnecessary assets contributes to overall improvement in the ROFA for any company.

Because maintenance is a core business process, it could benefit from benchmarking. The next question is what type of benchmarking should be used to gain the maximum benefits (Wireman, 2004). We have already touched on benchmarking but we develop the discussion in the following sections and explain how to set benchmarking in motion.

4.4.3 Types of Benchmarking

Types of benchmarking include the following:

1. Internal
2. Similar industry/competitive
3. Best practice

4.4.3.1 Internal Benchmarking

Internal benchmarking typically involves different departments or processes within a company. This type of benchmarking has some advantages in that data can be collected easily. It is also easier to compare data because many of the hidden factors (enablers) do not need to be closely checked. For example, the departments will have a similar culture, the organizational structure will likely be the same, and the skills of the personnel, labor relations, and management attitude will be similar. These similarities make data comparison quick and easy.

The greatest disadvantage of internal benchmarking is that it is unlikely to result in any major breakthrough. Nevertheless, it will lead to small, incremental improvements and should provide adequate ROI for any improvements that are implemented. The successes from internal benchmarking will very likely increase the desire for extensive external benchmarking (Wireman, 2004).

4.4.3.2 Similar Industry/Competitive Benchmarking

As the name suggests, similar industry or competitive benchmarking uses external partners in similar industries or processes. In many benchmarking projects, knowing competitors' practices can be very useful. This process may be difficult in some industries, but many companies are open to sharing information that is not proprietary.

Similar industry/competitive benchmarking tends to focus on organizational measures. In many cases, this necessitates meeting a numerical standard, rather than improving any specific business process. In competitive benchmarking, small or incremental improvements are noted, but paradigms for competitive businesses are similar. Thus, the improvement process will be slow (Wireman, 2004).

4.4.3.3 Best Practice Benchmarking

Best practice benchmarking focuses on finding the unarguable leader in the process being benchmarked. This search, which crosses industry sectors and geographical locations, provides the opportunity for an organization to develop breakthrough strategies. The organization studies business processes outside its industry, adapts or adopts superior business processes, and makes a quantum leap in performance. Being the early adaptor or adopter will give it an opportunity to lower costs or aggressively capture the market share.

One of the keys to being successful with best practice benchmarking is to define best practice. For example, does best mean the following:

- Most efficient?
- Most cost-effective?
- Most customer service–oriented?
- Most profitable?

Without this clear understanding, more resources will be needed to conduct a benchmarking project. Furthermore, the improvements will be mediocre at best.

In the United States, the Best Practices Committee for the General Services Administration (GSA) developed the following useful definition of best practices, applicable to a wide range of industries:

> Best Practices are good practices that have worked well elsewhere. They are proven and have produced successful results. They must focus on proven sources of best practices.

The Committee goes on to state as follows:

> They (organizations) should schedule frequent reviews of practices to determine if they are still effective and whether they should continue to be utilized.

This definition suggests best practices evolve over time. What was once a best practice in the past may only be a good practice now, and in the future it may even be a poor practice. Continuous improvement calls for movement.

When we are looking for best practice companies, we must understand that no single best practice company will be found. All companies have their strengths and weaknesses. There are no perfect companies. Because the processes requiring improvement by benchmarking will vary, the companies identified as the best will also vary. A company wanting to insure it is benchmarking its practices against the best has to adopt systematic and thorough planning procedures and data collection.

Of the three types of benchmarking, best practice is the superior one. It provides the opportunity to make the most significant improvement, as the companies being benchmarked are the best in the particular process. Best practice benchmarking provides the greatest opportunity to achieve the maximum ROI. Most importantly, it provides the greatest potential to design and use breakthrough strategies, resulting in an increase in the company's competitive position (Wireman, 2004).

4.4.3.4 Benchmarking Process

The following steps are necessary for a successful benchmarking project:

1. Conduct internal analysis.
2. Identify areas for improvement.
3. Find partners for benchmarkings.
4. Make contact, develop questionnaire, perform site visits.
5. Compile results.
6. Develop and implement improvements.
7. Do it again.

When conducting an internal analysis, it is important to use a structured format, for example, a survey. The goal is to identify weaknesses in the organization and find areas that need improvement. Using a survey, an organization can find where it has the greatest deviation from the average, and then begin its benchmarking project in that area.

Once the process areas needing improvement are identified, benchmarking partners who are markedly better in the process must be identified. The organizations must be contacted to see if they are willing to participate in benchmarking.

When the partners are willing to benchmark, a questionnaire should be developed, based on the earlier analysis. The questionnaire is sent to the partners; site visits are scheduled and conducted. The information gathered in this process is compiled and used to make recommendations for changes to improve the benchmarking process. Once the changes are implemented and improvements noted, the process starts over again.

An analysis should be conducted before each benchmarking exercise, instead of relying on the previous analysis. When one process is improved, it often generates improvements in other processes. These improvements would not be noted in an older analysis. Therefore, the newer benchmarking project will not produce the projected improvements and, in turn, the organization may stop viewing benchmarking as cost-effective.

Benchmarking is an evolutionary process. A company may start with internal partners and see incremental improvements. The process then extends to better practice partners, whether internal to the company or external. Based on the improvements made and any additional areas identified for the next round of improvements, the process is ultimately extended to benchmarking with best practice organizations.

The key to this evolution is always finding a partner who is measurably better in the process being benchmarked. Once process parity is achieved with the partner, a new partner must be found, one who is still measurably better in the process. The benchmarking process continues until the best is found and superiority over this partner's processes is achieved (Wireman, 2004).

4.5 Key Performance Indicators, Benchmarking, and Best Practices

Performance indicators, or measures, for best practices are misunderstood and misused in most companies. If properly used, performance indicators should highlight opportunities for improvement. Performance indicators should highlight a "soft spot" in a company, enable further analysis to find the problem causing the low indicator, and ultimately point to a solution.

Performance indicators are valuable tools in highlighting potential benchmarking areas. For example, if a certain set of performance indicators shows that a maintenance process, such as preventive maintenance (PM), needs to be improved, and the internal personnel for the company cannot identify the changes necessary to improve it, a benchmarking project may be the answer.

However, it is necessary to clarify that benchmarks are not performance indicators, and performance indicators are not benchmarks. Using performance indicators is an internal function for a company. A benchmark is an external goal recognized as an industry or process standard. The number in itself is meaningless, unless there is an understanding of how the benchmark is derived. Understanding the enablers and success factors behind the benchmark is important.

It must also be clearly understood that there is a difference between a benchmark and the process of benchmarking. The benchmark is a number, while benchmarking is a process of understanding a company's processes and practices, so they can be adapted or modified and then adopted (Wireman, 2004).

4.5.1 Continuous Improvement: The Key to Competitiveness

Since benchmarking is a continuous improvement tool, it should only be started if a company wants to make changes to improve. Companies cannot develop the attitude of "We have always done it this way." They must be willing to change to meet the challenges of increasing competitive pressure.

Benchmarking is a continuous improvement tool able to facilitate change. As best practice companies are examined and their processes understood, the gap between a company's present practices and best practices promotes the desire for change. When companies see, understand, and learn from best practice companies, they can identify what to change and how to make the changes to maximize their return on their investment. Best practices in another company or area provide a realistic and achievable picture of the desired future. However, this takes resources, both human and financial capital, to be successful. It is necessary to explore the tangible and intangible factors that combine to produce superior performance. It is also necessary to involve those people most directly connected with the business process being benchmarked, since they have to take ownership in the changed process (Wireman, 2004).

4.5.2 Benchmarking Goals

Before starting a benchmarking project, it is a good idea to review the goals of benchmarking. Benchmarking should enable the following:

1. Provide a measure for the benchmarked process to permit an "apples to apples" comparison.
2. Clearly describe the organization's performance gap when compared to the measure.

3. Clearly identify the best practices and enablers producing the superior performance observed during the benchmarking project.
4. Set performance improvement goals for the benchmarked processes and identify actions that must be taken to improve the process (Wireman, 2004).

Quantifying the organization's current performance, the best practice for the process, and the performance gap is vitally important. A management axiom says:

> If you don't measure it, you don't manage it.

This is true of benchmarking. Quantifiable measures are required if a clear improvement strategy is to be developed. This entails the following SMART requirements:

1. Specific—ensures the project is focused.
2. Measurable—requires quantifiable measures.
3. Achievable—ensures the project is within a business objective.
4. Realistic—ensures the project is focused on a business objective.
5. Time-framed—ensures the benchmarking project has a start and end date (Wireman, 2004).

4.5.3 Gap Analysis

Gap analysis (see Figure 4.9) is a key component of any benchmarking project and helps achieve the SMART objectives. Gap analysis is divided into the following three main phases:

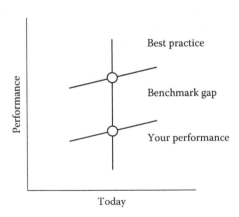

FIGURE 4.9
Gap analysis: best practice. (Adapted from Wireman, T., *Benchmarking Best Practices in Maintenance Management*, Industrial Press, Inc., New York, 2004, 256pp.)

1. Baseline—the foundation, or where a company is at present
2. Entitlement—the best a company can achieve with effective use of its current resources
3. Benchmark—the best practice performance of a truly optimized process

To use gap analysis effectively, the benchmarking project must be able to produce quantifiable results. All measures must be able to be expressed clearly and concisely so the improvement program can be quantified (Wireman, 2004).

The first step of gap analysis is to compare the company's process in quantifiable terms to the best practice results observed. In Figure 4.10, the gap between the observed best practice and the organization's current performance is plotted on the vertical axis. The horizontal axis shows the timeline. The chart highlights the need for the measures to be quantifiable if they are to be properly graphed.

The second part of gap analysis, shown in Figure 4.11, sets the time (T1) to reach what is called a current parity goal. This goal is focused on achieving the level of performance currently held by the best practice company. It also recognizes that the best practice company will have made improvements during this time period and will be at a higher level of performance.

The third step is to set a real-time parity goal. This level is reached when a company achieves parity on the benchmarked process with the best practice company, denoted in Figure 4.11 as T2.

The final step is the leadership position; at this point, a company's performance in the benchmarked process is recognized as having exceeded its partner's performance. This level is noted as T3 in Figure 4.11. At this point, the company will be recognized as the best practice company for the benchmarked process.

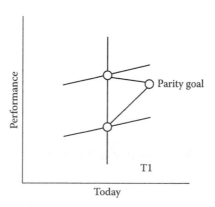

FIGURE 4.10
Gap analysis: parity goal. (Adapted from Wireman, T., *Benchmarking Best Practices in Maintenance Management*, Industrial Press, Inc., New York, 2004, 256pp.)

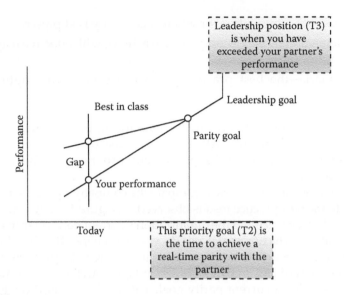

FIGURE 4.11
Gap analysis: reaching leadership position. (Adapted from Wireman, T., *Benchmarking Best Practices in Maintenance Management*, Industrial Press, Inc., New York, 2004, 256pp.)

Note that if a company is to effectively use gap analysis, all parameters must be quantifiable and time-framed. If not, gap analysis will be meaningless (Wireman, 2004).

4.5.4 Benchmarking Process

When the following checklist is used, a benchmarking process can be successful. If a disciplined approach is not followed, benchmarking is unlikely to produce any long-term results (Wireman, 2004).

A. Plan

B. Search

C. Observe

D. Analyze

E. Adapt

F. Improve

The checklist can be further expanded as follows.

4.5.4.1 A. Plan

1. What are our maintenance mission, goals, and objectives?
    a. Does everyone involved clearly understand the maintenance business function?

2. What is our maintenance process?

   a. What work flows, business process flows, and so on are involved?

3. How is our maintenance process measured?

   a. What are the current KPIs or performance indicators?

4. How is our maintenance process perceived as performing today?

   a. What is the level of satisfaction with the service performed by maintenance?

5. Who is the perceived customer for maintenance?

   a. Is the customer the operations department or shareholders/ owners? The answer to this question helps determine the level of understanding of maintenance within the organization.

6. What services are expected from the maintenance function?

   a. What service does maintenance perform? What is contracted externally? What isn't being done that needs to be done?

7. What services can maintenance deliver?

   a. Is maintenance capable of more? Are the staffing, skill levels, and so on at the correct level to perform the services?

8. What are the performance measures for the maintenance function?

   a. How does maintenance know if it is achieving its objectives?

9. How were these measures established?

   a. Were they negotiated or mandated?

10. What is the perception of our maintenance function compared to our competitors?

    a. *Are* internal perceptions worse than, as good as, or better than others?

4.5.4.2 B. Search

1. Which companies are better at maintenance processes than our company?

   a. *Suggestion*: search magazine articles and Internet sites.

2. Which companies are considered the best?

   a. *Suggestion*: consider the NAME Award: http://www.nameaward. com.

3. What can we learn if we benchmark with these companies?

   a. What are their best practices and how can they help our company?

4. Who should we contact to determine if it is a potential benchmarking partner?

   a. *Suggestion*: look for contacts in the articles or Internet sites searched previously.

4.5.4.3 C. Observe

1. What are their maintenance mission, goals, and objectives?
   a. How do they compare to our company's?
2. What are their performance measures?
   a. How do they compare to our company's?
3. How well does their maintenance strategy perform over time and at multiple locations?
   a. Are their current results an anomaly or are they sustainable?
4. How do they measure their maintenance performance?
   a. Are their measures different from our company's?
5. What enables their best practice performance in maintenance?
   a. Is it the plant manager, corporate culture, and so on?
6. What factors could prevent our company from adopting their maintenance policies and practices into our maintenance organization?
   a. How would we describe our culture, work rules, maintenance paradigm, and so on?

4.5.4.4 D. Analyze

1. What is the nature of the performance gap?
   a. *Suggestion*: compare their best practice to our practice.
2. What is the magnitude of the performance gap?
   a. How large is the benchmark gap?
3. What characteristics distinguish their processes as superior?
   a. *Suggestion*: detail the previously discussed enablers.
4. What activities do we need to change to achieve parity with their performance?
   a. What is the plan for change?

4.5.4.5 E. Adapt

1. How does the knowledge we have gained about their maintenance process enable us to make changes to improve our maintenance process?
   a. What do we need to do to improve?
2. Should we adjust, redefine, or completely change our performance measures based on the observed best practices?
   a. What are the differences and how can we benefit by the change?

3. What parts of their best practice maintenance processes would have to be changed or modified to be adapted into our maintenance process?

   a. *Note*: we need to be adaptors, not copycats.

4.5.4.6 F. Improve

1. What have we learned that will allow our company to achieve superiority in the benchmarked maintenance process?

   a. What can we change to eventually achieve the superiority position?

2. How can these changes be implemented into our maintenance process?

   a. *Suggestion*: develop an implementation plan.

3. How long should it take for our company to implement these changes?

   a. *Suggestion*: prepare a timeline for the implementation plan.

To gain maximum benefits from benchmarking, a company should conduct a benchmarking exercise only after it has attained some level of maturity in the core competency being benchmarked. Clearly, a company needs data on its own process before it can perform a meaningful comparison with another company (Wireman, 2004).

For example, in equipment maintenance management, the common benchmarks are the following:

1. Percentage of maintenance labor costs spent on reactive versus planned and scheduled activities.

2. Service level of the storeroom, that is, percentage of time spare parts are in the storeroom when needed.

3. Percentage of maintenance work completed as planned.

4. Maintenance cost as a percentage of the estimated replacement value of equipment.

5. Maintenance costs as a percentage of sales costs.

Without accurate and timely data and an understanding of how the data are used to compile the benchmark statistics, there will be little understanding of what is required to improve the maintenance process. This is true for whatever process is being benchmarked.

When partnering with companies considered the best in a certain aspect of a competency, it is important to have an example of an internal best practice to share with them. Benchmarking requires a true partnership, with mutual benefits. If we are only looking and asking during benchmarking visits, with no sharing, what is the benefit to the partner?

The final step to ensure benefits from benchmarking is to use the knowledge gained to make changes in the competency benchmarked. The knowledge should be detailed enough to develop a cost/benefit analysis for any recommended changes.

Benchmarking is an investment. The investment includes the time and money to do the ten steps described earlier. The increased revenue generated by the implemented improvements pays for the investment. For example, in equipment maintenance, the revenue may be produced through increased capacity (less downtime, higher throughput) or reduced expenses (efficiency improvements).

The revenue is plotted against the investment in the improvements to calculate the ROI. To ensure success, the ROI should be calculated for each benchmarking exercise (Wireman, 2004).

4.5.5 Benchmarking Code of Conduct

1. Keep it legal.
2. Be willing to give any information you get.
3. Respect confidentiality.
4. Keep the information internal.
5. Use benchmarking contacts.
6. Don't refer without permission.
7. Be prepared from the start.
8. Understand your expectations.
9. Act in accordance with your expectations.
10. Be honest.
11. Follow through with commitments.

While this list of suggestions may seem simply common sense, a surprising number of companies fail to apply them. This results in everything from minor disagreements between individuals to major legal battles. Recognizing that the other companies are our partners and treating them as such is key to successful benchmarking relationships (Wireman, 2004).

4.5.6 Traps in Benchmarking

When benchmarking is used properly, it can make a major contribution to the continuous improvement process. However, it can be devastating to a company's competitive position when used improperly. Some improper uses of benchmarking include the following:

1. *Using benchmarking data as a performance goal.* When companies benchmark their core competencies, they can easily fall into the trap of thinking a benchmark should be a performance indicator. For example, they focus all their efforts on cutting costs to reach a certain financial indicator, losing focus on the real goal.

   A company receives greater benefits when it understands the tools and techniques used by a partner to achieve a level of performance. This understanding allows the company not only to reach a certain number, but also to develop a vision of how to achieve an even more advanced goal.

   By focusing only on reaching a certain number, companies may change their organizations negatively (e.g., by downsizing or cutting expenses). They may also remove the infrastructure (people or information systems) and soon find that they are not able to sustain or improve the benchmark. In such cases, benchmarking becomes a curse.

2. *Premature benchmarking.* When a company attempts to benchmark before it is ready, it may not have the data to compare itself with its partners. It may end up making "guesstimates" that do it no good.

   The process of collecting data gives an organization an understanding of its core competencies and how it currently functions. Premature benchmarking will lead back to the first trap—simply wanting to reach a number. Companies who fall into this trap become "industrial tourists." They go to plants and see interesting things, but don't have enough of an understanding to apply what they see to their own businesses. The end results are reports that sit on shelves and never contribute to improved business processes.

3. *Copycat benchmarking.* Imitation benchmarking occurs when a company visits its partners and, rather than learning how the partners have changed their businesses, it concentrates on how to copy the partners' current activities. This practice may be detrimental to a company because it may not have the same business drivers as its benchmarking partners. There may be major constraints to implementing the partner's processes as well. Possible constraints include incompatible operations (7 days @ 24 h/day versus 5 days @ 12 h/day), different skill levels of the work force, differences in union agreements, different organizational structures, and different market conditions.

4. *Unethical benchmarking.* Sometimes a company will agree to benchmark with a competitor and then try to uncover proprietary information during site visits or through questionnaires. Clearly, this kind of behavior will lead to problems and ruin any chance

of conducting a successful benchmarking exercise at a later date. A second type of unethical benchmarking entails referring to or using the benchmarking partners' names or data in public without permission. This, too, will damage any chance for ongoing benchmarking between the companies. Even worse, the bad experience may prevent management from ever commissioning further benchmarking exercises with other partners (Wireman, 2004).

4.5.7 Other Pitfalls

Not every company is ready for benchmarking. That said, companies should not avoid benchmarking just because of a previous bad experience or because they have the attitude of "We are already the best" or "We are different than everyone else." Companies in which responsible individuals have such a mindset will have little chance of improving.

Once a company has the proper view of the benchmarking process, and disciplined guidelines are established and followed, desired improvements should follow. However, if a company does not benchmark for the right reasons, benchmarking efforts will fail (Wireman, 2004).

4.5.8 Procedural Review

Benchmarking opportunities are revealed when a company conducts an analysis of its current policies and practices. This should be a disciplined process, with 10 possible steps (also see Section 4.4.3.4 Benchmarking Process):

1. Conduct an internal audit of a process or processes.
   a. Education of key personnel in benchmarking processes is crucial at this point. They must fully understand and support the process.
2. Highlight potential areas for improvement.
   a. This requires understanding the cost of benchmarking compared to the financial benefits that will be derived. This should be presented as an ROI business case.
3. Do research to find three or four companies with superior processes in the areas identified for improvement.
4. Contact those companies and obtain their cooperation for benchmarking.
5. Develop a "pre-visit" questionnaire highlighting the identified areas for improvement (see step 2).
   a. This step requires a carefully planned approach to benchmarking and the discipline to adhere to the plan.
6. Make site visits to possible partners (see step 3).
   a. An interim report should be prepared after each visit and presented to the executive sponsor.

7. Perform a gap analysis on the data gathered by comparing them to current performance.
8. Develop a plan for implementing the improvements.
    a. The plan should include the changes required, personnel involved, and the timeline.
9. Facilitate the improvement plan.
    a. One or more members of the benchmarking project team should oversee the implementation of the plan to insure the changes are properly handled.
10. Start the benchmarking process over again.

Benchmarking helps companies find the opportunities for improvement that will give them a competitive advantage in their marketplaces. However, the real benefits do not accrue until the findings from the benchmarking project are implemented and improvements are realized (Wireman, 2004).

4.5.9 Final Points

1. It is necessary to explore the tangible and intangible factors that combine to produce superior performance and involve those people most directly concerned in the activity being examined.
2. Benchmarks are not the end-all. A benchmark performance does not remain a standard for long. Continuous improvement must be the goal (Wireman, 2004).

Table 4.1 shows benchmarks in many aspects of a company (financial, planning, maintenance, etc.) and suggests their importance. *Note*: this is merely a visual guideline intended to guide readers in the development of their own benchmarking processes (Mitchell, 2001).

4.6 Maturity of Maintenance as a Roadmap: Outcome of Benchmarking

The focus of the maintenance function is to insure all company assets meet and continue to meet the design function of the asset. Best practices, as adapted to the maintenance process, can be defined as follows:

> The maintenance practices that enable a company to achieve a competitive advantage over its competitors in the maintenance process. (Wireman, 2015)

TABLE 4.1

Maintenance Best Practice Benchmarks

Category	Benchmark
Yearly maintenance cost:	
Total maintenance cost/total manufacturing cost	<10%–15%
Maintenance cost/replacement asset value of the plant and equipment	<3%
Hourly maintenance workers as a percentage of total	15%
Planned maintenance:	
Planned maintenance/total maintenance	>85%
Planned and scheduled maintenance as a percentage of hours worked	~85%–95%
Unplanned downtime	~0%
Reactive maintenance	<15%
Run to fail (emergency + nonemergency)	<10%
Maintenance overtime:	
Maintenance overtime/total company overtime	<5%
Monthly maintenance rework:	
Work orders reworked/total work orders	~0%
Inventory turns:	
Turns ration of spare parts	>2–3
Training:	
For at least 90% of workers, hours/year	>80 h/year
Spending on worker training (% of pay roll)	~4%
Safety performance:	
OSHA recordable injuries per 200,000 labor hours	<2
Housekeeping	~96%
Monthly maintenance strategies:	
Preventive maintenance: total hours PM/total maintenance hours available	~20%
Predictive maintenance: total hours PM/total maintenance hours available	~50%
Planned reactive maintenance: total hours PRM/total maintenance hours available	~20%
Reactive emergency: total REM/total maintenance hours available	~2%
Reactive nonemergency: total REM/total maintenance hours available	–8%
Plant availability:	
Available time/maximum available time	>97%
Contractors:	
Contractors cost/total maintenance cost	35%–64%

Source: Adapted from Mitchell, J.S., *Handbook: Operating Equipment Asset Management*, 1st edn., Applied Research Laboratory, Pennsylvania State University, State College, PA, 2001.

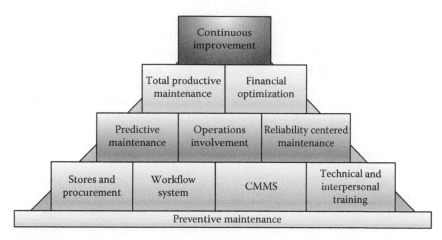

FIGURE 4.12
Maintenance management pyramid. (Adapted from Wireman, T., *Benchmarking Best Practices for Maintenance, Reliability and Asset Management: Updated for ISO 55000*, 3rd edn., Industrial Press, Inc., New York, 2015, 300pp.)

These practices (or processes) within maintenance fall into 11 categories:

1. Preventive maintenance
2. Inventory and procurement
3. Work flow and controls
4. Computerized maintenance management system (CMMS) usage
5. Technical and interpersonal training
6. Operational involvement
7. Predictive maintenance
8. Reliability-centered maintenance
9. Total productive maintenance
10. Financial optimization
11. Continuous improvement

Figure 4.12 illustrates the relationships, set up in the shape of a pyramid (Wireman, 2015).

4.6.1 Preventive Maintenance

A preventive maintenance (PM) program is the key to any attempt to improve the maintenance process. It reduces the amount of reactive maintenance to a level that allows other practices in the maintenance process to be effective. However, many companies have problems keeping the PM program focused. In fact, surveys show only 20% of US companies believe their PM programs are effective (Wireman, 2015).

Most companies need to focus on the basics of maintenance if they are to achieve any type of best-in-class status. Effective PM activities enable a company to achieve a ratio of 80% proactive maintenance to 20% (or less) reactive maintenance. Once the ratios are at this level, other practices in the maintenance process become more effective (Wireman, 2015).

4.6.2 Inventory (Stores) and Procurement

Note: for the purpose of this text, inventory and stores are used interchangeably.

The inventory and procurement programs must focus on providing the right parts at the right time. The goal is to have enough spare parts, without having too many spare parts. No inventory and procurement process can cost-effectively service a reactive maintenance process. However, if the majority of maintenance work is planned several weeks in advance, the inventory and procurement process can be optimized.

Many companies see service levels below 90%. As a result, more than 10% of requests face stock outs (i.e., an item is out of stock). This level of service leaves customers, in this case maintenance personnel, fending for themselves, stockpiling personal stores, and circumventing the standard procurement channels to obtain their materials. To prevent this situation, store controls are needed that will allow the service levels to reach 95%–97%, with 100% data accuracy. When this level of performance is achieved, the company can then start the next step toward improvement (Wireman, 2015).

4.6.3 Work Flows and Controls

This practice involves documenting and tracking the maintenance work that is performed. A work order system is used to initiate, track, and record all maintenance activities. The work may start as a request that needs approval. Once approved, the work is planned, scheduled, performed, and recorded. Unless the discipline is in place and enforced, data are lost, and true analysis can never be performed.

The system must be used comprehensively to record all maintenance activities. Unless the work is tracked from request through to completion, the data are fragmented and useless. If all maintenance activities are tracked through the work order system, effective planning and scheduling become a real possibility (Wireman, 2015).

Planning and scheduling requires someone to perform the following activities:

- Review the work submitted.
- Approve the work.

- Plan the work activities.
- Schedule the work activities.
- Record the completed work activities.

Unless a disciplined process is followed for these steps, productivity decreases and equipment downtime is reduced. At least 80% of all maintenance work should be planned on a weekly basis. Schedule compliance should be at least 90% on a weekly basis (Wireman, 2015).

4.6.4 Computerized Maintenance Management Systems Usage

In most companies, the maintenance function now stores data in computerized maintenance management systems (CMMS). CMMS software manages the functions already discussed.

Although CMMS has been used for almost a decade in some countries, results have been mixed. A recent survey in the United States shows the majority of companies use less than 50% of their CMMS capabilities. For CMMS to be effective, it must be used, and all data collected must have complete accuracy (Wireman, 2015).

4.6.5 Technical and Interpersonal Training

This function of maintenance insures technicians working on the equipment have the technical skills required to understand and maintain it. Additionally, those involved in the maintenance functions must have the interpersonal skills to be able to communicate with other departments in the company. They must be able to work as part of a team or in a work group environment. Without these skills, there is little possibility of maintaining the current status of the equipment, much less of making any improvements.

While there are exceptions, the majority of companies lack the technical skills within their organizations to maintain their equipment. In fact, studies show almost one-third of the adult population in the United States is functionally illiterate or just marginally better. When these figures are coupled with the lack of apprenticeship programs available to technicians, we see a workforce where the technology of the equipment has exceeded the skills of the technicians that operate or maintain it (Wireman, 2005).

4.6.6 Operational Involvement

The operations or production departments must take ownership of their equipment and be willing to support the maintenance department's efforts. Operational involvement, which varies from company to company, includes some of the following activities:

- Inspecting equipment prior to start-up
- Filling out work requests for maintenance
- Completing work orders for maintenance
- Recording breakdown or malfunction data for equipment
- Performing some basic equipment service, such as lubrication
- Performing routine adjustments on equipment
- Executing maintenance activities (supported by central maintenance)

The extent to which operations departments are involved in maintenance activities may depend on the complexity of the equipment, the skills of the operators, or even union agreements. The goal should always be to free up some maintenance resources to concentrate on more advanced maintenance techniques (Wireman, 2015).

4.6.7 Predictive Maintenance

Once maintenance resources are made available because the operations department has become involved, they should be refocused on the predictive technologies that apply to their assets. For example, rotating equipment is a natural fit for vibration analysis, electrical equipment a natural fit for thermography, and so on.

The focus should be on investigating and purchasing technology that solves or mitigates chronic equipment problems, not to purchase all of the technology available. Predictive maintenance (PDM) inspections should be planned and scheduled using the same techniques used to schedule preventive tasks. All data should be integrated into the CMMS (Wireman, 2015).

4.6.8 Reliability-Centered Maintenance

Reliability centered maintenance (RCM) techniques are applied to preventive and predictive efforts to optimize the maintenance programs. If a particular asset is environmentally sensitive, safety-related, or extremely critical to the operation, the appropriate PM/PDM techniques are chosen.

If an asset is going to restrict or impact the production or operational capacity of the company, another level of PM/PDM is applied with a cost ceiling in mind. If the asset is going to be allowed to fail and the cost to replace or rebuild the asset is expensive, yet another level of PM/PDM activities is specified. There is always the possibility that it is more economical to allow some assets to run to failure, and this option is considered in RCM.

RCM tools require data to be effective. For this reason, the RCM process is used after the organization has progressed to the point where it is compiling complete and accurate data on assets (Wireman, 2015).

4.6.9 Total Productive Maintenance

Total productive maintenance (TPM) is an operational philosophy whereby everyone in the company understands that his or her job performance impacts the capacity of the equipment in some way. For example, operations may understand the true capacity of the equipment and not run it beyond design specifications; they realize this could cause unnecessary breakdowns and respond accordingly.

TPM is like total quality management. The only difference is that companies focus on their assets, not their products. TPM can use the tools and techniques for implementing, sustaining, and improving the total quality effort (Wireman, 2015).

4.6.10 Financial Optimization

This statistical technique combines all of the relevant data about an asset, including downtime cost, maintenance cost, lost efficiency cost, and quality costs. It then balances those data against financially optimized decisions, such as when to take equipment off line for maintenance, whether to repair or replace an asset, how many critical spare parts to carry, and what the maximum-minimum levels on routine spare parts should be. Financial optimization requires accurate data; making these types of decisions incorrectly could have a devastating effect on a company's competitive position. When a company reaches a level of sophistication where this technique can be used, it is approaching best-in-class status (Wireman, 2015).

4.6.11 Continuous Improvement

Continuous improvement is best epitomized by the expression "best is the enemy of better." Continuous improvement in asset care requires an ongoing evaluation program that includes constantly looking for the "little things" that can make a company more competitive.

Benchmarking is one of the key tools for continuous improvement. As indicated in the previous section, the most successful type is best practice benchmarking. In maintenance, best practice benchmarking examines specific processes used in a company, compares them to companies that have mastered them, and maps changes to improve them. Benchmarking is a continuous improvement tool, a technique employed by mature organizations that are knowledgeable about the maintenance business process (Wireman, 2015).

4.7 Attempts at Standardization from Europe (EN) and the United States (ISO)

4.7.1 European Maintenance Standards

When an organization wants to compare maintenance and availability performance internally or externally, it needs a common platform of predefined indicators or metrics. Comparison of metrics when the bases of calculation are not the same is a frustrating non-value-added activity.

An organization should use standardized indicators or metrics for the following reasons (EN 15341):

- Maintenance managers can rely on a single set of predefined indicators supported by a glossary of terms and definitions.
- The use of predefined indicators makes it easier to compare maintenance and reliability performance across borders.
- When a company wants to construct a set of company indicators or scorecard, the development process is simplified if the indicators are predefined.
- The predefined indicators can be incorporated into various CMMS software and reports.
- The predefined metrics can be adopted and/or modified to fit a company's or a branch's specific requirements.
- The need for discussion and debate on indicator definitions has ended and uncertainties are eliminated.

The challenge was first met by the Society for Maintenance and Reliability Professionals (SMRP) and the European Federation of National Maintenance Societies (EFNMS). Since 2000, SMRP has defined 70 best practice metrics to measure maintenance and reliability performance. In 2000, EFNMS defined a set of indicators to measure maintenance performance. These indicators are now incorporated in the European standard EN 15341 Maintenance Key Performance Indicators, released in March 2007, and mentioned earlier in the chapter.

The performance indicators in EN 15341 were designed by the CEN Technical Committee 319 Working Group 6 (WG6). WG6 was set up by European experts in maintenance management to create an architecture of indicators to measure maintenance performance worldwide. The WG6 group looked at all maintenance indicators available in the literature and considered the guidelines, procedures, and experiences of many multinational industrial companies. From these, they selected three groups of KPIs: economic, technical, and organizational (Galar and Kumar, 2016).

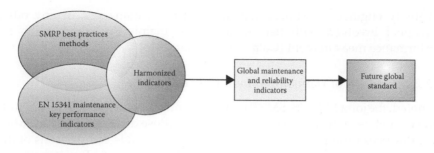

FIGURE 4.13
Process of the harmonized indicator project. (Adapted from Galar, D. and Kumar, U., *Maintenance Audits Handbook: A Performance Measurement Framework*, CRC Press, Taylor & Francis Group, April 6, 2016, 609pp.)

As a result, EN 15341 is a reliable reference, a worldwide standard that can measure and develop maintenance performance, considering and managing both external and internal influencing factors.

In 2006, SMRP and EFNMS conducted a joint effort to compare and document standard indicators for maintenance and reliability performance. The harmonization effort had the objective of documenting the similarities and differences between the SMRP metrics and the EN 15341 standards (Kahn and Gulati, 2006). The harmonization process is shown in Figure 4.13 (Kumar et al., 2013).

4.7.2 Objective of the Harmonized Indicator Document

With companies producing goods and supplying services on an international scale, the need for a common understanding of the indicators to measure maintenance and availability performance is paramount (Galar and Kumar, 2016). There will certainly be a global standard for maintenance indicators at some point in the near future.

The document produced by the harmonization project attempts to do precisely that. It offers the maintenance community a set of predefined indicators to measure maintenance and reliability performance on a global basis. The indicators can be used by all organizations with a need to measure, track, report, and compare maintenance and reliability performance (Galar and Kumar, 2016).

The document also gives a scale for measuring maintenance or reliability performance. The indicators or metrics are supported by a set of guidelines and examples of calculations (it is outside the scope of this document to give any ISBN 978-91-7439-379-8 23 recommended values or thresholds for the indicators, but see Galar et al., 2011). This provides maintenance professionals with an easy-to-use guide to indicators and their components.

The target group for the harmonized indicators document includes maintenance managers, asset managers, plant managers, operations managers, reliability engineers, technical managers, general managers, or any other personnel involved with benchmarking or maintenance and reliability performance measurement (Galar and Kumar, 2016).

4.7.2.1 Contents of Harmonized Indicators Document

As noted, the joint EFNMS–SMRP working group was intended to resolve differences between the EN 15341 indicators and those developed by the SMRP Best Practices Committee. The group made side-by-side comparisons of the indicator formulas and definitions of terms. The basis for the European terms includes EN 13306:2001 Maintenance Terminology and IEC 60050-191:1990 Dependability and Quality of Service. The SMRP definitions are contained within each indicator (metric) description and appear in an SMRP Glossary of Terms (SMRP Best Practices Committee, 2006). The comparisons resulted in two extensive lists, as there were terms or formulas common to both sets.

An indicator was defined as common if it had the same basic formula in both or could be universally applied. For these common indicators, the working group determined whether any differences could be eliminated without sacrificing the objective of the indicator. If differences could not be eliminated, they were qualified or explained. This is the essence of the harmonization process, graphically depicted in Figure 4.14.

It should be noted that the grouping of indicators is different. In EN 15341, the indicators are grouped into economic, technical, and organizational sets. The SMRP indicators are categorized in accordance with the five pillars of the SMRP maintenance and reliability body of knowledge: business and management, manufacturing process reliability, equipment reliability, organization and leadership and work management (SMRP Best Practices Committee, 2006).

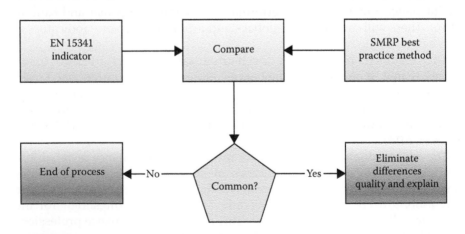

FIGURE 4.14

Harmonization process. (From Svantesson, T. and Olver, R., *Euromaintenance, SMRP-EFNMS Benchmarking Workshop*, Brussels, Belgium, April 8–10, 2008.)

The joint working group made good progress, announcing the first harmonization results in January 2007 and publishing the first edition of Global Maintenance and Reliability Indicators in April 2008 (Svantesson and Olver, 2008). To date, 29 metrics have been targeted for harmonization. When an indicator is harmonized, a statement to this effect is added to the SMRP metric description. Meanwhile, the SMRP metric is recommended by EFNMS for use as a guideline to calculate the EN 15341 indicators (CEN, 2007).

The harmonization work will continue until the complete list of SMRP indicators currently under development has been exhausted. Similar harmonization efforts are projected with other international maintenance organizations, such as the Technical Committee on Maintenance of the Pan American Federation of Engineering Societies (COPIMAN) or Maintenance Engineering Society of Australia (MESA). There are tentative plans to promulgate the use of these indicators as international standards. Discussions are ongoing with CEN/TC 319 to consider proposing the harmonized metrics as global standards or guidelines.

The calculation of indicators is exhaustive. Many maintenance managers read EN 15341 or SMRP Good Practices and extract valuable information, but they do not know the desired value of all the parameters, either partially or fully applied to their respective fields. In addition, in world-class maintenance, many figures have been proposed for many industries as a result of companies' experiences. If successful, the working group will lay the foundation for common methods for parameter calculation and the establishment of real numbers as targets (Kumar et al., 2013).

4.7.3 International Maintenance Standards

The most recent US proposal constitutes a series: ISO 55000, ISO 55001, and ISO 55002. Not just American but international, the ISO series constitutes a revolution in the traditional roadmap of maintenance performance (Sondalini, 2014).

4.7.3.1 Review of ISO 55000 Series

Simply stated, ISO 55001 is a standard for a management system for physical assets; it discusses the benefits of integrated asset management. ISO 55000 and ISO 55002 are companion documents: the former introduces the concepts, and the latter provides guidance on the application of ISO 55001 (Sondalini, 2014).

The series states that asset management is based on the following four fundamentals:

1. Bringing value for the organization and all its stakeholders in the use of its assets
2. Promoting alignment across the organization to achieve organizational goals

3. Providing managerial leadership with commitment, along with a culture of empowerment
4. Achieving assurance of required asset performance

Note that the four fundamentals are not principles of successful enterprise asset management (EAM); rather, they are outcomes of successful asset management.

It argues that six concepts drive asset management outcomes:

1. A physical asset is made of assemblies of individual components working as a system.
2. An organization is a system for doing the work valued by its owners through the use of knowledge and supporting processes.
3. Organizational processes and activities are series arrangements of sequential steps where successful outcomes are quality dependent.
4. Mistakes and errors in processes must be eliminated to get maximum value.
5. Probability and uncertainty exist everywhere, causing variation and creating risk.
6. A business aims to survive and profit, so short-term, long-term, and life cycle asset strategies must protect the business and make it flourish.

Harnessing these six principles is foundational to the successful use of physical assets and must be at the heart of all asset management system design, creation, and sustainability.

However, there is no acknowledgment in the ISO series that assets are made of systems of parts. Nor is there any recognition of the asset life cycle. Although there is strong promotion of integration between departments, nothing is said about preventing errors and defects. The influence of probability is noted for risk but not mentioned anywhere in respect to organizational process outcomes. There is no appreciation of the driving need for both business survival and financial success.

In short, the series does not incorporate all the important principles needed for world-class enterprise asset management performance. It is vital for companies to see the shortcomings and to correct the limitations so they do not waste time, money, and resources. In fact, satisfying the ISO standard should be of secondary importance to building an EAM system that contributes to the success of a company. Companies that build their EAM system solely on the basis of the ISO 55000 series have the wrong focus and purpose. As useful as it is to standardize asset management practices, a better solution is required for a company to become world class (Sondalini, 2014).

4.7.3.2 Contents of ISO 55001 and Its Companion Documents

The following subsections look more closely at the following divisions and their relation to asset management, as defined by the ISO 55001 and its companion documents:

- Context of the organization
- Leadership
- Support
- Operation
- Performance evaluation
- Improvement

4.7.3.3 Context of the Organization

The context refers to the internal and external environments, jurisdiction laws and regulations, the stakeholders within and without the organization, and the organization's purpose(s) and goals. These factors influence the development of a strategic asset management plan (SAMP) to achieve the goals. The SAMP is disseminated across the organization through individual asset management plans. It sets the activities undertaken in the workplace; when done correctly, these should deliver the asset management objectives and hence the organization's goals.

ISO 55001 does not mention the creation of a successful business, however, and the lack of focus on active production of business success curtails its strategic business value. It makes the asset management system principally a means to provide information and record events, get control over work done, and chart asset performance. That is obviously useful but it's not the focus a company needs.

All external and internal issues affecting both the achievement of the organizational goals and the asset management system should be identified. ISO 55002 lists the sorts of issues to appraise, but it does not explain the methodology for the comprehensive review. *Note*: an appropriate methodology is a business environmental scan.

To achieve the alignment of the asset management objectives and organizational objectives suggested in ISO 55001, it may be useful to document a clear progression from organizational objectives to asset management objectives, from asset management objectives to the SAMP, and from the SAMP to individual asset management plans (Sondalini, 2014).

4.7.3.3.1 Understanding Needs and Expectations of Stakeholders

The stakeholders are part of the context. All internal and external stakeholders who impact the asset management system must be identified. What do they want, and what do they need to know about the assets being managed? Their needs and expectations should be documented and taken into account when developing an asset management policy.

One approach is to compile a matrix showing all stakeholders and all their needs and expectations, including the ways each need and expectation is preferably addressed. ISO 55002 may be helpful here, as it provides lists of typical stakeholders in asset management. Companies should set criteria to determine priorities and resolve conflicts between stakeholders. The real aim of this requirement is to ensure proactive consultation and communication between relevant parties. Therefore, it is necessary to develop a communication plan explaining which stakeholders are involved and how they will interact about specific issues, events, or circumstances (Sondalini, 2014).

4.7.3.3.2 Determining the Scope of the Asset Management System

Scope is also part of context. This particular clause, called the "Scope" in ISO 55001, sets the boundaries for assets in the asset management system and defines the coverage of an asset management system. It requires an asset register listing all the assets covered by the asset management system; this is called the asset portfolio. All other assets not in the register are outside the scope.

The scope covers the whole organization's total involvement, resources, and interactions related to asset management in the widest sense. As well as covering internal processes and functions, it includes interfacing with outside suppliers, subcontracted services, regulators, and national and international considerations. It includes geographic locations and timescales of organizational involvement.

Scope is likely to be a written document with relevant subheaded sections or a spreadsheet with columns and rows titled with appropriate subheadings (Sondalini, 2014).

4.7.3.3.3 Asset Management System

A final element of context is the definition of every process used in asset management. This requires companies to create flowcharts for every process with all its interactions, and putting into place procedures of how to correctly conduct each phase/step of a process. Ultimately, a series of documents will cover all aspects of the asset management system and the SAMP. *Note*: the SAMP needs to be a stand-alone document that can be distributed (Sondalini, 2014).

4.7.3.4 Leadership

The upper management represents the leaders of asset management in an organization. Managers at this level of the company hierarchy are ultimately responsible for successful asset management. They set the vision and objectives. They align the organization with maintenance and provide the resources to achieve the goals. They put people into leadership roles in the organization and provide them with guidance and support to operate the required processes (Sondalini, 2014).

Company leaders should cover the following:

- Document the asset management policy and SAMP objectives.
- Show the company has integrated business processes that are collaborative, resourced, well-communicated, and effective.
- Prove the established outcomes are achieved.
- Ensure the company is continually improving.
- Practice risk management consistently.

4.7.3.4.1 Policy

Leadership must ensure the company's asset management policy justifies asset management as an organizational imperative and contains a set of principles for asset management in the organization (Sondalini, 2014).

4.7.3.4.2 Organizational Roles, Responsibilities, and Authority

The internal and outsourced roles needed to disseminate asset management in an organization and achieve the SAMP. All those involved need to be competent. All responsibilities for each role must be clearly specified so persons within and without the organization understand who is responsible for an activity. ISO 55001 specifically suggests responsibility and authority be provided to ensure the proper planning, functioning, performance, and upkeep of the asset management system. Typically, this is addressed with an organizational chart and the use of position descriptions in which responsibilities are clearly stated. It can also be addressed by specifying responsibilities and authority in procedures and work instructions. A matrix of roles listing responsibilities and authority is a useful way to show this information globally (Sondalini, 2014).

4.7.3.4.3 Planning

ISO 55001 says planning should address how to achieve the organization's asset management goals, address stakeholders' needs and requirements, and consider the risks and opportunities related to the asset, the asset's management, and the asset management system (Sondalini, 2014).

4.7.3.4.4 Addressing Risks and Opportunities

ISO 55002 states: "The overall purpose is to understand the cause, effect and likelihood of adverse events occurring, to manage such risks to an acceptable level, and to provide an audit trail for the management of risks. The intent is for the organization to ensure that the asset management system achieves its objectives, prevents or reduces undesired effects, identifies opportunities, and achieves continual improvement."

The SAMP explains how the asset management objectives are to be achieved. It cascades down to the plans for individual assets and the activities to address and control the risks. A company's planning will need to identify the processes used, the roles, the actions, and the addressed risks of an asset's management plan. The same applies, that is, process through to actions, for the asset's management, and for the effective use of the asset management system. As part of planning, the company also needs to specify how the plans, the activities, the asset's performance, and the asset management system's performance will be monitored and reported.

The most suitable methodology is to apply risk assessment and management to each asset, the asset's management, and the asset management system. Reliability centered maintenance (RCM) is a good option, as it applies failure mode effects analysis (FMEA) to assess and mitigate asset risks through an appropriate maintenance strategy and activities. To identify the risks to an asset's management and to the asset management system, process failure mode effects analysis (PFMEA), coupled with risk management, would be suitable (Sondalini, 2014).

4.7.3.4.5 Asset Management Objectives and Planning to Achieve Them

ISO 55000 defines objectives as the results to be achieved and asset management as the coordinated activity of an organization to realize value from assets. The asset management objectives are the specific results required from assets. Since achieving the organizational objectives is the purpose of the asset management objectives, there is a natural alignment between the organization's objectives and its asset management objectives.

The asset management objectives require planned actions to achieve them. These are called the asset management plans. The asset management plans require defined activities to be done on assets; they also require setting specific and measurable objectives, such as time frames, the resources to use, their costs, and the work quality to be performed to ensure asset risk management.

The asset management objectives must be allocated as roles and functions throughout the organization. The planned activities to achieve the objectives need to be derived and turned into jobs and tasks. Which is the best function for the responsibility? At what level in the organization should the objective be allocated? Answers to these questions depend on issues like the objective's importance, the risks for the organization, the effect on stakeholders, the difficulty involved, the information and knowledge needs, and the resourcing requirements.

ISO 55001 says asset management objectives work to achieve the following criteria: assess the extent of risk, be consistent with organizational objectives, have a clear means of achievement, monitor, measure, disseminate, and review results, and, finally, suggest updated procedures. The output of planning and setting the asset management objectives will be a tabular listing of specific, measurable, achievable, realistic, and time-bound (SMART) objectives, showing how each criterion is satisfied (Sondalini, 2014).

ISO 55001 stipulates the following specific requisites for asset management planning:

- Asset management planning must be integrated with the rest of an organization's planning.
- The output of an asset's management planning is the full scope of work addressing all the asset's risks during its use over its entire lifetime. The asset's management plan contains all subsequent risk mitigations and risk preventions; it identifies internal and external resources and any special tools or equipment required.
- Changing risk as the asset ages, or if there are changes in asset use, is a contingency to be factored into the asset's management plan.
- Risks with assets are to be considered as risks to the organization and addressed in organizational plans and contingency plans.

Identifying risk over the lifetime of an asset and determining the risk abatements to be implemented will require using a risk assessment and risk management methodology. Equipment criticality analysis can initially identify the full range of internal and external business risk implications from the ownership or use of an asset.

The asset's management planning can be done by asset type or class of asset if a company has many of the same design of asset used in similar ways on similar products, for example, an identical vehicle fleet.

ISO 55002 provides a structured approach to risk review and the identifying, analyzing, classifying, and eliminating of risk to assets, asset management, and asset management system. It advocates life cycle costing to identify the long-term financial implications of an asset's use (Sondalini, 2014).

4.7.3.5 Support

For an asset management system to function correctly, a company needs to know the processes, infrastructure, financing, knowledge, skills, information management, service delivery, and cultural environment that produce the intended asset performance (Sondalini, 2014).

4.7.3.5.1 Resources for Support

All resources needed to deliver the SAMP and the asset management plans must be identified. This clause in ISO 55001 addresses the lifetime of an asset and the commitments needed across the organization and throughout the asset's life cycle to get the desired performance from the asset. The company must identify and address all financial support, safety, human, equipment, tools, capital expenditure, for example, replacement, refurbishment, and so on, and internal and external resources.

ISO 55002 suggests using resourcing gap analysis to identify the resource gaps between an organization's capabilities and the requirements needed to achieve its asset management plans. Where there are constraints, the organization will need to prioritize its actions and the use of resources and consider going outside the organization to get needed competences. The identification of resources to implement asset management plans should be done when an asset's management plan is being developed. The required resources for asset management can then be accumulated and input into resourcing gap analysis (Sondalini, 2014).

4.7.3.5.2 Competence

An essential part of support is competence. Only competent people should be used in asset management–related roles, functions, and duties, including external suppliers, providers, and subcontractors. They must have the values and attitudes, awareness, knowledge, understanding, skills, and experience needed to correctly do the jobs given to them so that they deliver all necessary outcomes. (A competent person or organization consistently produces the desired outcome unsupervised.)

All competencies, that is, values and attitudes, awareness, knowledge, understanding, skills, and experience, needed for each position or function involved in an organization's asset management must be identified and documented. Proof of current competencies is required. Ongoing proof to confirm incumbents remain competent or have developed appropriate new competence is necessary. Typically, a training needs analysis and a gap analysis are done to confirm incumbents are competent. For those requiring further training and development, the shortfalls in capability become a short-term training plan. Typically, this clause of ISO 55001 is administered by a human resources group. ISO 55001 is clear: delivering the asset management objectives requires integration and cooperation of able and competent people within and without the organization. They must understand what needs to be done and work together effectively to achieve it (Sondalini, 2014).

4.7.3.5.3 Awareness

Awareness in ISO 55001 is very much about the extent to which a role incumbent (or a stakeholder) knows an organization's asset management system and its impacts and opportunities when working to deliver the asset management plan. In ISO 55002, awareness is implied to be far greater than a personal level of awareness; it is more about taking a business owner's perspective of risks, opportunities, and service delivery quality.

The *Oxford Dictionary* says being aware is having knowledge or perception of a situation or fact. To measure awareness, then, we might interview people using a set of targeted questions covering their understanding of asset management policy, how to effectively contribute to asset management success, how they view their contribution to achievement of the organization's asset management goals, and what more they can do to make a positive difference (Sondalini, 2014).

4.7.3.5.4 Communication

A strategy must consider how to get the right information to the right people, inside and outside the organization, at the right time. ISO 55000 recommends the use of a communication plan. The communication plan will cover the lifetime exchange of the information required by internal and external parties on individual assets, the assets' management, and the asset management system. All communication requirements related to the relevant assets, asset management, and the asset management system should be identified and addressed, including transfer and exchange of knowledge and intelligence for planning, execution, improvement, and monitoring the resulting performance (Sondalini, 2014).

4.7.3.5.5 Information Requirements

Support implies the availability of information. All information requirements related to the relevant assets, asset management, the asset management system, and the achievement of the organizational objectives must be identified and addressed. What type of information to keep will depend on its purpose. ISO 55002 helpfully lists a range of asset information requirements common to asset management and asset management systems.

ISO 55001 requires information on risks, roles, processes, stakeholders, and the content and worth of the information for decision-making. Companies need to address information collection, retention, management, consistency of terminology, and information traceability. This again lends itself to a systematic tabular listing of requirements documenting the analysis and decisions (Sondalini, 2014).

4.7.3.5.6 Documented Information

Documented information is defined in ISO 55000 as "information required to be controlled and maintained by an organization, and the medium on which it is contained." The ISO 55001 clause on documentation covers all media, every process in the asset management system, and all records of evidence.

The documents required by ISO 55001 include the following: those set by statute; documents that form evidence, usually known as records; internal and external documents used for the assets during their lifetime; and documents for running and improving the asset management system.

A company must have a structured and consistent way to develop, approve, and maintain these documents. They are considered controlled documents and become part of a document management system. They need to be identified, approved, available when needed, stored and preserved, and change-controlled; archival and disposition practices must also be specified.

Note: this entire clause is satisfied by the document management requirements of the earlier standard, ISO 9001. Instead of creating asset management–specific solutions for this clause, then, companies may simply incorporate asset information management into their existing ISO 9001 quality system, or adopt the ISO 9001 requirements to manage their asset management system's documented information (Sondalini, 2014).

4.7.3.6 Operation

ISO 55001 explains how to use an asset management system so that it delivers the outcomes it is meant to produce (Sondalini, 2014).

4.7.3.6.1 Operational Planning and Control

All processes, from beginning to end, have to be put into use and must operate as intended. The goal is to do exactly as planned to produce the intended results. The closer the actual results are to the planned results, the more effective the asset management system is. Companies must identify risks, use appropriate risk controls, and specify what evidence will be recorded, along with the performance measurements. The suggested methodology is process mapping (Sondalini, 2014).

4.7.3.6.2 Management of Change

An essential part of operational control is the management of internal or external changes, whether permanent or temporary. This needs to be evaluated, and suitable risk mitigations must be selected and approved before implementing any change. ISO 55001 makes it clear that changes are to be risk-assessed, with consideration of the impacts across the organization, including reviewing unintended consequences. The company must thoroughly plan implementation of change, and changes occur only after it has gone through the whole change management process. A comprehensive change management process and procedure must be created and communicated to all parties concerned. ISO 55002 provides a list of typical considerations to review when undertaking a change (Sondalini, 2014).

4.7.3.6.3 Outsourcing

In outsourcing, according to ISO 55000, companies "make an arrangement where an external organization performs part of an organization's function or process." The outsourcing of activities does not absolve the company of ownership, however. All work sent off-site or services subcontracted are still the company's responsibility; it is accountable for the successful completion of the work. This necessitates identifying the risks of outsourcing and proactively putting suitable controls and monitoring into place.

ISO 55001 requires a formal, documented process for outsourcing. Typically, legal contracts are first put into place between the organization and service provider. ISO 55001 asks companies to decide how to interact with the service provider, going so far as to ask for evidence of competence, awareness, establishing the processes for managing and monitoring the outsourced work and the processes for information exchange and document control. It is necessary to develop a documented process and procedures explaining how outsourcing is done and taking into account the complexity of the work and the risks it causes to the organization (Sondalini, 2014).

4.7.3.7 Performance Evaluation

Monitoring and evaluating the performance of assets, asset management, and the asset management system against their respective objectives is necessary to ensure the desired outcomes are being achieved. This provides confirmation of satisfactory or unsatisfactory performance and highlights opportunities for improvement (Sondalini, 2014).

4.7.3.7.1 Monitoring, Measurement, Analysis, and Evaluation

The information collected by monitoring, measurement, analysis, and evaluation is intended to improve the effectiveness of the processes used in the organization and ultimately to improve the organization's ability to meet owner and stakeholder needs.

To monitor, it is necessary to select measures and analyze results of measurement. ISO 55001 requires companies to determine the performance of their assets, asset management, and asset management system. Companies must determine what data and/or measures to use, what methods to use to monitor, measure, analyze, and evaluate, when to observe and check the process or procedure, when to report the findings and interpretations, and who receives the report. ISO 55002 provides a substantial list of the range of factors that can be observed through performance monitoring. ISO 55002 further advises: "A set of performance indicators should be developed to measure the asset management activity and its outcomes. Measurements can be either quantitative or qualitative, financial and non-financial. Indicators should provide useful information to determine both successes and areas requiring corrective action or improvement." ISO 55002 also makes the important point that asset management is actually about "asset risk management," so gauging the level of risk remaining to the organization is an important measure of asset management success.

Finally, the collecting of recorded evidence and its analysis, evaluation, and reporting constitutes documented information and must be controlled (Sondalini, 2014).

4.7.3.7.2 Internal Audit

On a regular basis, an organization must examine and report on its own asset management system effectiveness, the operation of its processes, the resulting performance against the respective standards it has set itself, and its compliance with the ISO 55001 standard. The suggested methodology is internal auditing (Sondalini, 2014).

4.7.3.7.3 Management Review

The ISO 9000 quality management system standard defines a review as an "activity undertaken to determine the suitability, adequacy and effectiveness of the subject matter to achieve established objectives." A management review, as required by ISO 55001, is a review of the suitability, adequacy, and effectiveness of the asset management, that is, the successfulness of the asset management system at delivering its objectives.

ISO 55001 lists a series of sub-clauses that must be addressed as part of a management review. The simplest way to do this is to create an agenda of items to be reported on at a meeting and ensure a record appears in the minutes of the meeting. A management review is an opportunity for upper management to get an overview of the organization's use and benefits from the asset management system. They ought to ensure the reports provided to them explore the full scope of the asset management system. It is wise to ask for an analysis of the organization's processes to identify ways to help it perform more effectively (Sondalini, 2014).

4.7.3.8 Improvement

It is expected that companies will want to improve the design and use of their asset management systems. Problems are simply opportunities to learn what more can be done to enhance the system. The best organizations are proactive at identifying potential risks and eliminating them and/or creating contingency plans to minimize the effects (Sondalini, 2014).

4.7.3.8.1 Nonconformity and Corrective Action

ISO 55001 says each incident must be immediately addressed to rectify the problem and control its effects, then analyzed to decide whether to eliminate the cause. Such analysis is called root cause failure analysis (RCFA). When changes are recommended, they enter the management of change process. The effectiveness of actions undertaken to address or prevent a problem need to be checked at appropriate times to confirm the actions are working. Further improvements are made if a solution is not effective.

Typically, this requirement is satisfied by developing a nonconformance management and corrective action procedure and communicating it to all parties concerned (Sondalini, 2014).

4.7.3.8.2 Preventive Action

ISO 55002 states: "Preventive actions, which may include predictive actions, are those taken to address the root cause(s) of potential failures or incidents, as a proactive measure, before such incidents occur. The organization should establish, implement, and maintain process(es) for initiating preventive or predictive action(s)." In other words, companies need to establish and maintain preventive action processes and procedures. ISO 55002 provides a list of things to consider when developing a preventive action procedure (Sondalini, 2014).

4.7.3.8.3 Continual Improvement

An asset management system and its operation must be continually improved by identifying opportunities to make it simpler, reduce its cost, do things faster, and deliver top-quality results. If the organization's assets, asset management, and asset management system perform better, all stakeholders benefit.

Companies must develop a clear methodology that delivers continual improvement. ISO 55002 provides a list of the issues to include in considerations of what can be done to encourage continual improvement of assets, asset management, and asset management systems (Sondalini, 2014).

4.8 Role of Maintenance Cost in a Maintenance Audit

Converting forecasts into concrete real numeric values requires extraordinary effort. As an alternative, the principles can be addressed by focusing on the system and its attributes rather than on specific outcomes. This is the "system audit approach." A maintenance audit is an examination of the maintenance system to verify maintenance management is carrying out its mission, meeting its goals and objectives, following proper procedures, and managing resources effectively and efficiently (Figure 4.15).

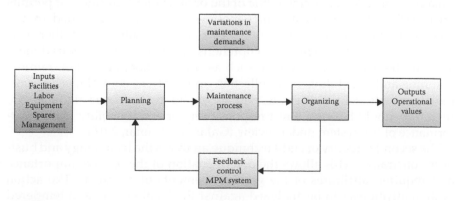

FIGURE 4.15
Input–output of maintenance transformation. (Adapted from Stenstrom, C., Maintenance performance measurement of railway infrastructure with focus on the Swedish network, Technical Report, Lulea University of Technology, Lulea, Sweden, 2012.)

This indicates a concentration on the maintenance system itself, as opposed to the quantification of its inputs and outputs. It is anticipated that using such an approach will yield a level of accuracy compatible with the information normally available on real performance. Subjectivity in performance measurement will not be overcome, but such subjectivity will become more visible (Galar and Kumar, 2016).

According to Kaiser (1991), De Groote (1995), The Institute of Internal Auditors (Stice et al., 1992), Mann (1983), and Duffuaa and Raouf (1996), an audit helps management achieve the following:

1. Ensure maintenance is carrying out its mission and meeting its objectives.
2. Establish a good organizational structure.
3. Manage and control resources effectively.
4. Identify problems and solve them.
5. Improve maintenance performance.
6. Increase the quality of the work.
7. Automate information systems to increase effectiveness and productivity.
8. Develop a culture of continuous improvement.

Auditing, as a general technique, can be divided into two categories. The first utilizes general audits based on a commonly assumed standard, designating what constitutes a "good system." System audits are a good tool for consultants, as they allow them to have a consistent standard of what a good maintenance system should be. Although this type of audit is normally isolated from a deep understanding of the organization's business, it permits the insertion of important attributes about which auditors have good knowledge, but whose importance varies from one organization to another. This kind of audit is a thorough and comprehensive review of the various dimensions in the maintenance system, such as organization, personnel training, planning and scheduling, data collection and analysis, control mechanisms, measurements, reward systems, and so on. To get unbiased findings, the reviewer should have no direct responsibility or accountability for the performance of the system under review (Galar and Kumar, 2016).

The second category of audit technique analyzes the technology and business constraints. This allows the determination of the relative importance and required attributes of the various elements of a system. The actual system attributes can be analyzed against the ideal system and tempered by the requirements for excellence in the particular activities making up the system.

4.8.1 Conducting an Audit

A system audit is usually conducted by using a questionnaire designed to provide a profile of the maintenance system. Typically, the questionnaire is structured to address specific key areas in the system to be audited. Responses to these questionnaires may take one of the following forms:

- Selecting either "yes" or "no"
- Choosing one or more of the available options
- Putting items on a Likert-type scale, for example, from 1 to 5, to indicate different degrees of agreement or lack thereof

Different weights may be assigned to different questions to reflect their relative contributions to system performance.

Even though they may use sophisticated assessment schemes, the underlying theory of system audits is obscure. Dwight (1994) suggests a procedure that relates the state of a system element, such as "feedback from operations," to its contribution to the system's overall performance. The overall performance of the maintenance system can be determined by aggregating the contributions to the business success of the observed states of all system elements with an influence on relevant asset failures. In this procedure, failure attributes affecting business success have to be identified. The same requirements apply to the system elements with an influence on failure attributes.

A more typical system audit focuses on conformity to a standard model, both in system design and execution. It is assumed that the standard can be universally applied to achieve superior performance. The maintenance system audit questionnaires by Westerkamp (1993) and Wireman (1990) are developed on the basis of this concept.

The audit process is usually done on site and includes the following steps: interviewing key people in the organization; conducting site inspections; reviewing process flows and mapping maintenance functions and control; reviewing relevant documentation; demonstrating system applications; attending key meetings; obtaining answers to structured questionnaires; and validating plant, equipment, and maintenance performance.

The results of the interviews and the answers to the structured questionnaires are analyzed to formulate action plans for improvement (Galar and Kumar, 2016).

Unfortunately, system audits fail to recognize that different organizations operate in different environments. Therefore, the second category of audit seeks to quantify the judgments of people with knowledge about the maintenance system, the organization's requirements, and the possible system elements in order to measure performance.

Products, technology, organizational culture, and the external environment are some key variables in an organization's operating environment, and they may be in a state of constant change. Superior performance can be achieved only if the internal states and processes of the organization fit perfectly into the specific operating environment. Sociotechnical systems (STS) analysis provides a methodology to design a system that will achieve this fit (Taylor and Felten, 1993). However, the basic assumption of a standard reference model implicit in the design of the typical audit questionnaire is problematic.

4.8.2 Historical Maintenance Audit Models

Authors are divided on the best way to audit and have come up with many different models. Westerkamp (1987) has developed an audit scheme covering 14 factors contributing to maintenance productivity and advocates automating the auditing process. The factors included in Westerkamp's audit are organization staffing and policy; management training; planner training; craft training; motivation; negotiation; management control; budget and cost; work order planning and scheduling; facilities, store, material, and tool control; PM and equipment history; engineering; work measurement; and data processing. He suggests obtaining information using a set of questions about each factor.

Kaiser (1991) has developed a maintenance management audit of key factors in the process of maintenance management. The basic components of Kaiser's audit include organization, workload identification, work planning, work accomplishment, and appraisal. Each of these components has six to eight factors. Using structured statements and weights, an overall score is obtained for the maintenance system. Improvements can be identified from the audit process.

Duffuaa and Raouf (1996) conducted a study on continuous maintenance productivity improvement using a structured audit; they propose a structured audit approach to improve maintenance systems. The factors in Duffuaa and Raouf's audit are organization and staffing; labor productivity; management training; planner training; craft training; motivation; management and budget control; work order planning and scheduling; facilities; stores, material, and tool control; PM and equipment history; engineering and condition monitoring; work measurement, incentives; and information systems. They suggest using an analytic hierarchy process (AHP) to determine the factors' weight and compute a maintenance audit index. They also propose root cause analysis to develop an improvement action program.

Duffuaa and Ben-Daya (1995) propose the use of statistical process control tools to improve maintenance quality, while Raouf and Ben-Daya (1995) suggest a total maintenance management (TMM) framework. An important component of TMM is a structured audit.

De Groote (1995) argues for the use of a maintenance performance evaluation approach based on a quality audit and on performance indicators of maintenance. The quality audit should be conducted in four stages: surveying the influencing parameters; analyzing collected data to make conclusions and recommendations; designing an improvement action plan; and justifying the proposed improvement plan based on cost–benefit. The evaluation should include the following five major factors: production equipment; organization and management of maintenance; material resources; human resources; and work environment.

Price Water House Coopers (PwC) (1999) has developed a questionnaire to evaluate maintenance programs, with the following 10 factors on the questionnaire: maintenance strategy; organization/human resources; employee empowerment; maintenance tactics; reliability analysis; performance measures/benchmarking; information technology; planning and scheduling; material management; and maintenance process reengineering. The questionnaire includes several statements about each factor, with each statement given a score ranging from 0 to 4.

Al-Zahrani (2001) reviewed various audit programs and surveyed managers and engineers in government and private organizations in Saudi Arabia to assess the factors affecting maintenance management auditing, with the aim of developing a suitable audit form for facilities maintenance. He proposes a form with six main components: organization and human resources; work identification and performance measures; work planning and scheduling; work accomplishment; information technology and appraisal; and material management. Each component has six to eight factors relevant to the performance of the maintenance system.

Five structured audit programs for maintenance systems are identified in the literature. These audits are developed by Westerkamp (1987), Kaiser (1991), Duffuaa and Raouf (1996), and Al-Zahrani (2001). An audit program consists of key elements in the maintenance systems examined through a set of statements or questions. Each statement or question has a score and a weight. Then based on the audit, a total weighted score is compiled and compared to an ideal score. Based on these scores, an action plan for improvement is formulated. The process is repeated periodically to ensure continuous improvement.

In addition to the balanced scorecard technique, Tsang et al. (1999) present a "systems audit" technique, based on STS analysis, to predict future maintenance performance, and data envelopment analysis (DEA), a nonparametric quantitative approach to benchmarking the organization's maintenance performance compared to its competitors. By defining the performance indicators in terms of ratios rather than absolute terms, De Groote (1995) develops a system for determining maintenance performance, using a four-stage quality audit approach.

Many authors argue for the necessity of obtaining both qualitative and quantitative results. Clarke (2002) suggests an audit must contribute a maintenance radar (graphic of spider Bells–Manson type) which includes all the economic aspects, human aspects, and so on of maintenance. He also proposes that a product of the audit must be good operative and technical practices in maintenance. Tavares (2002) makes use of the radar of maintenance in audits to represent the different areas of influence and dependency of the maintenance function. Many authors agree on generating these radars from massive surveys; in spite of their reliability, numeric data on the systems are not included in the radars; rather, there is a strong human factor (Papazoglou and Aneziris, 1999).

More recently, Galar and Kumar (2016) have developed an audit in this latter category. This model proposes a mixture of qualitative aspects (from questionnaires) and different weights obtained by maintenance KPIs. The KPIs should be strongly correlated with the answers to the questionnaires. This model shows the relationship between trends in questionnaires and indicators that validate the correlations or highlight the divergences, merging qualitative measures with quantitative ones (Galar and Kumar, 2016).

4.8.3 Fundamental Financial Information for Maintenance Audits

The key factor for excellence is the cost of maintenance with respect to the cost of the replacement of assets. Expensive maintenance occurs when the cost to maintain does not take into account the asset's productive capacity. The performance is directly proportional to the total cost of the asset.

$$EI1 = \frac{\text{Total maintenance costs}}{\text{Assets replacement value}} \quad (4.25)$$

This parameter, according to Peterson (2003), must be part of the financial direction, along with a long-term maintenance plan that optimizes the ROI of the acquired assets. The LCC must include the cost of acquisition, financial costs of maintenance, and so on. This information is crucial for replacement decisions.

The replacement value of assets should be calculated following the SMRP and harmonized. Weber and Thomas (2005) propose a corporate indicator of utility costs of maintenance with respect to sales in the manufacturing industry; in their view, a world-class reference for this indicator should be between 6% and 8%; the EI1 world-class reference is between 2% and 3%.

A relevant parameter is the overall equipment effectiveness (OEE). It is a key audit parameter, as proposed by Grencik and Legat (2007). They suggest auditing differences between global parameters and those in the various hierarchical levels of a company to get specific strategies.

$$OEE = OEEI1 \times OEEI2 \times OEEI3 \quad (4.26)$$

$$OEEI1 = \text{Availability} = \frac{\text{Total activity time}}{\text{Scheduled time}} \quad (4.27)$$

$$TI1 = \frac{\text{Total operation time}}{\text{Total operation time} + \text{Maintenance unavailability time}} \quad (4.28)$$

$$TI2 = \frac{\text{Availability time achieved during the required time}}{\text{Required time}} \quad (4.29)$$

The availability indicator is a main operating parameter. According to Olver and Kahn (2007), total unavailability can be close to 10% but the SMRP proposes setting limits at 90% availability achieved. The EFNMS is more ambitious and advises a 95% availability target for some sectors, such as pharmaceutical or food, agreeing with Olver at 90% in most industries.

$$OEEI2 = \text{Performance efficiency} = \frac{\text{Real Production to programed time}}{\text{Planned production to programed time}}$$
$$(4.30)$$

$$OEEI3 = \text{Quality rate} = \frac{\text{Total production} - \text{Defects}}{\text{Total production}} \quad (4.31)$$

Besides E1 and OEE, there is another financial indicator key for the preparation of the audit process:

$$EI3 = \frac{\text{Total maintenance costs}}{\text{Quantity produced}} \quad (4.32)$$

$$TPMI5 = \text{Decrease the unit production cost} \quad (4.33)$$

A parameter that handles the company's operations in the maintenance function can observe the cost reduction per unit.

Two parameters assess maintenance in many organizations: how many monetary units are loaded onto the finished product and how much this amount is reduced by applying new methods and new techniques. The EI3 parameter and percentage reduction are generally the only goals of maintenance when a financial perspective is expected to be achieved by this function.

In summary, E1, E3, and TPMI5 constitute the financial basis for maintenance audits while the classic OEE constitutes the effectiveness side. It is relevant to note that the OEE is extremely popular but does not address any financial achievement, and effectiveness is assessed independent of the cost (Galar and Kumar, 2016).

References

Al-Zahrani, A., 2001. Assessment of factors affecting building maintenance management auditing in Saudi Arabia. Master thesis. King Fahd University of Petroleum and Minerals, Dhahran, Saudi Arabia.

Cáceres, B., 2004. Cómo Incrementar la Competitividad del Negocio Mediante Estrategias para Gerenciar el Mantenimiento. In: *VI Congreso Panamericano de Ingeniería de Mantenimiento*, México City (Distrito Federal), México, September 23–24.

CEN, 2007. *EN 15341: Maintenance—Maintenance Key Performance Indicators*. European Committee for Standardization, Brussels, Belgium.

Clarke, P., 2002. Physical asset management strategy development and implementation. In: *International Conference of Maintenance Societies (ICOMS)*, Australasian Corrosion Association, Melbourne, Victoria, Australia, May 22–23, 2002, pp. 1–4.

Coelo, C., Brito, G., 2007. Proposta De Modelo para Controle de Custos De Manutenca Com Enfoque na Aplicacao De Indicadores Balanceados. *Boletim Tecnico Organizacao & Estrategia* 3(2), 137–157.

De Groote, P., 1995. Maintenance performance analysis: A practical approach. *Journal of Quality in Maintenance Engineering* 1(2), 4–24.

Dean, H., 1986. Integration of RCM analysis into the S-3A maintenance program. Calhoun: The NPS Institutional Archive, Dubley Knox Library, Monterey, CA, December 1986.

Duffuaa, S. O., Ben-Daya, M., 1995. Improving maintenance quality using SPC tools. *Journal of Quality in Maintenance Engineering* 1(2), 25–33.

Duffuaa, S. O., Raouf, A., 1996. Continuous maintenance productivity improvement using structured audit. *International Journal of Industrial Engineering* 3(3), 151–166.

Dwight, R. A., 1994. Performance indices: Do they help with decision-making? In: *Proceedings of International Conference of Maintenance Societies*, ICOMS-94, Sydney, New South Wales, Australia, pp. 1–9.

Eccles, R. G., 1995. The performance measurement manifesto. In: Holloway, J., Lewis, J., Mallory, G. (Eds.), *Performance Measurement and Evaluation*. Sage Publications, London, U.K., pp. 5–14.

Galar, D., Kumar, U., April 6, 2016. *Maintenance Audits Handbook: A Performance Measurement Framework*. CRC Press, Taylor & Francis Group, Boca Raton, FL, 609pp.

Galar, D., Parida, A., Schunnesson, H., Kumar, U., 2011. *Conference Proceedings on Maintenance, Performance, Measurement & Management (MPMM 2011)*. Luleå University of Technology, Lulea, Sweden, pp. 209–212.

Grencik, J., Legat, V., 2007. Maintenance audit and benchmarking—Search for evaluation criteria on global scale. *Eksploatacja i niezawodnoGü. Nr* 3(35), PNTTE, Warszawa, October 2007, 34–39.

Hansson, J., 2003. Total quality management—Aspects of implementation and performance. Doctoral thesis. Lulea University of Technology, Lulea, Sweden.

Kahn, J., Gulati, R., 2006. SMRP maintenance & reliability metrics development. In: *EuroMaintenance 2006—Third World Congress on Maintenance*, Congress Center, Basel, Switzerland, June 20–22.

Kaiser, H. H., 1991. *Maintenance Management Audit: A Step-by-Step Workbook to Better Your Facility's Bottom Line.* R.S. Means Co., Construction Consultants and Publishers, Kingston, MA.

Karlson, M., March 22, 2016. 29 popular financial KPIs you should add to your financial KPI dashboard. https://www.scoro.com/blog/financial-kpis-for-financial-kpi-dashboard/. Accessed September 13, 2016.

Kumar, U., Galar, D., Parida, A., Stenstrom, C., 2013. Maintenance performance metrics: A state-of-the-art review. *Journal of Quality in Maintenance Engineering* 19(3), 233–277.

Mann, L. J., 1983. *Maintenance Management.* Lexington Books, Lexington, MA.

Mitchell, J. S., 2001. *Handbook: Operating Equipment Asset Management,* 1st edn. Applied Research Laboratory, Pennsylvania State University, State College, PA.

Olver, R., Kahn, J., 2007. SMRP maintenance and reliability metrics development. In: *MARCON 2007,* Knoxville, TN, 2007.

Papazoglou, I. A., Aneziris, O., 1999. On the quantification of the effects of organizational and management factors in chemical installations. *Reliability Engineering & System Safety* 63(1), 33–45.

Peterson, S. B., 2003. The future of asset management. Strategic Asset Management Inc., Unionville, CT. http://www.plant-maintenance.com/articles/FutureofAsset-Management.pdf. Accessed October 10, 2016.

Price Water House Coopers (PwC), 1999. *Questionnaire of Auditing.* PricewaterhouseCoopers (PwC), Toronto, Ontario, Canada.

Raouf, A., Ben-Daya, M., 1995. Total maintenance management: A systematic approach. *Journal of Quality in Maintenance Engineering* 1(1), 6–14.

Smith, R., Mobley, R. K., March 31, 2011. *Rules of Thumb for Maintenance and Reliability Engineers.* Butterworth-Heinemann, Boston, MA, 336pp.

SMRP, 2006. *Best Practices Committee: SMRP Best Practice Metrics Glossary.* SMRP, McLean, VA.

Sondalini, M., 2014. How to build your ISO 55001 asset management system quickly and make ISO 55001 certification easy. Lifetime Reliability Solutions HQ. http://www.lifetime-reliability.com/home_pdfs/ISO_55001_standard_certification_Plant_Wellness_Way.pdf. Accessed September 10, 2016.

Stenstrom, C., 2012. Maintenance performance measurement of railway infrastructure with focus on the Swedish network. Technical Report, Lulea University of Technology, Lulea, Sweden.

Stice, J. D., Stocks, K. D., Albrecht, W. S., 1992. *A Common Body of Knowledge for Practice of Internal Auditing.* Institute of Internal Auditors, Orlando, FL.

Svantesson, T., Olver, R., 2008. *Euromaintenance, SMRP-EFNMS Benchmarking Workshop,* Brussels, Belgium, April 8–10.

Tavares, L., 2002. Administracio´n Moderna del Mantenimiento. Procesados por el Club de Mantenimiento & Mantenimiento Mundial. www.mantenimientomundial.com. Accessed December 10, 2016.

Taylor, J. C., Felten, D. F., 1993. *Performance by Design: Sociotechnical Systems in North America.* Prentice Hall, Englewood Cliffs, NJ.

Tsang, A. H. C., Jardine, A. K. S., Kolodny, H., 1999. Measuring maintenance performance: A holistic approach. *International Journal of Operations & Production Management* 19(7), 691–715.

Vergara, E., 2007. *Análisis de Confiabilidad, Disponibilidad y Mantenibilidad del Sistema de Crudo Diluido de Petrozuata*. Decanato de Estudios de Postgrado—Universidad Simón Bolívar, Caracas, Venezuela.

Weber, A., Thomas, R., November 2005. Key performance indicators. Measuring and managing the maintenance function. Ivara Corporation, Burlington, Ontario, Canada.

Westerkamp, T. A., 1987. Using computers to audit maintenance productivity. In: Hartmann, E. (Ed.), *Maintenance Management*. Industrial Engineering and Management Press, Atlanta, GA, pp. 87–94.

Westerkamp, T. A., 1993. *Maintenance Manager's Standard Manual*. Prentice Hall, Englewood Cliffs, NJ.

Wireman, T., 1990. *World Class Maintenance Management*. Industrial Press, New York.

Wireman, T., 2004. *Benchmarking Best Practices in Maintenance Management*. Industrial Press, Inc., New York, 256pp.

Wireman, T., 2005. *Developing Performance Indicators for Managing Maintenance*. Industrial Press, Inc., New York, 250pp., First printing February, 2005.

Wireman, T., 2015. *Benchmarking Best Practices for Maintenance, Reliability and Asset Management: Updated for ISO 55000*, 3rd edn. Industrial Press, Inc., New York, 300pp.

5

Consequential Maintenance Cost: A Problem Area

5.1 Economic Importance of Maintenance

Industrial plants increasingly use available front-end technologies, as they work toward total automation. Although they can expect more qualitative and quantitative production possibilities and better services, they must invest significantly in equipment.

As a result of the high cost of these investments, company managers try to optimize production capacities to maximize the return of investment. The shutdown of production due to a failure of important equipment compromises production, generating a loss of performance and increased expenses (Galar and Kumar, 2016).

Two types of costs can be distinguished in any company –variable and fixed.

1. Variable costs depend on the volume of production or sales, and include the following:
   a. Proportional costs of sales:
      i. Packing
      ii. Personnel
      iii. Shipping
      iv. Financiers (loans)
      v. Unpaid debts
   b. Proportional costs of manufacturing:
      i. Direct manpower
      ii. Materials
      iii. Energy
      iv. Corrective maintenance

2. Fixed costs are independent of the volume of sales, and include the following:

    a. Fixed costs of manufacturing:

        i. Indirect manpower

        ii. Amortization of production machinery and buildings

        iii. Preventive maintenance

        iv. Rental of equipment

        v. Insurance

    b. Fixed costs of administration:

        i. Materials

        ii. Personnel

        iii. Taxes

        iv. Amortization of office materials

From this initial cost breakdown into two maintenance costs, we can arrive at two conclusions:

1. Preventive maintenance must be considered a fixed manufacturing cost; an economic analysis can be counterproductive if preventive maintenance is not included.
2. Corrective maintenance, as a variable cost, influences the estimation and analysis of these costs and the gross margin of profit.

The maintenance function influences performance and revenues through the following:

- Cost of preventive maintenance (fixed)
- Cost of corrective maintenance (variable)
- Availability of equipment (Cost of failures)

These three parameters have an enormous influence on economic analysis; the simple cost model proposed here combines the three parameters, creating an easily understood way to perform such analysis (Galar and Kumar, 2016).

5.2 Classification of Maintenance Costs

Maintenance costs can be divided into four different areas that will appear on the scorecard of indicators proposed later in this chapter (Figure 5.1).

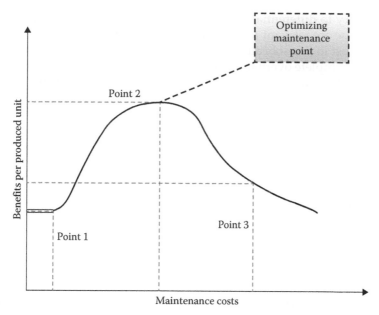

FIGURE 5.1
Benefits per produced unit versus maintenance costs. (Adapted from Galar, P.D. and Kumar, U., *Maintenance Audits Handbook: A Performance Measurement Framework*, CRC Press, Reference—609 p.—271 B/W Illustrations, February 25, 2016, ISBN: 9781466583917—CAT# K19014.)

The first division of maintenance costs comprises the actions of personnel, in both production and maintenance. *Note*: not all people performing maintenance actions belong to the maintenance department, especially when total productive maintenance (TPM) is implemented and the border between production and maintenance is blurred. Four cost subdivisions can be differentiated here: costs related to maintenance actions, purpose, accounting concepts, and negotiation with trade unions/professions.

The first division contains the costs related to maintenance actions which include the following:

- Cost of the daily maintenance: lubrications, painting, and so on
- Cost of the control and measurement of the degradation of the machinery through visual inspection, predictive maintenance, nondestructive testing
- Cost of the repair of the facility

A second subset of costs is related to the purpose (or type) of maintenance:

- Cost of corrective maintenance
- Cost of preventive maintenance
- Cost of modifying maintenance

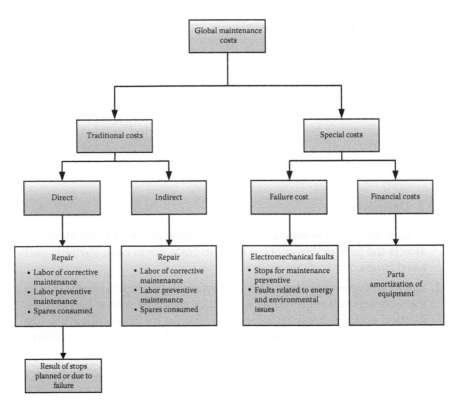

FIGURE 5.2
Breakdown of global cost into concepts. (Adapted from Galar, P.D. and Kumar, U., *Maintenance Audits Handbook: A Performance Measurement Framework*, CRC Press, Reference—609 p.—271 B/W Illustrations, February 25, 2016, ISBN: 9781466583917—CAT# K19014.)

Figure 5.2 gives a complete scheme of the global cost with the concepts defined above.

Lambán (2009) proposes a cost model based on activity-based costing (ABC), with a division into two large blocks referring to direct costs and indirect costs. As shown in Figure 5.3, the special costs are direct costs imputable directly to the equipment. The approach assumes the following:

1. The raw material of the maintenance process is the man hours used and the spare parts consumed, as well as smaller quantities of other accessories. This is particularly relevant in maintenance; in other processes, manpower is considered an operational cost. In this case, it is clearly a resource.

2. The operational costs derived from the execution of the tasks of maintenance (preventive, corrective, etc.) are associated with the cost of failure, that is to say, the repercussions to the product or service of the maintenance actions and the consequences of these actions.

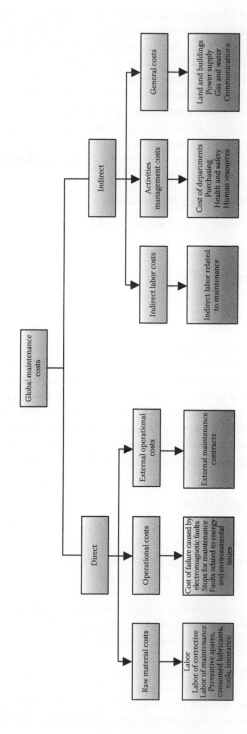

FIGURE 5.3

Costs of maintenance according to Lambán. ET model. (Adapted from Lambán, MP., et al., Modelo de gestión económica de la Cadena de Suministro, Primer Congreso de Logística y Gestión de la Cadena de Suministro, Zaragoza, Spain 2009.) (Adapted from Galar, P.D. and Kumar, U., *Maintenance Audits Handbook: A Performance Measurement Framework*, CRC Press, Reference—609 p.—271 B/W Illustrations, February 25, 2016, ISBN: 9781466583917—CAT# K19014.)

For example, an analysis of vibrations with a technician and a portable collector consumes direct resources of raw material (labor and equipment hours) but does not stop the machine; thus, the imputable operational costs will be zero. A relatively frequent example is halting production to take a machine to the manufacturer's workshop during its warranty period for an external overhaul. In this case, external operational costs will be incurred because the overhaul is subcontracted and there is an operational cost caused by the absence of the productive asset. However, the cost of internal maintenance will be zero for the raw material used, that is to say, spares parts or manpower, since actions are performed by contractors (Galar and Kumar, 2016).

5.2.1.2 Imputation of Global Cost of Maintenance

The global cost can be imputed, as previously noted, at the following levels:

- Machine level
- Set of equal machines/group level
- Section level
- Plant level

Costs are commonly imputed to the machine level or to a group of machines (e.g., a room containing centrifugal pumps). Grouped machines are installed in sections; thus, the section costs can be obtained by summation.

The collection of data will depend on the level at which the financial indicators are required, bearing in mind the bureaucratic complexity and administrative and document management in case users want to obtain detailed machine indicators in plants with hundreds of assets, numerous subsets, factories with multiple locations, and so on (Galar and Kumar, 2016).

5.2.1.3 Optimization of Global Cost

In terms of mathematics, and at first glance, when some fixed and variable costs are increased, even financial ones, the global cost should also increase, but this does not happen because the failure cost is the inverse of the other costs; that is to say, it tends to decrease when these grow (Figures 5.4 and 5.5).

If the cost of preventive maintenance is increased (indirect manpower, programmed shutdowns, the consumption of lubricants, tools, etc.), the cost of corrective maintenance will drop (direct manpower, consumption of spare parts, repairs by external companies, etc.).

The increases and reductions in maintenance can stem from attempts to find the proper balance of planned and unplanned maintenance.

Corrective maintenance can be programmed (as a result of preventive maintenance) or reactive/unprogrammed (as a result of a shutdown due

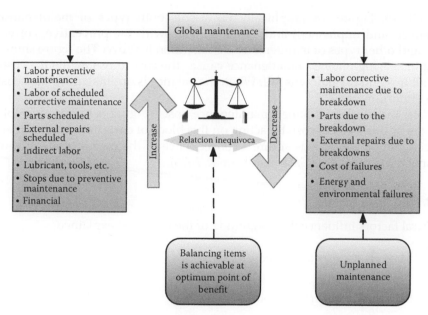

FIGURE 5.4
Relationship between the amount of preventive maintenance and medium repair. (Adapted from Galar, P.D. and Kumar, U., *Maintenance Audits Handbook: A Performance Measurement Framework*, CRC Press, Reference—609 p.—271 B/W Illustrations, February 25, 2016, ISBN: 9781466583917—CAT# K19014.)

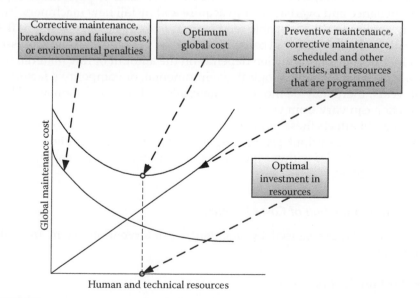

FIGURE 5.5
Global maintenance cost versus Human technical. (Adapted from Galar, P.D. and Kumar, U., *Maintenance Audits Handbook: A Performance Measurement Framework*, CRC Press, Reference—609 p.—271 B/W Illustrations, February 25, 2016, ISBN: 9781466583917—CAT# K19014.)

to failure). Figure 5.8 graphically represents both types of maintenance (planned and unplanned) and its relationship with the preventive, corrective and other types of maintenance that mention it above. The figure shows that increasing planned maintenance causes the cost to rise, while the costs of failures and breakdowns fall. When the balance is optimal, the global cost is a minimum value.

Note: the block of planned maintenance costs (Figure 5.4) is responsible for the optimal global cost. By acting on this block of costs, failures and their effects can be controlled—a direct consequence of the excess or absence of planned maintenance (Galar and Kumar, 2016).

5.2.1.4 Other Factors

Several factors influence the magnitude of the costs just explained:

- Markets for manufactured products
- Production rate
- Seasonality of certain products

These factors determine whether a shutdown due to failure will cause an economic loss and the associated size of that loss.

In the food industry, for example, production is often concentrated in 2 or 3 months of the year, with the rest of the time available to perform all types of maintenance and overhauls, or to acquire and install new machinery. This usually applies to the perishable food industry, like fruit, vegetables, or fish, materials that are available at certain times of the year and must be processed quickly. This type of production depends on the amount of raw material that is available. Due to meteorological, environmental, or competitive factors, in some years, there may be more raw materials and/or more demand. Thus, production can vary from year to year.

The market affects these industries as well. For example, stringent parameters of quality can limit production and/or increase costs. The export of products to countries without specific regulations will circumvent this (Galar and Kumar, 2016).

5.2.1.5 Determination of Cost of Failure

The costs of failure caused by maintenance interventions stem from the following:

- Lost production costs
- Costs of accidents
- Costs of loss of image, sales, reputation
- Costs of quality rejections

At the beginning of every year, the rate of lost monetary units to machine/ hours of nonavailability must be fixed.

All work orders (WOs) must show if a failure has resulted in nonavailability, in an increase in energy consumption, or in environmental problems. As a result of nonavailability, one or all of the following may occur:

- Energy savings because the equipment is not being used: annual savings/12 = monthly rate or monthly discount.

- Increased power consumption, for example, an energy failure during the entire work order, from the initial request to closure, increase the costs.

- Environmental deterioration: costs determined by fines or sanctions imposed (Galar and Kumar, 2016).

5.2.1.6 Cost Breakdown on Shop Floor

Most companies differentiate between maintenance performed in the field and in the workshop; the corresponding costs are also separated. For that reason, the manpower used for corrective maintenance, the consumed spare parts, and so on can be broken down to determine which areas (workshop or field) result in greater cost (Galar and Kumar, 2016).

5.3 Costs after Downtime

Downtime (DT) caused by the nonavailability of equipment and equipment breakdown is among the most common unanticipated factors having a nontrivial impact on equipment productivity and project and organizational performance (Hanna and Heale, 1994; Elazouni and Basha, 1996).

DT is any time that will increase the total time in cycle component (i.e., time from the beginning to the end of a specific process). In general, break times and lunch times are not considered DT as they are already included in the total scheduled runtime of equipment (ICONICS Inc., 2011).

Despite its significance, few companies pay attention to the impact of DT or take managerial action to reduce it. To address this issue, several researchers have made significant efforts to minimize DT by providing theoretical frameworks and models. Vorster and Sears (1987) suggest the concept of failure cost profiles (FCPs). Vorster and De La Garza (1990) refine the FCP concept by developing a cost model designed to quantify the intangible (consequential) costs associated with lack of availability (LAD) and DT. Tsimberdonis and Murphree (1994) offer operational failure cost (OFC) profiles and charts to be used as decision support tools. More recently,

Edwards et al. (2002) developed a model to predict the hourly cost of DT for tracked hydraulic excavators operating in the UK opencast mining industry. All have contributed to establishing theoretical bases for DT and quantifying DT costs, and the reader is referred to their work for more information.

There has been little attention paid to the less-tangible costs of DT, however (Edwards et al., 1998). Previous research has not provided sufficient details of the causes and consequences of DT. It is known that DT can have a number of consequences that vary according to the nature of the project, activity, and equipment. Some research in Nepal has attempted to explore the impact of DT on projects (Kirmani, 1988). However, the existing DT models cannot be applied to equipment management in developing countries where the contractors face a number of problems related to equipment management, such as a lack of expertise and generally poor cost accounting practices. These equipment management-related factors, coupled with the working environment, result in expensive repair and DT costs (Kirmani, 1988).

5.3.1 Reasons for Downtime

There are many possible reasons for DT, including the following:

- *Equipment fault*: The control system detects an undesirable condition that could cause equipment, part, or personnel damage or injury and shuts down the process.
- *Equipment over cycle*: Equipment has taken longer to complete a cycle than configured. Usually this is due to operator input, such as part loading and unloading.
- *Changeover*: Equipment is converting from one part type to another; this usually requires tooling or fixture changes. The control system is notified of the change and the system is shut down.
- *In-process maintenance*: Components or tooling used within the process need replacement or attention due to wear or other criteria.
- *Scheduled maintenance*: Equipment is taken out of production for servicing. Usually this is for preventative maintenance, such as lubrication, or replacement or repair of worn or defective components.
- *Starved*: Equipment has run out of raw materials or subcomponents and is waiting for these; equipment can also be waiting for a part.
- *Blocked*: Equipment is unable to release the current part. There may be no more room in the finished part container or it is absent; possibly the next station has not released its part and there is no buffer between them.
- *Production meeting*: The supervisor calls a meeting, usually for quality alerts or scheduling changes.

The reasons for DT should be entered into the system (CMMS) before restarting so the monitoring application can log the reason as the equipment goes back into production (ICONICS Inc., 2011).

Note: some entries are automatic and some are made by operators.

5.3.2 Impact of Downtime

This section considers the quantitative impact of DT. Previous research has reported that factors related to plant and equipment breakdown, particularly from a management perspective, must be considered in assessing the impact of DT (Edwards et al., 1998). Therefore, we first identify the generic factors and the related processes (see Figure 5.6), some of which are incorporated into causal loop diagrams. Then, using the diagrams, we analyze the dynamic consequences of DT (Nepal and Park, 2004).

5.3.3 Downtime Factor Analysis

5.3.3.1 Site-Related Factors

As shown in Figure 5.6, examples of site-related DT factors include poor working conditions, uncertainties during equipment operation, and the location of the site. All three factors may affect the performance of equipment. Difficult working conditions (for example, in the mining or construction industries) may cause equipment to deteriorate rapidly, triggering sudden failure. Proactive action can have a significant effect, but in some cases, managers may not have enough knowledge of the site conditions, especially in the construction industry. In addition, the uncertainty of operation (operating in different environmental conditions) causes a greater risk of equipment breakdown (Arditi et al., 1997; Edwards et al., 1998). The location of the site may limit the type and size of equipment that can be transported to it (Day and Benjamin, 1991). Finally, the remoteness of a site may affect the repair time of equipment by affecting communication and the prompt procurement of parts (Nepal and Park, 2004).

5.3.3.2 Equipment-Related Factors

Factors related to equipment include its age, type, quality, complexity of operation, and degree of usage. A company's procedures, policies, and site management actions can have significant influence on the selection, use, and operation of equipment. The risk of equipment breakdown is often related to the complexity and sophistication of a piece of equipment (Elazouni and Basha, 1996; Arditi et al., 1997). It is, therefore, important to have proper knowledge about the capacity, complexity, and technical suitability for the use of the equipment under the given conditions.

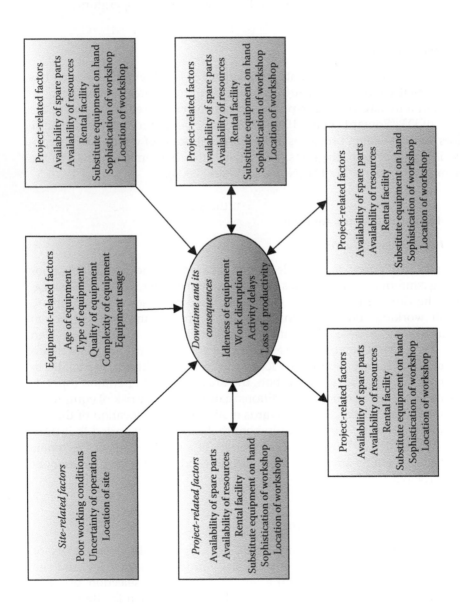

FIGURE 5.6
Downtime factor analysis. (Adapted from Nepal, M.P. and Park, M., *Eng. Constr. Archit. Manage.*, 11(3), 199, 2004.)

5.3.3.3 Crew-Level Factors

Human crews are involved in the equipment maintenance, operation, and production process. The factors in this category include the skill level of operators and mechanics, fatigue, morale, and motivation. An operator's skill is one of the most important factors; it affects his or her performance and the direct cost of DT through job efficiency (Elazouni and Basha, 1996; Arditi et al., 1997; Edwards et al., 2002). In addition, misuse of equipment because of operator negligence or lack of proper training, not to mention a lack of know-how on the part of the equipment supervisor, may result in increased frequency and cost of DT (Pathmanathan, 1980). Another important consideration is the morale, motivation, and fatigue of the crew. For example, management may attempt to increase the work rate though extensive overtime and increased pressure on crews to avoid the impact of DT. Above a certain threshold level, both factors can have negative effects on productivity by causing fatigue and lowering morale (Roberts and Alfred, 1974; Cooper, 1994). Furthermore, when the job context—such as supervision, resource availability, worker compensation, and work environment—is degraded, a lack of worker motivation can result in a loss of productivity (Maloney and McFillen, 1986).

5.3.3.4 Force Majeure

This category includes events that are unanticipated by project participants, particularly those related to natural calamities. Examples include floods, landslides, and accidents. Such events may result in delays in equipment maintenance and affect project performance. Maintenance planners should anticipate some events, particularly seasonal ones that may affect equipment (e.g., the effect of winter storms on construction), and take the necessary precautions to reduce their likely impact on DT. Adopting proper safety practices and increasing security measures can control other events, such as vandalism and accidents (Pathmanathan, 1980).

5.3.3.5 Company Procedures and Policies

This category includes the company's standard procedures and policies about equipment management and may include factors such as maintenance policies, replacement decisions, inventory management and control, standby repair and maintenance facilities, and procurement systems. The equipment policies of a construction a company reflect the priorities set by top management and are significant in terms of resource allocation and strategic planning (Sözen and Giritli, 1987). Not all companies can justify, for example, the costs of carrying an inventory of spare parts. Note that this may be influenced by the number of available jobs. A company's policies may also reflect the corporate-level strategy and existing market conditions. Furthermore, maintaining a fleet of equipment can be strategically important to a company if the awarding of a contract is based on the condition and availability of equipment. However, as equipment management procedures and policies can vary companywide, they can have different implications for DT (Nepal and Park, 2004).

5.3.3.6 Project-Level Factors

Project-level factors, such as the availability of spare parts, resources, and rental facilities, substitute equipment on hand, the location and sophistication of a workshop, and other project-specific requirements, vary considerably and are related to DT. They are mostly influenced by such factors as a company's action plan and procedures, local and national market conditions, requirements of the project owner, and, to some extent, site-related factors. This means maintenance managers may have little control; for instance, they may find it difficult to get spare parts and materials to repair equipment. In addition, any delay in the time required for skilled mechanics to arrive on site may paralyze the work. Receiving substitute equipment on time is a major challenge to projects located in remote areas. Finally, the availability, location, and sophistication of a workshop can have a considerable influence on DT (Nepal and Park, 2004).

5.3.3.7 Shop-Level Management Actions

Maintenance managers may influence DT in a number of ways, including substituting failed equipment, waiting for failed equipment to be repaired, adding or changing resources, accelerating activities, transferring crews to other operations or sites, and changing the sequence of work. Each of these actions, when implemented properly, may reduce the impact of DT. If the selected course of action is not appropriate or is implemented in an improper way, however, it may exacerbate the situation. For example, extended use of overtime to accelerate work without improving the work environment may erode the motivation level of crews; it may also increase fatigue and, thus, induce more errors and rework (Thomas and Raynar, 1997; Eden et al., 2000). Consequently, a project may suffer loss of productivity.

5.3.3.8 Consequences of Downtime

Some of the important consequences of DT include idle equipment and crews, work disruptions, activity delays, and loss of productivity. Each of these consequences may, in turn, interact with the actions of maintenance managers, the company's procedures and policies, project-level factors, and crew-level factors, as indicated by the two-way arrows in Figure 5.7.

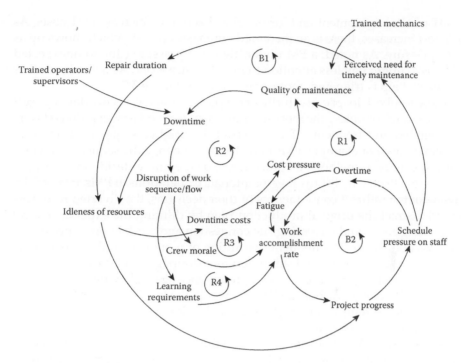

FIGURE 5.7
Dynamics of equipment downtime. (Adapted from Nepal, M.P. and Park, M., *Eng. Constr. Archit. Manage.*, 11(3), 199, 2004.)

In the construction industry, for example, projects are primarily "solution-driven" and mostly focus on minimizing costs and limiting immediate consequences (Mitropoulos and Tatum, 1999). Thus, it is possible that construction site managers may underestimate the actual impact of DT that may be triggered by their actions. They need to understand DT and its possible impact on project performance in a systematic way (Nepal and Park, 2004).

5.3.4 Dynamics of Downtime

DT factors related to crews and site management actions have been incorporated into causal loop diagrams (see Figure 5.7). The first and most noticeable effect of DT is idle resources, which, if it lasts for extended periods, will slow down the progress of a project. Slow project progress increases schedule pressure (Neil, 1989). When project managers (PMs) are under schedule pressures, they might become distracted from proper supervision and resort to hasty maintenance. This produces low quality of maintenance, which then increases DT, which, in turn, increases schedule pressure, generating the vicious reinforcing loop denoted as R1 in Figure 5.7 (Nepal and Park, 2004).

DT of vital equipment and/or of critical activities increases DT costs. As DT cost increases, there is an increased emphasis on cost, which shows up as cost pressure. As noted, a PM under high cost pressure (due to unexpected DT costs) might pay less attention to maintenance work. As a result, another feedback loop is triggered, represented as R2 in Figure 5.7.

These feedback loops cause further cost pressures and can slow down project progress until, or unless, the root causes are identified and proper action is taken.

Extended and frequent DT can disrupt the original sequence of work in several ways. For example, PMs may decide to change the sequence of work, introducing new methods or procedures, or they may decide to divert the resources affected by DT to other maintenance operations. If PMs are not fully aware of the indirect consequences of their decisions, the diverted resources could distract the original production plan by diluting the experience level of existing crews and increasing site congestion and work interference (Piper and Vachon, 2001). The frequent disruption of work can erode crew morale (Eden et al., 2000). This can lead to frequent stoppages and the imposition of additional learning requirements for crews, once again slowing down project progress (Piper and Vachon, 2001). As a result, the feedback effects caused by two additional reinforcing loops, indicated as R3 and R4 in Figure 5.7, also affect the production process.

A PM may seek options to relieve schedule pressure. For example, schedule pressure can be reduced by timely maintenance of equipment by project staff, as indicated by balancing loop B1. Other common managerial actions to avoid the impact of DT on project progress are the use of overtime and the placing of pressure on staff to increase the work completion rate. Overtime can facilitate the progress of construction by increasing working hours, as conceptualized with balancing loop B2 in Figure 5.7, but as it continues, it can lower productivity by causing fatigue in workers (Cooper, 1994; Thomas and Raynar, 1997).

In summary, DT has a number of possible ramifications on operations of a construction project. Managers need to understand how DT and subsequent managerial decisions can affect project performance. Not all variables are quantifiable, however, and many subjective issues may arise. Our suggested framework for evaluating the impact of DT in terms of the time and cost of a particular project (Nepal and Park, 2004) can be expressed as follows:

$$\text{The DT percentage} = \left(\frac{\text{Total DT hours}}{\text{Total planned working hours}} \right) * 100. \quad (5.1)$$

DT costs are the monetary value of idle equipment and the time when it is unavailable (Pathmanathan, 1980). DT costs can be categorized into tangible and intangible costs (Vorster and De La Garza, 1990). Tangible costs include the costs of labor, materials, and other resources required to repair the equipment, operators' wages, and the loss of production because of equipment

DT. Intangible costs include costs accrued as a result of the loss of production, loss of labor productivity, extended overhead costs, and, in some cases, liquidated damages and late-completion charges (Pathmanathan, 1980; Tsimberdonis and Murphree, 1994).

5.3.4.1 Repair Cost

The cost of repairing failed equipment consists of three items: (1) labor (mechanics/helpers), (2) materials, and (3) spare parts. As it is not appropriate to charge major repairs and complete overhauls to a project (Tsimberdonis and Murphree, 1994), these items are not included in this cost category. In other words, it includes the costs applicable to the failed equipment, but not to its periodic and scheduled maintenance work.

5.3.4.2 Cost of Idle Time for Laborers, Operators, and Supervisors

This cost includes the cost incurred for idle time of human resources because of DT associated with failed equipment or equipment that is forced to be idle as a result of breakdown of another piece of equipment. As equipment fails, wages will continue to be paid to laborers, operators, and supervisors who are idle. The effect is more pronounced when equipment works in conjunction with large crews and the DT causes them to be idle (Selinger, 1983). This effect tends to continue unless substitute equipment is mobilized or crews are transferred to other operations.

5.3.4.3 Cost of Idle Time of Equipment

This cost category includes the cost of failed equipment and equipment, if any, that remains idle because of the breakdown. The main consideration in financing equipment is that it should be used fully and productively and should earn adequate revenues to recover its investment cost. Thus, some sort of penalty costs should be levied on idle equipment because of the expectation that, as far as possible, productive assets should be kept in good condition (Vorster and De La Garza, 1990). The cost, which is calculated as an expected rent charged for the equipment, reflects the opportunity cost that would be earned if the equipment was not failed.

5.3.4.4 Cost of Substitute Equipment

This category of cost occurs only when management decides to substitute for the failed equipment by drawing from a company fleet or from an outside agency. This cost normally occurs only when a breakdown lasts for an extended period of time and the DT may affect critical activities (Nepal and Park, 2004).

5.3.4.5 *Project-Associated Costs*

Project-associated costs are related to the contractual obligations and clauses agreed upon for a particular project (Tsimberdonis and Murphree, 1994). Costs such as liquidated damages, additional claims, and late completion penalties belong in this category. Sometimes, additional costs may be incurred when DT in a particular work area disrupts other work (Nepal and Park, 2004).

5.3.4.6 *Loss of Labor Productivity*

Loss of labor productivity is caused by DT due to work disruptions, crowding of workers, extended overtime, accelerated working, learning curve effects, and so on. The effects of these factors are well recognized in the literature (Halligan et al., 1994; Schwartzkopf, 1995; Eden et al., 2000). Thomas (2000) says that the economic consequences of the loss of labor productivity are quite severe.

5.3.4.7 *Other Costs*

This category includes other indirect costs, such as overtime costs, cost of accelerations, incentives paid to crews, and miscellaneous petty expenses attributable to the DT event. Based on these cost categorizations, the percentage of DT cost in any project is calculated in terms of the budgeted cost for the period and expressed as follows:

$$\text{The percentage of DT cost for any project} = \left(\frac{\text{Total DT cost}}{\text{Budgeted project cost}} \right) * 100. \tag{5.2}$$

Lastly, the DT cost impact, which represents the cost per hour of the breakdown of particular equipment, is calculated as follows:

$$\text{DT cost impact} = \left(\frac{\text{Total DT cost}}{\text{Total DT hours}} \right) * 100. \tag{5.3}$$

5.4 Intangible Aspects of Maintenance Costs and Uncertainty

Intangible costs are not easily measured. Common intangible costs include a drop in employee morale, dissatisfaction with working conditions, or customer disappointment with a decline in service or product quality.

Intangible costs have an identifiable source are often not predicted. They may occur after a new practice or policy is put into effect, such as a cut in staffing levels or in employee benefits. Managers can try to estimate intangible costs as soon as they see a pattern of loss. This estimate will be the basis for a decision to either change or continue a practice that frustrates employees or customers. If a new procedure has injured an employee, the company may need to act quickly to avoid government fines and inspections (Reed Newsome, 2016).

5.4.1 Sources of Intangible Costs

As noted above, intangible costs are not always foreseen. For example, when corporate management puts a new program or policy into place that is not appropriate for a given location, unintended intangible costs may ensue because what works well at a work site in one place may clash with the employee work culture at another. For example, managed labor systems, which measure productivity automatically and chart it according to a preset standard, may improve productivity at one facility but harm performance at another. A site where the employees take extended lunches and unauthorized breaks could benefit from this type of automated monitoring. The new system could actually improve the morale among conscientious employees who resent their coworkers' lack of effort. The same managed labor system could be a disaster at a site where employees work as a team and already closely track their departments' speed and productivity. Workers may become anxious and confused, wondering how the new system will affect their pay raises or continued employment. They may refuse to assist their coworkers, fearing that being off-task will hurt their own productivity (Reed Newsome, 2016).

5.4.2 Addressing Intangible Costs

After intangible costs are incurred, management must decide how to address them. In general, the company will either decide to absorb the cost or act to eliminate its source. This decision will be based on the best estimate of the intangible cost that management can make, to take decisions. The cost of training new employees, after long-time employees have left for other opportunities is one variable used to estimate intangible costs. If a company decides to continue an unpopular policy, it may invite employees to informational meetings to reduce employee confusion and discontent, thereby reducing the intangible costs incurred. A change that has lowered the quality of customer service may require a public relations outreach to keep customer goodwill, or it may require the company to come up with some other customer benefit to replace what was lost (Reed Newsome, 2016).

5.4.3 Iceberg Effect and Intangible Costs

Some think that intangible costs are arbitrary and cannot really be controlled (something like predicting the future). They perceive it as much like an iceberg: a small part is perceived but the larger part lies hidden under the surface. The submerged part of the iceberg represents the hidden costs, which are less tangible than visible monetary ones, perhaps, but equally deadly. The problem of cost visibility as a consequence of the "iceberg effect" is shown in Figure 5.8. To avoid the iceberg effect, cost analysis must tackle all aspects of the life cycle cost (LCC).

An intangible cost of maintenance is located in the development opportunities which may be postponed or lost because available resources are devoted to other maintenance tasks. Other intangible costs of maintenance are the following:

- Customer dissatisfaction when a request for repair or modification cannot be handled within a reasonable period of time.
- Hidden errors introduced by changing the asset configuration/design during maintenance reduce overall product quality.
- Worker prejudice.

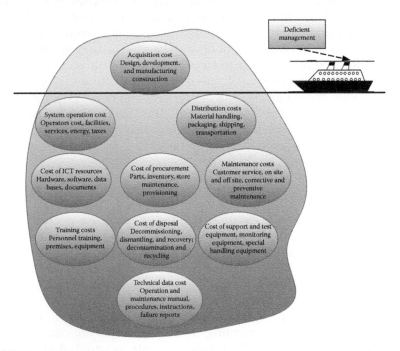

FIGURE 5.8
Iceberg effect in costs associated with service life. (Adapted from Galar, P.D. and Kumar, U., *Maintenance Audits Handbook: A Performance Measurement Framework*, CRC Press, Reference—609 p.—271 B/W Illustrations, February 25, 2016, ISBN: 9781466583917—CAT# K19014.)

A final cost is a reduction in productivity when alternative methods (which are not usually optimal) must be used (Galar and Kumar, 2016).

Intangible costs are subtle; in some cases, they are hidden in larger budgets, buried in ordinary operating expenses, or camouflaged by existing activities. Characteristically, they affect individuals and society in the long run. For example, the time it takes to do clerical and technical tasks, such as ordering, installing, and securing hardware and software, usually supersedes the time it takes to do existing tasks (Western Transportation Institute, 2009).

5.4.4 Analysis of Benefits and Intangible Costs

In what follows, we use the guidelines set by the US Department of Transportation's Federal Highway Administration (FHWA) for the benefit–cost analysis of agency costs, user costs and benefits, and nonuser impacts as investment costs, operational costs and benefits, and externalities, respectively. We think this case study is especially interesting because of FHWA's use of a maintenance decision support system (MDSS) with an integrative human machine interface (HMI) (Western Transportation Institute, 2009). Although we draw on the case of the FHWA, the definitions and concepts have a much broader applicability; hence, their inclusion and discussion here.

5.4.4.1 Intangible Benefits and Costs Defined

Intangible benefits and costs are theorized to be present based on various logical arguments, observations, and experiences. They are often qualitative and result from loose or overlapping connections (spin-offs). Attaching monetary values to intangibles is difficult for the following reasons:

- They lack the common units of measurement applied to tangible costs.
- They are often described in terms of value, the estimation of which is a key source of inaccurate analysis.
- Though intangible, they still affect customer choice and satisfaction.
- Ultimately, an agency must rely on its corporate culture to assess the value of intangible benefits.

Keen (1981) defines decision support systems as interactive systems "designed to help improve the effectiveness and productivity of managers and professionals." He identifies a range of functional areas and types of tasks including the following:

- "They are non-routine and involve frequent ad hoc analysis, fast access to data, and generation of non-standard reports."
- "They often address 'what if?' questions."
- "They have no obvious correct answers; the manager has to make qualitative trade-offs and take into account situational factors."

In the simplest sense, decision support systems provide fundamentally intangible benefits. Keen provides a set of decision support system benefits frequently cited in earlier case studies:

- Increase in number of alternatives examined
- Better understanding of the business
- Fast response to unexpected situations
- Ability to carry out ad hoc analysis
- New insights and learning
- Improved communication
- Control
- Cost savings
- Better decisions
- More effective teamwork
- Time savings
- Better use of data resources

Among these, only cost and time savings can be tracked to a straightforward benefit–cost analysis.

FHWA's use of performance measures is an attempt to quantify the outcomes of winter maintenance activities, and its maintenance decision support system (MDSS) is an attempt to quantify the application of rules of practice to weather prediction. Thus, representative quantitative performance measures can be adopted in the analysis of tangible benefit as a consequence of intangible cost reduction. However, if we take the definition of intangible as an aspect of the product or outcome of a service offering that has a value but is difficult to see or quantify, we need different measures, using interviews to elicit information, for example.

As the following discussion of intangible benefits and costs will show, many of the intangible benefits gained through the application of this MDSS result from an integrative human machine interface (HMI), that is, the incorporation of a quantitative technology platform with the ability to compare activity alternatives, and the required collection, documentation, and inputs of quantitative performance definitions and metrics (Western Transportation Institute, 2009).

5.4.4.2 MDSS Function Analysis

The stated objective of an MDSS is to provide support (through software systems) for proactive maintenance decision-making before and during adverse weather events, resulting in a higher level of service, reduced operational costs, and/or safer highway conditions.

The functions performed by this MDSS to meet the described objective form the framework for the analysis of operational intangibles. The aim is to use a benefit–cost analysis to determine whether benefits outweigh costs at an acceptable ratio. Another way to put this is to determine whether costs are too high for the function proposed to supply the customer with what the customer desires.

The first step in identifying functions is to determine the purpose of the system. MDSS has the ability to merge weather, maintenance, and RWIS data into a unified visualization and decision support tool to realize maximum benefit.

The functions of the resulting hybrid MDSS have three tiers: global, primary, and secondary. MDSS integrates several functions essential to maintenance in a single suite relating them in ways not previously accomplished. These integrated functions can be primary or secondary. Primary functions are created as part of the MDSS development process, such as the road treatment module. Secondary functions are or can be accomplished by existing systems, such as road weather forecasts.

The function analysis system technique (FAST) diagram shown in Figure 5.9 provides a helpful explanation of the relationship of the functions of a developed MDSS (Kaufman and Woodhead, 2006).

In the FAST diagram, two outputs or objectives culminate in the product (the intangible costs and benefits, indicated as "LOS") and are shown to the left of the scope line (Figure 5.9). In this case, the objectives are as follows:

- To produce real-time assessments of current and future conditions
- To make real-time maintenance recommendations

FAST defines only basic and supporting functions. The global, primary, and secondary terms described earlier are used in this FAST diagram. Portray is the global basic function of the MDSS. Suggest, Predict, and Integrate are the primary basic functions. Together, the four represent the purpose or mission of the MDSS. Model Pavement Conditions, Initialize Conditions, and Track Treatments are the primary supporting functions. Track Conditions is a secondary supporting function.

Darker lines indicate primary connections along the critical path. The forked path to the left of the function Record Resources indicates that both Equipment and Materials are supporting functions. The forked path to the left of the function Track Treatments represents Manual Data Entry or Mobile Data Collection as supporting functions. The dashed line boxes at the top represent specifications or particular parameters that must be achieved to satisfy the functions (Western Transportation Institute, 2009).

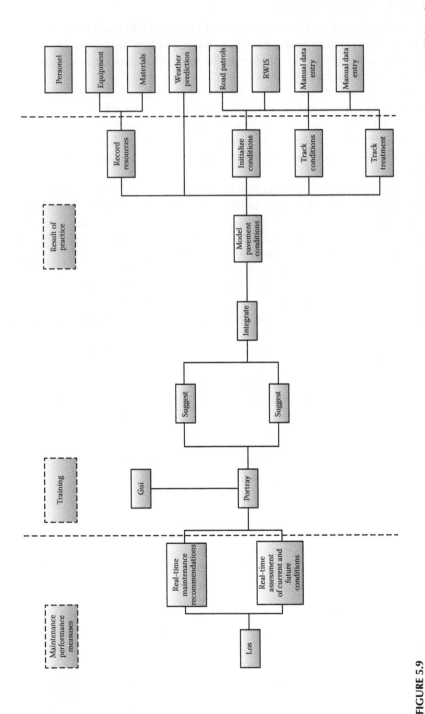

FIGURE 5.9
FAST diagram of MDSS functions. (Adapted from Western Transportation Institute, Analysis of Maintenance Decision Support System (MDSS) benefits & costs, Study SD2006-10, Final Report, Bozeman, MT, Iteris, Inc., Boise, ID, May 2009.)

5.4.4.3 Intangible Benefits of MDSS Function

Interestingly, a study shows that when interviewed, the supervisory and operational personnel directly related to the deployment of an MDSS associated with highest-order functions (closer to the left of the FAST diagram) focus on what is familiar (e.g., road weather information system (RWIS)); on improved access (Internet-distributed and in-vehicle); and on unrealized supporting functions (e.g., mobile data collection (MDC)). The Portray function is regularly considered the primary function in descriptions of the MDSS use and potential. This function uses a graphical user interface (GUI) to present the results of the other basic functions, the portrayal of the treatment alternatives, and the prediction of the associated pavement conditions (Western Transportation Institute, 2009).

5.5 Combining Tangible and Intangible Maintenance Costs and Benefits

Integrating intangible costs into monetary assessments has always been a challenging task, given the heterogeneity of decision-relevant factors, as well as the diversity of possible investment scenarios. In general, it is not possible to use one type of approach for all types of investment (Andresen, 2001). Therefore, we suggest selecting a technique based on the investment's main focus and motivation (Lucas, 1999). The following investment types can be distinguished in the asset management domain:

- *Direct returns as the main investment focus*: In this case, for example, if a new technology is required to accelerate processes, the financial assessment is the key to the management's investment decision. The solution is the one with the highest calculated return. The intangible assessment is secondary and mainly focuses on the assessment of risks, which must not exceed risk limits predefined in the company's policy.

- *Indirect returns as the main investment focus*: Here, the return on investment (ROI) is not reliably quantifiable. Examples include the implementation of technologies in manufacturing to improve data accuracy. This may streamline processes and analyses; this, in turn, could improve a company's reputation with its suppliers and customers. In this type of investment, the management should allocate a budget for the envisaged implementation, look for implementations meeting these budget constraints, and select the implementation with the maximum intangible evaluation score as determined by a decision tree.

- *Strategic investments opening up new opportunities*: As strategic aspects are very relevant, the key factor in the investment decision is the strategy score determined for the intangible assessment. Preferably, the alternative with the best score is selected, as long as the investment meets budgetary constraints, and the risks are deemed manageable. If alternative solutions have significantly different risk scores, the management is advised to do a trade-off analysis between strategic impact and risk.

- *Transformational investments*: These investments facilitate a complete reorganization of processes. In such investments, all tangible and intangible parameters may be relevant. As a result, management needs to define minimum thresholds for all criteria. In a first step, all investment alternatives not meeting the thresholds are discarded from further evaluation. For the assessment of remaining solutions, decision-makers may use a modified balanced scorecard approach combining the financial perspective, the operational perspective, the strategic perspective, and the risks perspective.

- *A technology as unique solution to implement functionality*: The selected technology may be the only possible solution to achieve certain functionality (e.g., to identify products that are reliable in harsh or dirty environments). Here, the key issue is how much management is willing to pay for the enabled functionality. Therefore, in a first step, management defines target costs not to be exceeded. In a second step, all implementations meeting the defined thresholds are assessed from an intangible perspective. Finally, management does a trade-off analysis of the remaining solutions' intangible scores and their calculated financial returns.

- *Mandatory investments*: These are investments required by law or contracts. If the investment is mandatory, intangible aspects play a secondary role (as the investment is required anyway) and the focus of managers will be on cost reduction. In terms of the intangibles, managers will primarily look at the risks of the proposed solution and make sure these do not exceed predefined, critical values. Out of the solutions meeting intangible risk requirements, the least expensive one is selected, unless there are very large differences in the intangible score (Ivantysynova et al., 2009).

5.5.1 Maintenance Practices Influencing Maintenance Costs: Causal Relations between Various Cost Factors

The expanding practice of outsourcing manufacturing operations has created a greater need to identify and reduce costs. One of the largest controllable manufacturing costs is in maintenance operations. But what is the true cost of maintenance systems? The first step in determining the true cost of maintenance efforts is to determine and understand the type of maintenance system.

The first thing to ask is what result is the maintenance system designed for? Is it producing predominantly reactive activities? If so, it should likely be redesigned. When maintenance systems are mostly reactive, up to 50% or more of maintenance spending can be eliminated. Reactive maintenance systems should simply be converted into proactive ones.

Statistics show that the cost of a typical repair is 5–15 times greater than the cost of the (proactive maintenance) effort that would have prevented the failure from occurring. When the cost of the loss of product, business opportunity, client rapport, and similar indirect costs are added to the cost of the repair, we see how the true cost of reactive maintenance practices can quickly add up.

Compounding this already costly situation is the fact that—especially in reactive maintenance operations—many maintenance systems do not identify root causes of failure and, as a result, suffer the same failures repeatedly. This escalates maintenance costs and results in a downward spiral of system and equipment reliability. Ultimately, the viability of the business itself is in question.

Many executives and senior managers believe the maintenance operation is solely responsible for a good behavior and function of an asset during the manufacturing operation. Maintenance, of course, plays a major and leading role in assets operation but the maintenance process alone cannot deliver an optimum reliability.

High levels of equipment reliability can only be established through the deployment of plant-wide, disciplined and integrated processes.

The elements essential to establishing optimum levels of reliability are the following:

- Appropriate specification and design practices
- Professional purchasing practices
- Appropriate storage facilities
- Precise installation methods
- Well-defined and consistent start-up and commissioning procedures
- Consistent operating practices
- Proactive maintenance processes

Note: Defects resulting from deficiencies in each of these elements are not necessarily equal. Defects introduced during design, start-up, and commissioning, and the operating process have significantly more influence on reliability than the remaining elements (see Figure 5.10) (Dabbs, 2004).

5.5.2 Estimation of Cost Elements

Operation and support costs in the exploitation phase are based on activities planned for the life cycle phase of operation and logistical support and are normally difficult to determine.

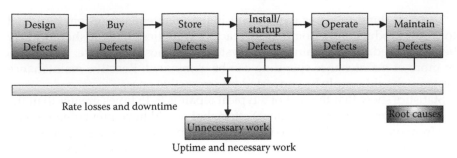

FIGURE 5.10
Introduction of defects and effect on reliability. (Adapted from Dabbs, T., The True Cost of Maintenance, CMRP, Life Cycle Engineering (LCE), As appeared in the April 2004 issue of *Pump & Systems*, 2004.)

Operation costs are a function of the requirements of the system or product and its utilization.

Support costs are a function of the inherent reliability and maintainability of the system's design, and the logistic requirements to support all the actions of maintenance, programmed and not programmed, throughout the projected service life. Logistical support includes maintenance, personnel and their training, supply management (spare parts and inventories), support and test equipment, transport and invoicing, facilities, information and communication technology (ICT) resources, and certain technical and engineering data. Individualized estimations of the cost of support to the system are based on the anticipated frequency of maintenance actions or the average value of maintenance and on the resources required during maintenance activities. The costs are deduced from reliability and maintainability predictions, from the analysis of logistical support, and from any other information on support obtained from the design data associated with the system or product.

Proposals, catalogs, reports on design, and studies by suppliers (or potential suppliers) can be a source of data for cost estimation. Potential suppliers will present/display proposals that include forecasts of the cost of acquisition and service life. If the data from the supplier are used, the analyst must be aware of what is included and not included to avoid omissions and repetition of costs.

During the final phases of the development and production of the system, when the system or product is tested or put into operation, the obtained experience represents the best data source for analysis and definitive evaluation. These data are collected and used to analyze the cost of the service life. The operation data are also used as much as possible to evaluate the impact on the cost of the service life of any proposed modification to the equipment or its logistical support elements.

The data must be coherent and comparable to be useful in the estimation process. The inability to decipher data frequently stems from a lack of comprehension or different interpretations of the definitions, errors in the production volumes, absence of certain necessary elements of cost, inflation, and so on.

Organizations register their costs in various ways; accountancy data are often submitted to public authorities in categories different from those used internally for LCC estimations. Categories also change from time to time. Because of these differences in definitions, the first stage of the cost estimation must fit all the data to the definition used. Consistency in the definition of the physical characteristics and the performance of assets is needed as well. When the cost data are obtained from several sources, it is important to arrive at suitable definitions of the physical characteristics, the performance of assets, and the cost elements. Integration of disparate data sources leads to more accurate estimations.

Estimations of the production cost often consider both recurrent and non-recurring costs. The two costs must be clearly distinguished and carefully defined.

Recurrent costs are those that are repetitive and occur when an organization produces similar goods or services on an ongoing basis. Variable costs are also recurrent costs, as they are repeated with each production unit. Because recurrent costs are not limited to variable costs. A fixed cost that is paid on a repeatable basis is a recurring cost. For example, in an organization that provides architectural and engineering services, renting office space that is a fixed cost is also a recurring cost.

Non-recurring costs, then, are all those that are not repetitive, although total expenditure can be cumulative in a relatively short period. Typically, non-recurring costs imply developing or establishing an ability or ability to operate. For example, the cost of acquiring the real estate in which a plant will be built is a non-recurring cost, as is the cost of building the plant itself.

If research and development (R&D) of a new product is considered a cost of current production, it will not be reflected in the costs of the new product. In this case, the cost to introduce a new product will be underestimated, and the cost of existing products overestimated. Recurrent and nonrecurring costs must be separated, lowering the costs of current production, adjusting the costs of existing products, and transferring the impact of R&D to the production of the new product, not penalizing the current one.

There are many large recurring costs associated with operation, maintenance, and removal. Energy, fuel, lubricants, spare parts, training, maintenance manpower, and logistical support during the service life are some of these costs. When the life of many complex systems is estimated and recurring costs of this type are considered, the total figure can be high, even higher than the initial cost. Often operation and maintenance (O&M) costs are given little attention, and the initial investment cost takes precedence. This is a mistake, the cost of the life cycle of a system, considering O&M, is the real system cost. In fact, O&M requires proper LCC estimation for optimization.

purposes and LAD groups. As with any classification system, the definition of LAD groups necessitates striking a balance between a large number of small, fine-tuned groups and a smaller number of groups whose membership may not be precisely similar.

Machines in a given LAD group will probably perform a number of tasks (loaders can load trucks, blend material, or do general cleanup work) and different amounts of time will probably be allocated to each task (Vorster and De La Garza, 1990).

5.5.4.2 Scenarios

Each LAD group requires one or more scenarios (i.e., tasks) under which its items operate. Each scenario requires the following:

1. An identifying code.
2. A textual description (optional).
3. The percentage of operating time spent on the task by each member of the LAD group. A logical extension of the model would allow each item to have its own percentage assignments in its LAD group's scenario.

Each scenario requires entering cost-estimating parameters for one or more of the four LAD cost classifications mentioned earlier (Fuerst et al., 1991).

The ability to estimate LAD costs depends on being able to describe what is likely to happen when a failure occurs. It is necessary to articulate scenarios describing the task and what happens when a member of a given LAD group breaks down. This is done to focus thinking and provide a static background against which the cost effects of the described failure can be assessed.

The equipment and vehicles in a particular LAD group frequently perform more than one task and frequently fail under different circumstances. Therefore, more than one failure scenario may be applicable to a given LAD group, and the percentage of time spent or work done by members of a LAD group under a given scenario will need to be assessed. This makes it possible to determine a weighting factor for each scenario so the LAD costs can be weighted and added at a later stage. It must be stressed that the scenarios play no role in computing LAD costs; they only provide a predefined description of a situation as a background for the estimating process (Vorster and De La Garza, 1990).

5.5.4.3 LAD Cost Model

The LAD Cost Model is a tool to help estimate the LAD costs over a given period for a particular machine and for a LAD group as a whole. The definition of LAD groups, the description of the scenarios, and the input parameters needed to estimate each LAD cost category create an estimating environment that draws on data unique to a particular machine and on the operating conditions unique to each group of machines. The model can be implemented on a microcomputer using any of several popular programmable database products. As in most computer programs, successful implementation relies on the ability of programmers to create an intuitive environment for users.

In what follows we describe the data necessary to support the model, including the algorithms for calculating LAD costs from the data (Fuerst et al., 1991).

5.5.4.3.1 LAD Group Information Requirements

Each LAD group is described as follows:

1. An identifying code (no more than 5 characters)
2. An optional longer description (around 50 characters)

5.5.4.3.2 Equipment Item Definitions

Equipment items are defined and assigned to each LAD group; each item is described by the following:

1. An identifying code such as a serial number or license plate number.
   a. Any optional information to enhance the description, such as make, manufacturer, or year. These data are not essential to the operation of the model and are used solely to enhance outputs and reports.
   b. The LAD group to which the item belongs.
2. Performance data. Data describing each item's previous or projected future performance will include the following:

   W_i = number of hours the ith item is in use during the study period

   D_i = number of hours during the study period the ith item is unavailable due to breakdowns

   V_i = number of times the ith item breaks down during the study period

Note: if the in-use hours, hours unavailable due to breakdown, and number of breakdowns reflect values for some previous time period, the output from the model evaluates the LAD costs for that period. If these items are projections for some future time period, the model's output predicts LAD costs for that period (Fuerst et al., 1991).

5.5.5 Associated Resource Impact Cost Procedure

This procedure is used to assist in estimating the parameters needed to calculate the ARI or associated resource investment costs.

The parameters apply to a given LAD group working under a given scenario. ARI costs for a particular machine in the LAD group are calculated from performance data unique to the machine and the ARI parameters for the machine's LAD group (Fuerst et al., 1991).

ARI costs are defined as costs that occur as a direct result of a failure. Figure 5.12 shows the accumulation of ARI costs along a timeline stretching from the point when the failure occurs and normal operations cease (C) to the point when normal operations resume (R). Each associated resource is affected differently by a failure and, thus, will have its own:

1. *Impact lag (profile CL in Figure 5.12)*: This is the period from the failure to the start of the impact on the resource, as shown by the line segment CL in Figure 5.12. For certain types of resources, such as the driver of a failed truck, this lag period will be very short. For other resources, this lag will be relatively long, as when a bulldozer fails and affects a loader that is loading material stockpiled by the bulldozer.

2. *Impact duration*: This is the time from the failure to the end of the impact on the resource, shown by the line segment CD in Figure 5.12. The impact duration can be equal to the total duration of the impact (CR) if re-planning is not possible. It will be substantially shorter if resources can be reassigned during the period affected by the failure.

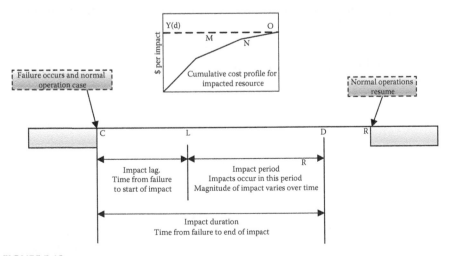

FIGURE 5.12

Timeline of ARI cost. (Adapted from Vorster, M.C. and De La Garza, J.M., *J. Constr. Eng. Manage.*, 116(4), 656, 1990.)

3. *Impact period*: This is the time during which ARI costs accumulate. The impact period, the period in which the impacts and their associated costs actually occur, is given by line segment LD in Figure 5.12.

4. *Cumulative cost profile*: This defines how the accumulated impact cost on an associated resource grows as the impact period increases. The rate at which the impact cost of an associated resource accumulates with the increasing failure duration is shown in the rectangle LMNO above the timeline in Figure 5.12. It also varies from impacted resource to impacted resource; thus, it is necessary to define a cost accumulation method for each of the impacted resources. The cumulative cost profile (LMNO) reflects the way the cumulative cost of the impact on a particular resource grows over time. LMNO in Figure 5.12 shows that an impact of duration LD yields a cost of $Y(d)$.

The ARI cost for a given machine in a given period is determined by multiplying the total ARI cost per impact by the number of failures suffered by the machine in the period (Fuerst et al., 1991).

Each of the associated resources impacted by a failure is affected differently, so each has its own impact lag (CL) to represent the period from the time of the failure to the start of the impact on the resource.

For certain types of resources, such as the driver of a failed truck, the lag period in this case may be very short; for another ones, it may be relatively long, as well as when a bulldozer fails and impacts a loader loading material it has stockpiled.

The cost accumulation curve depends on the number of all associated resources affected during each portion of the impact period, and the extent to which they are affected. Thus, the cost per hour per associated resource when working, the cost per hour per associated resource when idle, and the number of these resources affected are all required. Once these are specified, the program completes a table or screen similar to Figure 5.13 (Fuerst et al., 1991).

The answers for two questions help establish the impact of duration over the costs:

1. Will re-planning eliminate the impact of failure on this resource within the day, if failure occurs early enough within the day? In this case, an estimate of the impact lag and the impact period (or duration) is required. The cumulative cost profile covers the period from the start to the end of the estimated impact period. In this situation, it is possible to bring in a replacement for the failed item relatively quickly. The expected ARI cost for a failure under this scenario is calculated by assuming that the failure is equally likely to occur at any hour of the workday.

   If the answer to this question is no, the next question to ask is

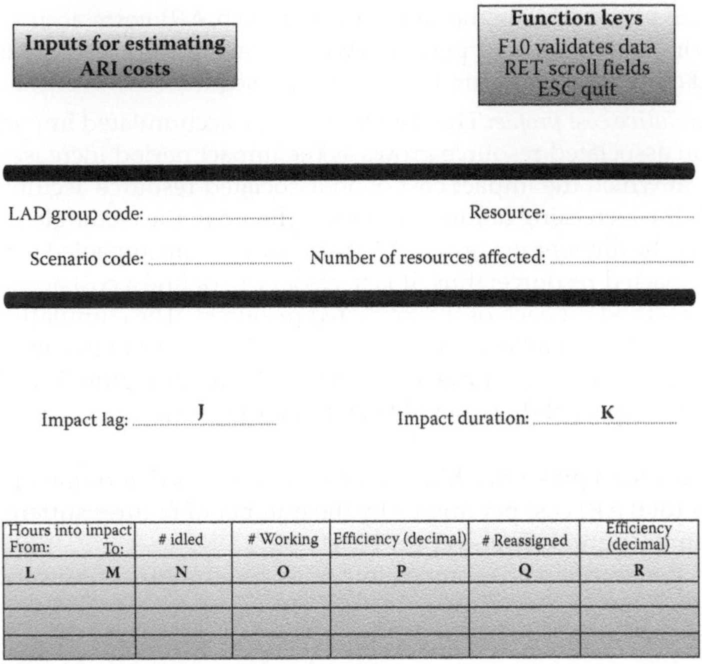

Inputs for estimating ARI costs		Function keys
		F10 validates data
		RET scroll fields
		ESC quit

LAD group code: ....   Resource: ....

  Scenario code: ....   Number of resources affected: ....

       Impact lag: ....J................   Impact duration: ....K................

Hours into impact From: To:		# idled	# Working	Efficiency (decimal)	# Reassigned	Efficiency (decimal)
L	M	N	O	P	Q	R

FIGURE 5.13
Input parameters for estimating ARI cost. (Adapted from Fuerst, M.J. et al., A model for calculating cost of equipment downtime and lack of availability in directorates of engineering and housing, USACERL Technical Report P-91/16, Corps DEH Equipment Maintenance Management System of Engineers, Champaign, IL, March 1991.)

2. Will the impact of the failure on this resource terminate at the end of the day? If yes, normal operations can be resumed at the start of the next working day. In this case, the cumulative cost profile starts at the end of any impact lag and continues to the end of the workday. The expected ARI cost for a failure under this scenario also assumes the failure is equally likely to occur at any hour of the workday.

   If the answer to this question is no, the effects of the failure can last beyond the end of the workday, and the following workday estimates are required:

   a. The impact lag

   b. The most optimistic duration (i.e., the shortest it will ever be), designated t_1

   c. The most likely duration, designated t_2

   d. The most pessimistic duration (i.e., the longest it will ever be), designated t_3

The three estimates for the duration are used to calculate the parameters of a generalized beta distribution. First, normalize the three estimates according to Hahn and Shapiro, 1967:

$$u_i = \left(\frac{t_i - t}{t_3 - t_1}\right) \Rightarrow u_i = O; \quad u_3 = 1 \tag{5.4}$$

Now, use these normalized estimates to calculate estimates for the mean (μ) and variance (σ^2) of the resulting beta distribution for the normalized variables:

$$\mu = \frac{(u_1 + 4u_2 + u_3)}{6} \tag{5.5}$$

$$\sigma^2 = \frac{(u_3 - u_1)}{6} \Rightarrow \sigma^2 = \frac{1}{36} \tag{5.6}$$

These estimates of the mean and variance of the beta distribution, used in most program evaluation review technique (PERT) applications, assume the standard deviation is one-sixth of the range between the maximum and minimum durations (Swanson and Pazer, 1969; Antill and Woodland, 1982).

Next, calculate the parameters of the beta distribution for the normalized variables:

$$\alpha = \frac{(1-\mu)\left(\mu*(1-\mu) - \sigma^2\right)}{\sigma^2} \tag{5.7}$$

$$\beta = \frac{\mu * \alpha}{1-\mu} \tag{5.8}$$

Numerically integrating the product of the resulting generalized beta distribution with the cumulative cost profile produces the expected ARI costs for a single failure of an item for the designated LAD group. Multiplying the cost of a single failure by the number of failures for an item produces the ARI cost for the item of the LAD group during the study period. *Note:* other distributions can be used depending on the available information (Fuerst et al., 1991).

5.5.6 Lack of Readiness Costs

The costs occurring when an item of equipment or a vehicle fails can be divided into two broad categories. The first includes the tangible cost of the labor, materials, and other resources needed to repair the machine. The second includes all the intangible, or consequential, costs that arise from the failure and impact the organization as a whole. Tangible costs are easy to record and estimate using normal cost accounting methods. Intangible costs present a different problem in that they cannot be assessed with any degree of certainty except under very rigid, well-defined circumstances (Fuerst et al., 1991).

The search for an effective or at least consistent methodology to access the dollar value of intangible costs is important because success will bring rigor to many aspects of equipment management that still remain subjective despite advances in recording and processing data on tangible costs.

Quantifying intangible costs with a reasonable degree of accuracy can influence equipment decision-making in three ways.

First, these costs can represent the impact that less-than-perfect performance in a particular machine has on the organization as a whole. They can be used to compare one machine with another and identify members of a fleet that merit special attention (Vorster and De La Garza, 1990). The basic trade-off in equipment management is between capital costs and operating inferiority, where the latter is defined to include both the direct costs of repair and the intangible costs arising from the failure (Terborgh, 1949). The annual cost of interruption caused by component failure has been defined as the product of the annual frequency of failure, the average duration of failure, and the DT cost per unit (Cox, 1971). This definition is suited to situations where the equipment working on a particular task is configured as a single rigid system and where failure in one component causes the whole system to go down (Fuerst et al., 1991).

Intangible costs can also be assigned to a particular year of equipment life on the basis of an estimated percentage of DT multiplied by the planned hours of operation for the machine and the hourly cost of a replacement or rental machine (Nunnally, 1977). This approach focuses on the failed machine alone, however, and disregards any effect the failure may have on the system as a whole (Fuerst et al., 1991).

Second, intangible costs can be used to assess the effectiveness of maintenance policies and procedures. Effective maintenance operations should keep the mechanical quality of equipment at a high level, thereby ensuring low intangible costs. This means the balance between maintenance expenditures and intangible costs is a good measure of maintenance effectiveness.

The third intangible costs can be used as an input to an economic replacement model. Under these conditions they are added to normal owning and operating costs to give a better assessment of economic life. These costs can play an important part in economic life studies, because they highlight the fact that neither costs nor economic life is independent of the impact of DT and the lack of availability (Vorster and De La Garza, 1990).

5.5.6.1 Lack of Readiness Cost Procedure

Lack of readiness or LOR costs are the penalty costs that could or should be levied because of the expectation that resources representing capital investments in assets should be kept in a ready condition as much as possible. They are based on the concept that there should be some penalty mechanism that motivates managers to ensure that as many of the fleet as possible are ready for deployment when needed (Vorster and De La Garza, 1990). In many ways, these costs are analogous to ongoing depreciation and interest charges.

The methodology used to quantify LOR costs for a machine belonging to a particular LAD group is essentially similar to that used for ARI costs, and is set out in Figure 5.14. Point C identifies the time when failures occur and normal operations cease. The impact lag reflects the fact that the penalty should only be applied after a reasonable period defined by the time CL. The impact duration is given by CD, which, in this case, equals the period of the DT. Penalties should stop when the machine is repaired and can resume work, whether it is needed or not. Normal operations resume at point R.

The cumulative cost profile is straightforward, in that LOR costs relate only to units in the LAD group under study and have nothing to do with any other resources. The profile starts at point L and has a uniform slope proportional to the penalty cost per hour (Vorster and De La Garza, 1990).

The fact that the cumulative cost profile is linear makes it possible to calculate the LOR costs on a monthly basis using the following form (Vorster and De La Garza, 1990):

$$LOR = P * \left[D - (V - L) \right] \qquad (5.9)$$

where
  LOR is the lack-of-readiness costs for a machine in a month
  P is the lack-of-readiness penalty cost in \$/h
  D is the number of hours a particular unit breaks down and is unable to respond to operational demands in a month
  V is the number of times a machine breaks down and disrupts planned operations in the month
  L is the impact lag in hours

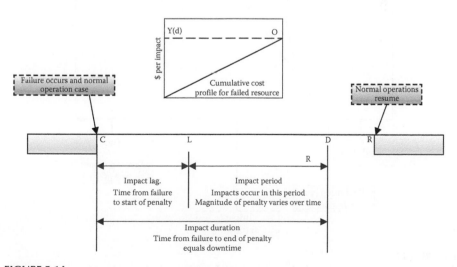

FIGURE 5.14
Timeline of LOR costs. (Adapted from Vorster, M.C. and De La Garza, J.M., *J. Constr. Eng. Manage.*, 116(4), 656, 1990.)

5.5.7 Service-Level Impact Costs

Service-level impact (SLI) costs occur when, in a pool of similar equipment performing a certain service, lack of reliability in one or more pool members causes the others to work in a more costly manner to maintain the required level of service. SLI costing is shown in Figure 5.15.

The common pool of resources from which a certain level of service is demanded corresponds to an LAD group. When quantifying SLI costs for one member of the group, consider the following three points:

1. The number of pieces of equipment needed to satisfy operational demands under normal conditions
2. The probability that a certain number of pieces of equipment will be available in any single day given the overall availability of each member in the LAD group
3. The costs of the actions taken to ensure that the service level is maintained when the number of items in service falls below that required to satisfy operational demands (Fuerst et al., 1991)

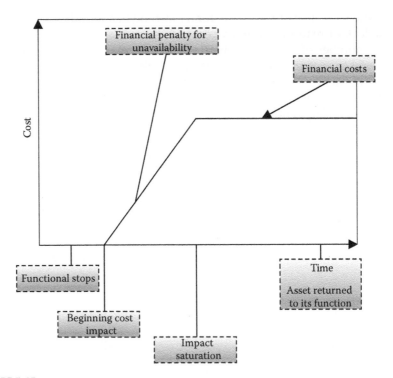

FIGURE 5.15
Financial costs in SLI costing. (Adapted from Galar, P.D. and Kumar, U., *Maintenance Audits Handbook: A Performance Measurement Framework*. CRC Press, Boca Raton, FL, Reference—609 p.—271 B/W Illustrations, February 25, 2016, ISBN: 9781466583917—CAT# K19014.)

5.5.7.1 SLI Cost Procedure

The common pool of resources from which a certain level of service is demanded corresponds to an LAD group. The problem of quantifying SLI costs for a member of the group must take the following three factors into account:

1. The operational demands placed on the LAD group in terms of the number of pieces of equipment needed to satisfy operational demands under normal conditions
2. The overall work capacity of the LAD group, as defined by the probability that a certain number of pieces of equipment will be available in any one day given the overall availability of each member in the LAD group
3. The cost of the action that will be taken to ensure the service level is maintained when the work capacity of the LAD group falls below the number required to satisfy operational demands

This extremely complex problem has been addressed by developing a Monte Carlo simulation model that performs the following five functions (Vorster and De La Garza, 1990):

1. The down ratio for each of the members of the LAD group listed as units $X = A, B, C, …, N$ is calculated for the month under study using the following form:

$$Z = \frac{D}{D+W} \tag{5.10}$$

where
   Z is the down ratio
   D is the number of hours a particular unit breaks down and is unable to respond to operational demands in a month
   W is the hours worked by the machine in the month

Quantifying these costs depends on the following factors:

   a. Level of demand of the service versus the available pieces of equipment to fulfil the demand.
   b. Group ability to work is defined by the condition of a given number of pieces of equipment and is based on the availability of each member of the group (Galar and Kumar, 2016).
2. The down ratio of each individual machine in the LAD group (machines $X = A, B, C, … N$) is used in a simulation model to produce the following two results: the probability $P_{(q)}$ of having $q = 0, 1, 2, 3, … m$ units

in the LAD group down and incapable of working in any one day; the frequency with which unit $X = A, B, C, \ldots N$ is listed as down on the days when the number of units down equals $q = 0, 1, 2, 3, \ldots m$.

3. The two results of the simulation are used to calculate the joint probability $P_{(X,q)}$ that q units in the LAD group are down in a given day and unit $X = A, B, C, \ldots N$ will be included among the down units.

4. A monthly charge reflecting the additional expenditure needed to maintain the service level if $q = 0, 1, 2, 3, \ldots m$ units are down on a particular day is calculated from a series of user inputs.

5. The SLI costs for the particular machine in a month are calculated by multiplying the monthly charge by the joint probabilities $P_{(X,q)}$ for the machine and summing over all values of q.

The DT ratios of the individual machines in the LAD group are used to simulate two results (Fuerst et al., 1991):

1. $P_{(x)}$, the probability of having $x = 1 \ldots n$ items in the LAD group down and incapable of working in any day.

2. $P_{(i|x)}$, the conditional probability of item i being down, given that x items are down that day. For example, $P_{(2|4)} = 0.4$ means when four items are down, item 2 will be down 40% of the time. In other words, when four items are down, 40% of item 2 is down. The sum of the $P_{(i|x)}$ over for a given x equals x.

The expected SLI cost for item i on a day when x items are down is expressed as follows:

$$SLI_{ix} = \frac{c_x * P(i|x)}{x} \tag{5.11}$$

where
C_x = Cost per Day if only x > n-m items are down
m = The number of items required to maintain normal service

The expected daily SLI cost for item i is therefore given by

$$SLI_i = \sum SLI_{ix} * P(x) \tag{5.12}$$

5.5.8 Alternate Method Impact Costs

Alternate method impact (AMI) costs occur when the failure and continuing DT of a given machine force a change from an optimum to a less-than-optimum method, thereby causing the organization to incur additional cost. An example is the use of a loader and truck in place of a more efficient motor scraper because the scraper is not available. The use of standard vehicles rather than customized, more efficient vehicles to collect refuse is another. AMI costs normally occur only after an extended period of DT, and they frequently involve specific expenditures associated with mobilizing and demobilizing the resources needed for the alternative method (Vorster and De La Garza, 1990).

AMI costs appear when the failure of a piece of equipment in the group forces a change in the method (operation or manufacturing) and the organization undergoes an additional failure cost, proportional to the differential cost between the applied methodologies (optimized and alternative). The evolution of this type of cost once failure starts can be seen in Figure 5.16. In this case:

1. The new method implies a vertical jump in cost at the beginning of using the alternative; this represents the cost to configure the new method.
2. Cost per unit of time during the use of the new method is the difference between the costs per unit of time between the original method and the alternative method.
3. A second vertical jump at the end of the use of the alternative method reflects the cost to return to the original method.

The graph in Figure 5.16 shows the three sections of costs implied in the use of alternative methods while the repair of the original equipment takes place. The first stage indicates the configuration of the new method shortly after the activity of the original equipment ends and a decision is made to use the alternative method.

After it is configured and deployed, the alternative method usually performs more poorly than the original method, thus generating a cost. Once the original equipment is repaired and the original method can be restored, there is an additional and final cost related to the restitution of the original method and the disassembly of the alternative one (Galar and Kumar, 2016).

5.5.8.1 AMI Cost Procedure

AMI costs occur when the failure and continuing DT of a machine in an LAD group forces a change in the method used to carry out the work described in the scenario. The change is assumed to be from an optimum to a less-than-optimum method; thus, the organization suffers a consequential cost proportional to the cost differential between the methods and the quantity of work done under the less favorable circumstances.

The rationale developed to quantify ARI costs in Figure 5.12 is used for the third time, as can be seen in Figure 5.17. C and R again represent the points when normal operations cease and resume; CL shows the lag from failure to, in this case, the introduction of the alternative method, CD shows the impact duration, and LD shows the impact period. The cumulative cost profile is

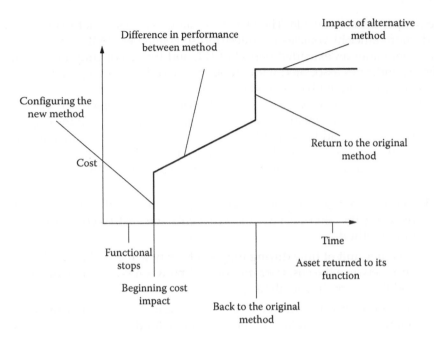

FIGURE 5.16
Cost of use of alternative methods. (Adapted from Galar, P.D. and Kumar, U., *Maintenance Audits Handbook: A Performance Measurement Framework*. CRC Press, Boca Raton, FL, Reference—609 p.—271 B/W Illustrations, February 25, 2016, ISBN: 9781466583917—CAT# K19014.)

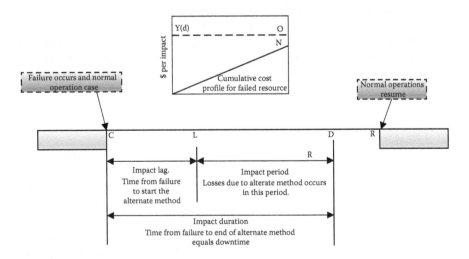

FIGURE 5.17
Timeline of AMI costs. (Adapted from Vorster, M.C. and De La Garza, J.M., *J. Constr. Eng. Manage.*, 116(4), 656, 1990.)

essentially the same as that for the LOR cost module with the following three exceptions (Vorster and De La Garza, 1990).

A vertical step (LM) appears initially to reflect the setup costs associated with mobilizing the new method.

1. The slope of the profile in the range M to N is proportional to the cost and production differential between the methods.
2. A second vertical step (NO) is included at the end to reflect the cost of breaking down or demobilizing the new method.

In practice, the mobilization and demobilization of an alternate method occur only in the case of severe failures. Thus, a *mobilization percentage* is used to reflect the proportion of failures relative to all failures for which mobilization and demobilization occur (Fuerst et al., 1991).

The cumulative cost profile and simplicity of the concept hide a critical problem; the AMI cost for a given machine in a given period cannot be obtained by multiplying the total AMI cost per impact $Y(d)$ by the number of failures the machine experiences in the period, because the mobilization and demobilization costs are incurred in a limited number of severe failures. It is necessary to define a mobilization percentage that reflects the proportion of severe failures relative to all failures and use it to scale down the effect of the mobilization and demobilization estimates (Vorster and De La Garza, 1990).

The linear nature of the cumulative cost profile between M and N and the use of a mobilization percentage make it possible to calculate AMI costs on a monthly basis using the following form:

$$AMI_i = S * Q * \left[Di - \left(L_i * V_i \right) \right] + V_i * M_p * \left(M_z + D_z \right) \qquad (5.13)$$

where
    AMI_i is the alternative method impact costs for item i in the study period
    S is the cost surcharge in \$/unit caused by the alternative method
    D_i is the number of hours during the study period the ith item is unavailable due to breakdowns
    Q is the quantity produced in units per hour by the alternative method
    V_i is the number of times the ith item breaks down during the study period
    H is the impact lag in hours
    M_p is the mobilization percentage
    M_z is the cost of mobilization
    D_z is the cost of demobilization

Because the model provides the mechanisms needed to quantify several forms of consequential costs, it can model various situations. However, it is complex to get a good trade-off between accuracy and complexity as Figure 5.18 illustrates. In developing the model, every increase in complexity must be checked to ensure meaningful and relevant improvement (Fuerst et al., 1991).

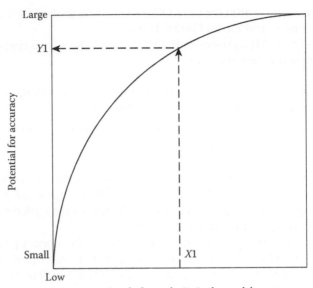

FIGURE 5.18

Relationship of complexity and potential for accuracy. (Adapted from Fuerst, M.J. et al., A model for calculating cost of equipment downtime and lack of availability in directorates of engineering and housing, USACERL Technical Report P-91/16, Corps DEH Equipment Maintenance Management System of Engineers, Champaign, IL, March 1991.)

References

Andresen, J. L., 2001. A framework for selecting an it evaluation method—In the context of construction. PhD thesis. DTU, Lyngby, Denmark.

Antill, J. M., Woodland, R. W., 1982. *Critical Path Methods in Construction Practice*, 3rd edn. John Wiley & Sons, Canada, pp. 301–302.

Arditi, D., Kale, S., Tangkar, M., 1997. Innovation in construction equipment and its flow into the construction industry. *Journal of Construction Engineering and Management* 123(4), 371–378.

Cooper, K. G., 1994. The $2,000 hour: How managers influence project performance through the rework cycle. *IEEE Engineering Management Review* 22(4), 12–23.

Cox, E. A., 1971. Equipment economics. In: Havers, J. A., Stubbs, Jr, F. W. (Eds.), *Handbook of Heavy Construction*, 2nd edn. McGraw-Hill Book Co., New York, pp. 7–15.

Dabbs, T., 2004. The True Cost of Maintenance, CMRP. Life Cycle Engineering (LCE). As appeared in the April 2004 issue of *Pump & Systems*. Charleston, Carolina del Sur.

Day, D. A., Benjamin, N. B. H., 1991. *Construction Equipment Guide*, 2nd edn. John Wiley & Sons, Inc., New York.

Eden, C., Williams, T., Howick, S., 2000. The role of feedback dynamics in disruption and delay on the nature of disruption and delay in major projects. *Journal of the Operational Research Society* 51, 291–300.

Edwards, D. J., Holt, G. D., Harris, F. C., 1998. Financial management of construction plant: Conceptualizing cost prediction. *Journal of Financial Management of Property and Construction* 3(2), 59–73.

Edwards, D. J., Holt, G. D., Harris, F. C., 2002. Predicting downtime costs of tracked hydraulic excavators operating in the UK opencast mining industry. *Construction Management and Economics* 20, 581–591.

Elazouni, A. M., Basha, I. M., 1996. Evaluating the performance of construction equipment operations in Egypt. *Journal of Construction Engineering and Management* 122(2), 109–114.

Fuerst, M. J., Vorster, M. C., Hicks, D. K., 1991. A model for calculating cost of equipment downtime and lack of availability in directorates of engineering and housing. USACERL Technical Report P-91/16. March 1991. Corps DEH Equipment Maintenance Management System of Engineers, Champaign, IL.

Galar, P. D., Kumar, U., February 25, 2016. *Maintenance Audits Handbook: A Performance Measurement Framework.* CRC Press, Boca Raton, FL. Reference—609pp., 271 B/W Illustrations. ISBN: 9781466583917—CAT# K19014.

Hahn, G. J., Shapiro, S. S., 1967. *Statistical Models in Engineering.* John Wiley & Sons, New York, pp. 91–96.

Halligan, D. W., Demsetz, L. A., Brown, J. D., Pace, C. B., 1994. Action–response model and loss of productivity in construction. *Journal of Construction Engineering and Management* 120(1), 47–64.

Hanna, A. S., Heale, D. G., 1994. Factors affecting construction productivity: Newfoundland versus rest of Canada. *Canadian Journal of Civil Engineering* 21, 663–673.

Ivantysynova, L., Klafft, M., Ziekow, H., Gunther, O., Kara, S., 2009. RFID in manufacturing: The investment decision. In: *Proceedings of the Pacific Asia Conference on Information Systems (PACIS) 2009.* Association for Information Systems. Posted at AIS Electronic Library (AISeL).

ICONICS, Inc., 2011. *KPI, OEE and Downtime Analytics,* An ICONICS White paper. ICONICS, Inc., Foxborough, MA.

Kaufman, J. J., Woodhead, R., 2006. *Stimulating Innovation in Products and Services with Function Analysis and Mapping.* John Wiley & Sons, Inc., Hoboken, NJ.

Keen, P. G. W., March 1981. Value analysis: Justifying decision support systems. *MIS Quarterly* 5(1), 1–15.

Kirmani, S. S., 1988. The construction industry in development: Issues and options. Policy planning and research staff, infrastructure and urban development department. Report INU 10, World Bank, Washington, DC.

Lambán, M. P., 2010. Determinación de Costos de Procesos de la Cadena de Suministros e Influencias de Factores Productivos y Logísticos. PhD thesis. Universidad de Zaragoza, Zaragoza, AR, Spain.

Lambán, M.P., et al., 2009. Modelo de gestión económica de la Cadena de Suministro, Primer Congreso de Logistica y Gestion de la Cadena de Suministro, Zaragoza, Spain.

Lucas, H. C., 1999. *Information Technology and the Productivity Paradox—Assessing the Value of Investing in IT.* Oxford University Press, Oxford, U.K.

Maloney, W. F., McFillen, J. M., 1986. Motivational implications of construction work. *Journal of Construction Engineering and Management* 112(1), 137–151.

Mitropoulos, P., Tatum, C. B., 1999. Technology adoption decisions in construction organizations. *Journal of Construction Engineering and Management* 125(5), 330–338.

Neil, J. M., 1989. Concepts and methods of schedule compression. *Transactions of the American Association of Cost Engineers* S9, 71–77.

Nepal, M. P., Park, M., 2004. Downtime model development for construction equipment management. *Engineering, Construction and Architectural Management* 11(3), 199–210.

Nunnally, S. W., 1977. *Managing Construction Equipment*. Prentice-Hall, Englewood Cliffs, NJ, p. 226.

Pathmanathan, V., 1980. Construction equipment downtime costs. *Journal of Construction Division* 106(4), 604–607.

Piper, C. J., Vachon, S., 2001. Accounting for productivity losses in aggregate planning. *International Journal of Production Research* 39(17), 4001–4012.

Reed Newsome, B., 2016. What are tangible costs & intangible costs? Powered by studioD. http://smallbusiness.chron.com/tangible-costs-intangible-costs-51412. html. Septiembre 2016.

Roberts, E. B., Alfred, P., 1974. A simple model of R & D project dynamics. *R & D Management* 5(1), 1–15.

Schwartzkopf, W., 1995. *Calculating Lost Labour Productivity in Construction Claims*. Wiley Law Publications, John Wiley & Sons, Inc., New York.

Selinger, S., 1983. Economic service life of building construction equipment. *Journal of Construction Engineering and Management* 109(4), 398–405.

Sözen, Z., Giritli, H., 1987. Equipment policy as one of the factors affecting construction productivity: A comparative study. In: Lansley, P. R. Harlow, P. A. (Eds.), *Managing Construction Worldwide: Productivity and Human Factors in Construction. Fifth International Symposium, CIOB, CIB*, London, U.K., pp. 691–696.

Swanson, L. A., Pazer, H., 1969. *PERTSIM Test and Simulation*. International Textbook Company, Scranton, PA, p. 11.

Terborgh, G., 1949. *Dynamic Equipment Policy: A MAPI Study*. Machinery and Allied Products Institute and Council for Technological Advancement, p. 27.

Thomas, H. R., 2000. Schedule acceleration, work flow, and labour productivity. *Journal of Construction Engineering and Management* 126(4), 261–267.

Thomas, H. R., Raynar, K. A., 1997. Scheduled overtime and labour productivity: Quantitative analysis. *Journal of Construction Engineering and Management* 123(2), 181–188.

Tsimberdonis, A. I., Murphree, E. L., Jr., 1994. Equipment management through operational failure costs. *Journal of Construction Engineering and Management* 120(4), 522–535.

Vorster, M. C., De La Garza, J. M., 1990. Consequential equipment costs associated with lack of availability and DT. *Journal of Construction Engineering and Management* 116(4), 656–669.

Vorster, M. C., Sears, G. A., 1987. A model for retiring, replacing or reassigning construction equipment. *Journal of Construction Engineering and Management* 113(1), 125–137.

Western Transportation Institute, 2009. Analysis of Maintenance Decision Support System (MDSS) benefits & costs. Study SD2006-10, Final Report. Bozeman, MT, Iteris, Inc., Boise, ID, May 2009.

6

Maintenance Services and New Business Models: A New Way to Consider Costs

6.1 New Maintenance Service Providers

The competition in international markets has led to a situation where industrial maintenance services are increasingly outsourced. Many regional clusters have fragmented into global value chains, in which geographically scattered companies specialize in particular activities. At the same time, the competition has changed from being between companies to being between networks (or chains). As a result, more attention has to be paid to sharing the value between all the network members and keeping the whole network competitive. Most of the existing management tools are not directly suitable for networks, because they have been designed for individual companies. There is a need for practical tools to manage maintenance service networks to increase the value of the cooperation of the network members.

The current global transformation is visible in several ways. One is that companies that have previously focused completely on manufacturing are now interested in transferring at least some part of their focus to the production of services. Thus, they are integrating service offerings with their manufacturing activities. All over the world, the share of services in total production is already nearly 70%, with the outsourcing trend generating an increase in the number of companies focusing entirely on providing industrial maintenance services.

The proliferation of maintenance outsourcing indicates that companies firmly believe they can achieve added value from a new kind of network model. In addition, previous studies have proved that companies can improve their profitability and overall competitiveness by incorporating new services systematically into their business. The reasons for this include, for example, better predictability of company sales and cash flows, reaching new markets, protecting the company's actual core product, and improved customer satisfaction and cooperation.

With automation, lean production philosophies, and tightening competition, the significance of industrial production equipment availability to business performance is increasingly emphasized. This has led to a rapid diversification of industrial services. Accordingly, the provision of service is becoming an important part of the global competitiveness of industry. Many equipment providers have tried to extend their business by focusing their investments on the development of industrial services. A large installed base of equipment offers significant possibilities to equipment providers to develop their business in a new direction, especially in the field of maintenance. At the same time, they must retain their traditional business, so creating new service offerings is a challenge.

In short, there is a clear demand for independent maintenance companies. In the best case scenario, these maintenance companies will integrate large arrays of services, and provide their customers with comprehensive maintenance contracts and solutions both regionally and internationally.

6.1.1 New Maintenance Network

Traditionally, the maintenance research has taken a technical point of view, and the role of the service provider has been emphasized. Only recently have aspects of costs and profitability, as well as the perspectives of the service buyers and equipment providers with service offerings, been considered.

The role of maintenance is significant and will continue to grow. All equipment grows old, but there is an increasing lack of maintenance within companies. For one thing, it is expensive: the annual investment in maintenance is a significant item in the budget. Recently, more companies have turned to outsourcing in an attempt to reduce costs. But the outsourcing trend has created new kinds of decision-making problems. These require increased cost awareness, and decision-makers must consider value production throughout the whole life cycle of a plant, not just an asset, as well as asset-level maintenance services in multi-operator networks.

With today's economic situation, companies may be unwilling to make large investments in production equipment, causing maintenance costs to rise. The increasing maintenance costs can partly be compensated for by turning to active networks and cooperation between companies. In the very near future, one of the largest challenges in industrial maintenance will be to make the profitability information and value thinking transparent on a network level. However, this transparency will provide the tools required for effective maintenance and proactive maintenance management.

6.1.2 Performance Measurement of Maintenance Services

The measurement of diverse parameters associated with maintenance services plays an important role in assessing, for example, the quality of customer service or the effectiveness and availability of the service. The measurement and assessment of services can also be used in business management and control. Finally, measurement can be used for the identification of customer value created by a service, using, for example, life cycle costing.

Measuring service production is traditionally based on measuring service quality (e.g., errors), time (e.g., availability and accessibility of the service), and costs (e.g., remaining at a specific cost level) (Cameron and Duckworth, 1995). Service production is a complex and multidimensional entity, making its measurement difficult; the direct output of the service is often insufficient as an end result. In addition, the benefits and effects which create customer value have to be taken into account (Lönnqvist et al., 2010).

Vargo and Lusch (2004) emphasize service-oriented thinking, suggesting the use of a progressive and desirable model with more advanced indicators. Service-oriented thinking emphasizes the value-in-use received by the customer who uses the service (e.g., the customer receives a solution for a specific problem), instead of the nominal value-in-exchange (e.g., the price for purchasing the service). The measurement of maintenance services and their value has been seen as a significant challenge (Grönroos and Ojasalo, 2004; Berry and Bendapudi, 2007). The general challenges of measurement also hold in the service production context, but the service context creates the following extra challenges for measurement (Lönnqvist et al., 2010):

- The output is difficult to define because services are intangible, and the importance of qualitative factors is emphasized.
- The produced services vary in service content and quality. An overhaul operation, for example, can be extremely customized, making it difficult to compare produced services.
- The content of services is not stable but changes over time. It is difficult to develop indicators which take this change into account.
- The role of the customer may be very important to the contribution of resources. For example, the production unit's personnel may participate actively in a maintenance operation.
- The customer must appreciate the usefulness of the outputs. The direct output of the service itself is not enough, as the services must also cause the desired effect.

Some researchers have given practical advice on measuring service production. McLaughlin and Coffey (1990) recommend dividing services into subsets and exploring the quality, time, and costs of the different subsets. Jääskeläinen (2010) proposes a framework which can be applied to developing a measurement system for service production. The framework includes the main contingencies and provides the user with a list of the issues to be covered in the construction of a measurement system.

Before constructing a measurement system, the main value elements of the provided service should be identified to make the development of the indicators easier. This way, the main elements of customer value can be measured. If proper measures are developed, the service will be high quality and will meet the customer's expectations.

Because of their intangibility and high customer orientation, services cannot be measured through numerical values only. Instead, qualitative and subjective assessment methods have to be adopted, for example, to identify the customer value and assess the quality of the service process (Bititci et al., 2012).

6.1.3 Impact of Servitization on Life Cycle

Servitization has a huge impact on the life cycle. The term "life cycle" has been used extensively in the previous chapter. Life cycle costs include all the cost items of an object from design to decommissioning. The focus is often on life cycle cost (LCC) assessment, with life cycle profits (LCP) excluded from the calculations. However, adding maintenance costs to the assessment becomes justified through aspects like shortened idle time or improved dependability (Kärri et al., 2011). Jonker and Haarman (2006) emphasize the importance of creating value in maintenance by using a life cycle perspective.

The equipment provider, maintenance company, and customer have traditionally conducted life cycle calculations from their own points of view. According to the theories of business networks, however, life cycle modeling should include all perspectives to ensure benefits for all network members.

Life cycle thinking is typically essential in decision-making situations like pricing, purchasing decisions, product design, and replacement of production equipment. Life cycle accounting can be a resource for the equipment provider seeking to sell a product, and it can help the customer decide to buy. The most interesting situations are those in which a service provider and a customer use life cycle models together to search for the most appropriate and economical purchasing entities and service packages, considering the whole life cycle period.

Life cycle thinking is also present in the general financing of large investment projects, such as public–private partnerships. In this financing model, the service entity provided by the supplier includes, for example, the building of the object, the object's ownership and operation, and the transfer of control. In this kind of business model, the role of the equipment provider is more significant than in traditional business (Pekkarinen et al., 2012). Today, individual equipment providers are interested not just in delivering equipment and related maintenance services, but also in conducting operation services for their customers.

A common challenge of asset owners and maintenance professionals is to keep production equipment and assets available and productive. In other words, they have to stop the increase in the lack of maintenance and decrease the accrued need for maintenance operations. To reverse the trend, investments should be allocated to areas where the value and effectiveness of the investment will be the highest. Life cycle thinking should be adopted more extensively, especially given today's financial, social, and environmental concerns. Finally, maintenance networks should exploit the perspectives of life cycle accounting.

6.1.4 Life Cycle Accounting in Business Networks

The discussion of value-based pricing has recently contributed to the research field of maintenance service pricing. As competition and cost awareness increase, customers are increasingly questioning the fixed pricing of services. Even though the fixed prices have meant high-cost controllability from the customer's point of view, customers are now interested in the value-creating ability of the purchased maintenance service. Value-based pricing can be based, for example, on the reliability of production equipment, the realized volume of production, or the actual usage of the service. Value-based pricing is, however, challenging for the service provider. The value elements of the service can vary remarkably depending on the customer company. In addition, value-based pricing can include some additional risks to the service provider. This emphasizes the need to observe the situation constantly, so that the value, as well as the risk, created by the service can be divided fairly between network members.

A service offering includes all the value elements that the company is able to deliver to its customers. In product-oriented business, the service offering has generally been built around products and their supportive services (e.g., Anderson and Narus, 1999). In the service business, services are classified more specifically in different classes. For example, the concept proposed by Grönroos (2000) is based on consumer marketing; and it includes a core service, assistant services, and support services. Boyt and Harvey (1997) categorize industrial services into three groups based on the activity of the seller–customer relationship: basic services, average-level services, and complex services. In project business, an increasingly studied topic, the service offering is considered to consist of four elements: core product support services, customer support services, services supporting the customer's other functions, and services supporting the customer's network (Cova and Salle, 2000). This reflects the high level of networking in industrial markets, and it takes the role of the networks into account while planning the service offering. The service offering should contain all the value elements that the service provider is able to offer both to its customer and to the whole network. In some novel business models, the role of financial elements (e.g., financial models) has increased in the construction of service offerings. A modern, solution-oriented offering may include product elements, service elements, and financial elements (Pekkarinen et al., 2012).

The main goal in these new scenarios is to study how the value of a maintenance service can be identified and evidenced to each of the business network members (customer, maintenance company, and equipment provider),

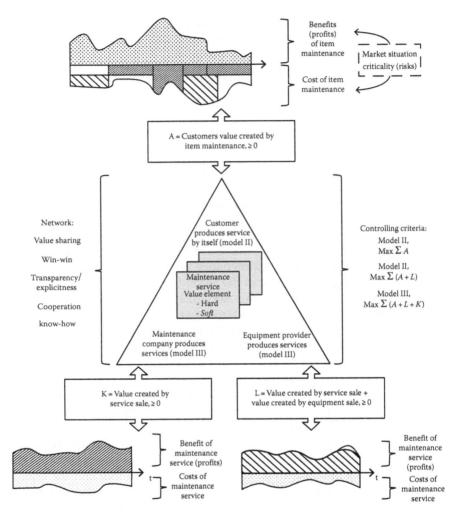

FIGURE 6.2
Value-based life cycle model of maintenance services. (Adapted from Sinkkonen, T. et al., *Int. J. Ind. Syst. Eng.*, 2016.)

Finally, the independent maintenance company can be responsible for producing services on either the asset level or the level of the whole factory. The role of the maintenance company in the network is justified if the company is able to create additional value for the maintenance service packages, higher than what the customer and the equipment provider can create either alone or in cooperation. Figure 6.2 shows this increased value schematically.

From the perspective of managing the whole network, it is a question of using life cycle thinking to gain a better overall view of the creation of value and the distribution of the value between various network members. Each network member may aspire to examine the created value only from his/her own point of view, but if the customer works in close cooperation with the equipment provider, together they can create value.

The most interesting situation is the one in which all the network members try to create value collaboratively, using their potential know-how to improve the competitiveness of the whole network (Sinkkonen et al., 2016).

6.1.5 Managing Value in the Network

Measurement is a significant tool of network management. The essential parts of developing a measurement system are identifying the essential value elements of measurement and management from the perspective of managing maintenance services, and determining how the measurements can be executed in practice.

Hence, the measurement system to be developed should integrate different value elements and diverse measurement practices, while helping to manage maintenance service networks. Both hard, ratio-based measurements and soft assessments based on subjective surveys and/or mapping should be included in the measurement system.

Several analytical models have been developed to examine how different network members are capable of creating value from the perspective of profitability asset management (AM). Benchmarking studies between networks can be conducted with these models, and they incorporate the measurement system into the management of the maintenance networks.

This results in the diverse management of a network, with all the important aspects of added value and profitability taken into account. The models emphasize the importance of flexible AM as the demand changes.

In the models, AM is seen as an optimization problem of both fixed assets and working capital. Previous research concludes that maintenance companies can achieve a significant competitive advantage by keeping their fixed assets small and by managing their working capital efficiently (Marttonen et al., 2013).

So far, the models have not been applied in the network context, where asset optimization could become a mutual goal of all companies/providers in the network. This will undoubtedly become a common scenario and business configuration over the next few years, as companies and providers seek flexible AM options with mutually benefiting decision-making situations.

6.1.6 Mutation of Maintenance toward Asset Management

The maintenance world is evolving just like anything else. New methods, processes, and techniques are constantly being developed, tested, and implemented. Every company wants to produce as much as possible, at the lowest cost, with the highest return, at the best efficiency rate, and, of course, without running their assets into the ground. The current trend is to slowly eliminate maintenance management and replace it with AM.*

---

* http://www.reliableplant.com/Read/8801/maintenance (accessed June 19, 2016).

Maintenance today contributes to the aim of sustainable development, including environmental and energy-saving aspects, safety aspects, and economic aspects. Automation and integrated production have resulted in larger technical systems, which are more difficult to control, and more sensitive and vulnerable to breakdowns. Advanced maintenance has a critical role if a company is to remain competitive, particularly when safety and availability are important.

Reliability, availability, and lifetime planning were first advanced in the nuclear energy industry. The aerospace industry quickly followed, developing methods to assure reliability by distributing and duplicating the crucial features of reliability. Reliability analyses have now been adopted in most industrial fields.

However, existing methods are not always easily applicable to power plants, or to the process and metal industries, where availability is often a more important criterion than reliability. In other words, downtime is more important than a small probability of failure. Failure can be acceptable if the repair and restarting times are short. Maintainability and maintenance support performance are extremely important in such cases.

Traditionally, the manufacturer guaranteed the faultless action of a product for a certain warranty period, but life cycle profit (LCP) planning is gaining popularity. LCP is based on the reliability of a product during its whole lifetime. Statistically defined failure frequency, availability, and product lifetime can now be used to ensure competitiveness.

Today's product design methods are mainly based on optimizing the performance of products, with little attention given to reliability and lifetime estimations. Few design tools emphasize reliability and availability. Figure 6.3 shows the influence of a wide range of technological advances over the last two decades. Because of the great variety of different techniques, there is a need to approach reliability and maintainability problems from a more holistic point of view, starting with the customer and ending with the user (Holmberg et al., 2010).

The Technical Research Centre of Finland (VTT) has developed a systematic approach (Holmberg, 2001; Holmberg and Helle, 2008) to improve synergistic interactions between various fields of expertise. In its logical and comprehensive structure, each expert can find his or her place and see the connections to experts from other fields, all working with the same aim of a satisfied end user, as shown in Figure 6.4 (Holmberg et al., 2010).

The probability of personnel, equipment, and environmental damage can be analyzed and the accident consequences estimated by systematic study risk control. The critical parts are identified, the probability of system failure and lifetime are calculated, and the operability costs are estimated by statistically based techniques of reliability control. When the critical parts of the production system requiring improvement have been identified, the right techniques and tools for improvement actions can be located in the fields of

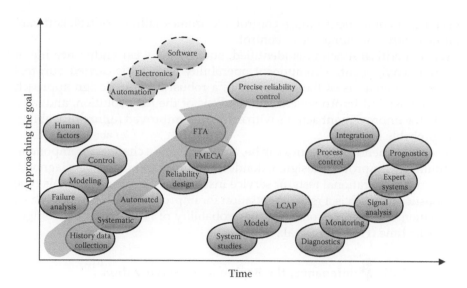

FIGURE 6.3
Fusion of advanced maintenance technologies. (Adapted from Holmberg, K. et al., Chapter 2: Maintenance today and future trends, in: *e-Maintenance*. Springer, XXV, 2010, 511p., Hardcover, http://www.springer.com/978-1-84996-204-9.)

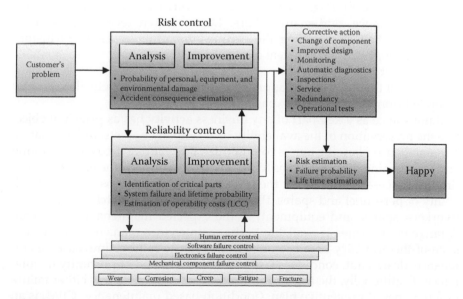

FIGURE 6.4
Holistic approach to maintenance integration. (Adapted from Holmberg, K. et al., Chapter 2: Maintenance today and future trends, in: *e-Maintenance*. Springer, XXV, 2010, 511p., Hardcover, http://www.springer.com/978-1-84996-204-9.)

mechanical component failure control, electronics failure control, software failure control, or human error control.

When a critical function is identified, such as the wear endurance life of a certain component, a component operability analysis is carried out; this includes an analysis of the old solution, a robust lifetime design approach to recommended improvements, an analysis of the new solution, and, as a result, the improvement actions with estimated improved failure probability and probable lifetime.

The recommended measures can be, among others, a change of component, redundancy, improved design, extended monitoring, automatic diagnosis, inspections, operational tests, or service instructions. The output of this type of holistic approach is a recommendation for improvement, together with an estimation of the effects on risks, the probability of failure, and the component's lifetime (Holmberg et al., 2010).

6.1.7 Holistic Maintenance, the Beginning of Servitization

The concept of holistic maintenance has been developed extensively during recent decades. As noted earlier, a significant driver was the reliability of the nuclear power and aircraft industries, but the competiveness of the manufacturing industry, coupled with improvements in transport areas, has led to a range of integrated program philosophies and holistic planning processes. These include reliability-centered maintenance (RCM), total productive maintenance (TPM), total quality maintenance (TQM), lean maintenance, and many others. Data-oriented techniques such as proportional hazard modeling allow some integration of the asset's history (Jardine et al., 1998). A good deal of proprietary know-how is also available in the market.

The aim of these integrated approaches is to offer a complete plan, usually compiled from the strategies expanded here.

Maintenance as a key part of any business activity has as principal objective the preservation of the availability of a company's assets. In formulating a maintenance plan, the aim is to minimize the combined cost of operating and maintaining the asset. Plans for maintenance consider the need for a continuous process of operations, the possibility of breakdowns, and the availability of personnel and spares. Planning is required to match the resources (workers, spares, and equipment) to the expected maintenance workload. A range of strategies is available to the maintenance manager. The current state-of-the-art policy uses a combination of run-to-failure, time-based maintenance, design out, condition-based maintenance, and opportunity maintenance. Traditionally, the trigger for initiating maintenance was either failure or a time-based preventive plan. Condition-based maintenance (CBM) is an improved method of preventing failures, based on detection of asset deterioration (Holmberg et al., 2010).

6.2 Impact of Business and Technological Environment on Maintenance Costs

Maintenance is generally considered a cost center, not a business opportunity. However, the understanding of maintenance business is developing, as more companies recognize that maintenance processes have more margin for optimization than operation processes. Financial indicators considering the return on investment (ROI)(e.g., return on inversion, return on net assets (RONA), overall equipment effectiveness (OEE)) (EN-15341, 2007) can be a good starting point to understand the three main areas of productivity: availability, quality, and craft (human) effectiveness.

Cost efficiency in carrying out maintenance operations is still emphasized even though reliability and availability have improved considerably in most companies (AEM, 2005). One indicator used to measure this efficiency is "wrench time." This indicator measures overall craft/human effectiveness (Peters, 2003) and represents the percentage of time an employee applies physical effort to a tool, equipment, or materials in the execution of assigned work. It is used to define how well organized the company is at planning, scheduling, and performing work. According to Wireman (2003), a representative wrench time in many organizations in the United States is 35% of the total maintenance time.

An increasing number of companies recognize maintenance as a natural way to extend operation and productive processes, and are looking for an opportunity to move toward service business. The main idea is to become less vulnerable to market changes. The trend toward services is part of a trend toward more sustainable business models, with environmental aspects becoming economically significant. This new model looks for lower costs for raw materials, energy, and recycling, as shown by Takata et al. (2004). The trend in cost reduction in information and communications technologies (ICT) enables their use in an increasing number of scenarios. This is in line with reductions in the price of commodities (Cashin and McDermott, 2002) and multiple products that facilitate capture, transmit and analyze data and information in fast, distributed, and economical ways. New standards have appeared that facilitate a way to understand how different technologies may interoperate (Thurston and Lebold, 2001).

All this new technology available for maintenance and operations activities is affecting how operators do their work. Many studies show it is beneficial to use e-Maintenance and CBM, and even small steps can be economically justified. The transformation from corrective to predictive maintenance strategies will lead to a decrease in breakdowns causing production stoppages and an increase in availability. Whereas condition-monitoring systems have conventionally concentrated on minimizing the accidents caused by ineffective maintenance. The capability of equipment to produce quality products, for example, products that satisfy customer requirements, is greatly affected by maintenance effectiveness.

Maintenance also affects the technical performance of the production department. Technical performance can be assessed by determining overall equipment effectiveness (OEE) in TPM or by using a modified version of OEE, overall process effectiveness (OPE). Efficient maintenance adds value through better resource utilization (higher output), enhanced product quality, and reduced reworking and scrapping (lower input production costs). In addition, it avoids the need for additional investment in capital and people by expanding the capacity of existing resources (Alsyouf, 2004). A company's capital investments are influenced by maintenance-related factors, such as equipment/component useful life, equipment redundancy, extra spare parts inventory, buffer inventory, damage to equipment due to breakdown, extra energy consumption, and so on.

Finally, maintenance can have an impact on customers, society, and shareholders. Deardeen et al. (1999) says firms try to capture new customers, satisfy them, and retain them by giving them assurance of supply on time. This depends on adequate production capacity with minimum disturbances and high-quality products. The impact of maintenance on society is reflected in its effects on safety, the environment, and ecology (Alsyouf, 2004). And the impact of maintenance on shareholders can be traced by analyzing the effect of maintenance on generated profit, usually measured by indexes such as return on investment (ROI) percentage (Ahlmann, 1984, 1998).

In other words, there is a need to know how to assess the impact of maintenance practices on a number of performance outcomes (Mitchell et al., 2002).

6.2.3 Industry 4.0 and Maintenance 4.0

Assets are complex mixes of complex systems. Each system is built from components which, over time, may fail. Previously, the diagnosis of problems occurring in systems was performed by experienced personnel with in-depth training and experience (Chiang et al., 2001). Computer-based systems are now being used to automatically diagnose problems to overcome some of the disadvantages associated with relying on experienced personnel (Price and Price, 1999). Asset management (AM) enables the realization of value from assets throughout the full life cycle. It involves coordinated and optimized planning, asset selection, acquisition/development, utilization, care (maintenance), and ultimate disposal or renewal of the appropriate assets and asset systems. The layered structure of AM is presented in Figure 6.6.

ISO 55000 is an international standard for asset management launched in 2014. It is based on PAS 55 (Publicly Available Specification) published by British Standard Institution in 2004, which gives guidance for good practices in physical asset management. ISO 55000 extends the definition of assets as it covers any asset that creates value for the organization, not just physical production assets.

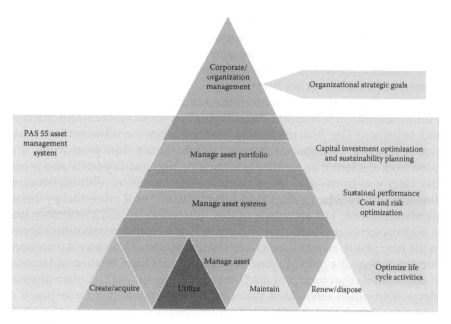

FIGURE 6.6
Layered structure of asset management, based on PAS55. (Adapted from Woodhouse, J., Setting a Good Standard in Asset Management, 2013, pp. 27–30.)

The standard provides a framework for developing an AM system and enables organizations to achieve business objectives through the effective and efficient management of their assets (ISO 2014) (Koskinen et al., 2016).

Data relevant to AM are gathered, produced, and processed on different levels by different IT systems, for example, enterprise resource planning (ERP) for business functions; supervisory control and data acquisition (SCADA) for monitoring process and controlling the asset; computerized maintenance management system (CMMS) and condition monitoring (CM) for maintenance functions; and safety instrumented systems (SIS) for safety-related functions. The challenge is to provide intelligent tools to monitor and manage assets (machines, plants, products, etc.) proactively through ICT, focusing on health degradation monitoring and prognosis instead of fault detection and diagnostics. Maintenance effectiveness depends on the quality, timeliness, accuracy, and completeness of information related to machine degradation state. This translates into the following key requirements: prevention of data overload, ability to differentiate and prioritize data (during collection as well as reporting), and prevention, as far as possible, of the occurrence of information islands (Koskinen et al., 2016).

While the use of a good version of either technology can help to achieve the maintenance goals, combining the different data sources into one seamless system can exponentially increase the positive effects to the operations and maintenance (O&M) group's performance. Only a few years ago, the

6.2.4.1 CBM, the First Step to the Fourth Industrial Revolution in Maintenance

The main and first outcome of the analysis of huge amounts of maintenance data is usually reliability prediction based on failure information and maintenance actions. Such preventive maintenance schemes as block replacement and age replacement are usually time-based and do not consider the current health state of the assets, making them inefficient and less valuable (Lee et al., 2006).

In contrast, condition based maintenance aims to avoid failure by detecting early deterioration and spotting hidden or potential failures. CBM initiates maintenance when machine condition deteriorates. It uses degradation analysis to investigate the evolution of the physical characteristics, or performance measures, of an asset leading up to its failure. The component or equipment is usually replaced or repaired as soon as the monitored level of the performance measure exceeds the normal level. CBM offers the following benefits (Holmberg et al., 2010):

- Better planning of repairs is possible, reducing out-of-production time.
- Inconvenient breakdowns and expensive consequential damage are avoided.
- Failure rate is reduced, improving asset availability and reliability.
- Spares inventory is reduced.
- Unnecessary work is avoided, keeping the repair team small but highly skilled.

CBM methods and practices have continued to improve over recent decades and are used in a wide range of industries. For example, a major manufacturer of elevators for high-rise buildings will continuously monitor the braking systems and acceleration and deceleration of elevators globally to meet high safety requirements.

The main idea of CBM is to use the product degradation information extracted and identified from online sensing techniques to minimize the system downtime by balancing the risk of failure with the achievable profits. The decision-making in CBM focuses on predictive maintenance, with many diagnostic tools and methods now available. Sensor fusion techniques are commonly used because of the inherent benefits of mutual information from multiple sensors (Hansen et al., 1994; Reichard et al., 2000; Roemer et al., 2001). Because of their excellent capacity for describing machine performance, vibration signature analysis and oil analysis have been successfully employed for prognostics for a long time (Kemerait, 1987; Wilson, 1999). Alternative approaches using time/stress, temperature, acoustic emissions (Goodenow et al., 2000), and ultrasound are popular as well. However, there has been no major breakthrough in prognostics. Many methods are still based on traditional signal-processing methods (Hardman et al., 2000).

Generally speaking, current prognostic approaches can be classified into three basic groups: model-based prognostics, data-driven prognostics, and hybrid prognostics.

- An example of a characteristic model-based prognostics application includes data collected from model-based simulations under normal and degraded conditions. Prognostic models are constructed based on different random load conditions, or modes.

- In the absence of a reliable or accurate system model, another approach to determine the remaining useful life of an asset is to monitor the trajectory of a developing fault and predict the amount of time until it reaches a predetermined level requiring action. This is the data-driven prognostic method. The alpha–beta–gamma tracking filter and the Kalman filter, two well-known tracking and prediction tools, have been applied to gearbox prognostics (McClintic, 1998; Ferlez and Lang, 1998). Both filters have been investigated for their ability to track and smooth features from gearbox vibration data. The literature mentions the use of Kalman filters to track changes in features like vibration levels, mode frequencies, or other waveform signature features; they can also be used to estimate future hazard rate, probability of survival, and remaining useful life (Swanson, 2000, 2001).

- A hybrid method fuses the model-based information and sensor-based information and takes advantage of both model- and data-driven methods (Hansen et al., 1994). This guarantees a more reliable and accurate prognostic result.

The goal of the raw data analysis is to identify information that does not represent the normal asset working conditions and/or outliers that cannot be obtained because of problems in the condition-monitoring system.

Based on these ideas and considering that the monitored variables, that is, the fingerprint features, are available, all that is needed to perform a measurement-based strategy is to define the maintenance threshold, that is, the asset state or condition in which the predictive maintenance action is conducted. In this case, the maintenance threshold is equal to the fingerprint feature threshold because the maintenance action will only depend on the asset health condition. As an example, a predictive maintenance action should be performed every time the fingerprint feature reaches its threshold. The maintenance action can be expected to recover totally or at least partially the fingerprint feature value, and the asset will be restored to its normal health condition.

The machine operational condition can be monitored as well; this allows the identification of machine usage profiles based on the accumulated power consumption (calculated KPI).

6.2.4.2 PHM, One Step Ahead of CBM

Maintenance combines various methods, tools, and techniques in a bid to reduce maintenance costs while increasing the reliability, availability, and security of equipment. The most common type of maintenance is corrective maintenance (also called unplanned maintenance, or run-to-failure maintenance), which takes place only at breakdowns. Time-based preventive maintenance (also called planned maintenance) sets a periodic interval to perform preventive maintenance regardless of the health status of a physical asset.*

With the rapid development of modern technology, products have become increasingly complex, while better quality and higher reliability are required. This raises the cost of preventive maintenance. Preventive maintenance has become a major expense of many industrial companies, leading them to implement CBM (Chawla and Kumar, 2013).

CBM is designed to avoid unnecessary maintenance tasks. A very important element of the maintenance strategy, especially when deploying CBM, is called prognostics. Prognostics is of great interest to both industry and research centers as it can significantly improve the efficiency of a CBM program. Prognostics basically tries to predict how much time remains before a fault or failure will occur, given the current situation of the asset and the operation. In other words, prognostics is based on predicting the remaining useful life (RUL) of a system before a failure occurs.

With prognostics, industries can potentially reduce the costs of both preventive and corrective maintenance. If they can predict when a machine or asset may fail, they will be able to prevent or limit maintenance activities. Prognostics is mainly based on mathematical models for predicting the remaining useful life. These models are built using one of four approaches: experience-based, data-based, model-based, or hybrid methodology (Hernandez and Galar, 2014).

Figure 6.7 illustrates the hierarchy of possible approaches in relation to their applicability and relative costs.

A typical example is the use of models for modeling fatigue initiation and propagation of cracks in structural components. Other examples include the study of the estimation of RUL for bridge structures using a system based on

---

* https://www.maintenanceassistant.com/run-to-failure-maintenance/ (accessed June 21, 2016).

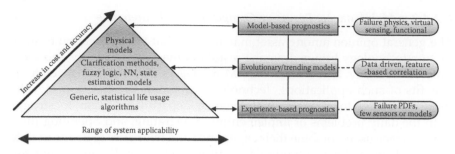

FIGURE 6.7

Hierarchy of prognostics approaches. (Adapted from Hernandez, A. and Galar, D., Techniques of prognostics for condition-based maintenance in different types of assets, Technical report, Department of Civil, Environmental and Natural Resources Engineering, Operation, Maintenance and Acoustics, Luleå University of Technology, Luleå, Sweden, 2014, ISSN: 1402-1536.)

experience or a study of vibrations in rotating machines using a data-driven approach (Hernandez and Galar, 2014).

PHM is focused on the detection of incipient faults, the assessment of the current health of the system, and the prediction of the remaining useful life in a component. This approach has been accelerated by the fact that many systems, within their design, now include devices (sensors, controllers, software, and intelligent algorithms) that enable the development of a methodology for maintenance scheduling with a certain level of automation and dynamism in terms of decision-making and with a better assessment of actual operating conditions.

The PHM concept and structure have been developed using well-known maintenance models, diagnosis and prediction techniques, such as preventive maintenance, reliability-centered maintenance (RCM), and condition based maintenance (CBM). The PHM should avoid preventive maintenance actions with high and unnecessary costs, optimizing the maintenance plan with a consequent reduction in the need for material for unscheduled substitutions. Once the most important failure modes are detected, the variables used to provide information about the presence of these possible failure modes are identified too. They can be used to develop normal behavior models (reference models), which will be later used for the estimation of behavior performance.

The detected, incipient fault condition should be monitored, trended from a small fault as it progresses to a larger fault, until it warrants some maintenance action and/or replacement. By employing such a system, the health of a machine, component, or system can be known at any point in time, and the eventual occurrence of a failure can be predicted and prevented, enabling the achievement of near-zero downtime (Lee et al., 2014).

6.2.4.3 e-Maintenance: Pervasive Computing in Maintenance

The general opinion among asset managers is that the application of information technology brings dramatic results in machine reliability and O&M process efficiency; however, few O&M managers can show or calculate the benefits of such applications. Technology providers are trying to develop more advanced tools, while maintenance departments seem to struggle with the daily problems of implementing, integrating, and operating them. Maintenance users combine their experience and heuristics to define maintenance policies and to use condition monitoring systems. The resulting maintenance systems are a heterogeneous combination of methods and systems in which the integrating factor between the information and business processes is the personnel. Information about the assets is human-based, forming an organizational information system and creating a high reliance on the expertise of the AM staff (Koskinen et al., 2016).

The contribution of O&M data to the different asset management stages has been triggered by the emergence of intelligent sensors for measuring and monitoring the health state of a component, the gradual implementation of ICT, and the conceptualization and implementation of e-maintenance. Two main systems are deployed in maintenance departments today: computer maintenance management systems (CMMS) are the core of traditional maintenance record-keeping practices and condition-monitoring systems (CM) directly monitor asset component parameters. Attempts to link observed CMMS events to CM sensor measurements have been fairly limited in their approach and scalability. Moreover, information from SCADA (mainly used in operation for supervisory purposes) is seldom fused with other information (Stene-erik et al., 2013).

During the last decade, global competition and the advancement of ICT have forced production and process industries to continuously transform and improve. The business scenario is focusing on e-business intelligence to perform transactions with an emphasis on customers' needs for enhanced value and improvement in AM. Such business requirements compel organizations to minimize production and service downtime by reducing machine performance degradation. The organizational requirements stated in ISO 55000 necessitate the development of proactive maintenance strategies to provide optimized and continuous process performance with minimized system breakdowns and maintenance. With these changing systems of the business world in the twenty-first century, a new era of data e-services coming from asset data has emerged (Koskinen et al., 2016).

A top-shelf CMMS can perform a wide variety of functions to improve maintenance performance. CMMS is primarily designed to facilitate a shift in emphasis from reactive to preventive maintenance (PM) by allowing maintenance professionals to set up an automatic PM work order generation. CMMS can also provide historical information, which is then used to adjust the PM system setup over time to minimize repairs that are unnecessary, while still avoiding run-to-failure repairs. PM for a given piece of equipment can be set up on a calendar schedule or a schedule that uses meter readings. A fully featured CMMS also includes inventory tracking, workforce management, and purchasing in a single package that stresses database integrity to safeguard vital information. The final result is optimized equipment uptime, lower maintenance costs, and better overall plant efficiency. In summary, CMMS contains asset hierarchy and is an excellent starting point for taxonomy but not ontology (Galar et al., 2015).

In contrast, CM and SCADA systems accurately monitor the real-time equipment performance and condition in order to alert operators and maintenance professionals of any changes in performance or behavior trends. A CM and SCADA package can track a variety of measurements, and the very best CMs are expert systems that can analyze measurements like vibration and diagnose machine faults. Such expert system analysis puts maintenance procedures on hold until absolutely necessary, thus extracting maximum equipment uptime. In addition, the best expert systems offer diagnostic fault trending where individual machine fault severity can be observed over time. SCADA systems provide information on the performance of the asset and screenshots of a number of performance indicators at a time (now commonly called casting). This information, together with a number of alarms and incidences linked to sensor thresholds, has been historically used for process monitoring but never fused with other information to get immediate benefits (Koskinen et al., 2016).

SCADA, CMMS, and CM systems have strong benefits that make them indispensable to O&M improvements. CMMS is a great organizational tool but cannot directly monitor equipment conditions, whereas a CM system excels in monitoring those conditions but is not suited for organizing overall maintenance operations (Stene-erik et al., 2013). Finally, SCADA collects huge amounts of data for supervision of the process but is not integrated in a holistic way for AM. The logical conclusion is to combine these technologies into a seamless system to avoid catastrophic breakdowns and eliminate needless repairs to equipment that is running satisfactorily, while at the same time optimizing the operation (Koskinen et al., 2016).

So far, integration has been addressed largely from the viewpoint of representing the collected information to the end user (operator or manager) in an effective manner, that is, bridging the gap between information platforms and increasing the popularity of the e-maintenance concept, which uses such integration (Galar et al., 2016).

The awareness and use of e-maintenance has grown during the last few years. The hardware and software have developed rapidly, and it is increasingly possible to access Internet wirelessly. At the same time, manufacturers are moving toward taking care of the machinery they have produced all the way through its life cycle. For this, they need cost-effective tools, and this is where e-maintenance can help. Accordingly, researchers are focusing on e-maintenance (Jantunen et al., 2011). e-Maintenance advantages can be slotted into the three following categories:

- Maintenance types and strategies
- Maintenance support and tools
- Maintenance activities

6.2.4.3.1 Improvement in Maintenance Types and Strategies Using e-Maintenance
Potential improvements can be summarized as follows:

- *Remote maintenance*: By leveraging wireless (e.g., Bluetooth) and Internet technologies, users may log in from anywhere and with any kind of device as soon as they get an Internet connection and a browser. Any operator, manager, or expert has the ability to remotely link to a factory's equipment through the Internet, allowing them to take remote actions, such as setup, control, configuration, diagnosis, debugging/fixing, performance monitoring, and data collection and analysis. Consequently, the manpower retained at the customer's site can be reduced. In addition, there are facilities to diagnose problems when an error occurs. Finally, preventive maintenance can be improved by machine performance monitoring (Muller et al., 2008).

One of the greatest advantages of e-maintenance is the ability to connect field systems with expertise centers located at distant geographical sites, allowing a remote maintenance decision-making that adds value, trims expenses, and reduces waste. This makes the development of an asset information management network a sound investment (Baldwin, 2004).

Web enablement of computerized maintenance management systems (e-CMMS) and remote condition monitoring or diagnostics (e-CBM) avoid the expense of software maintenance, security, and hardware upgrade. Computer science experts can add new features and/or migrations without users noticing (Muller et al., 2008).

- *Cooperative/collaborative maintenance*: e-Maintenance represents the opportunity to implement an information infrastructure connecting geographically dispersed subsystems and actors (e.g., suppliers with clients and machinery with engineers) on the basis of existing Internet networks. The resulting platform facilitates cooperation between different human actors, different enterprise areas (production, maintenance, purchasing, etc.), and different companies (suppliers, customers, machine manufacturers, etc.).

An e-maintenance platform introduces an unprecedented level of transparency and efficiency (Figure 6.8), and it can support business process integration. As a result, there is the chance to radically reduce interfaces between

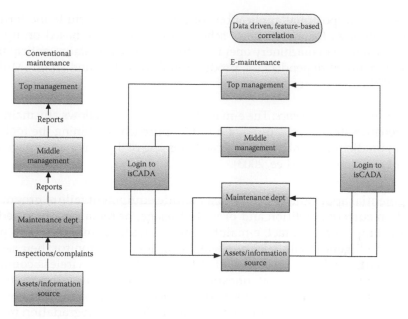

FIGURE 6.8
Implementing e-maintenance. (Adapted from Muller, A. et al., *Reliab. Eng. Syst. Saf.*, 93, 1165, 2008.)

personnel, departments, or even IT systems. The integration of business processes significantly contributes to the acceleration of total processes, to an easier design (lean processes), and to the synchronization of maintenance with production, maximizing process throughput and minimizing downtime costs. In general, this leads to fewer process errors, improved communication processes, shorter feedback cycles, and improved quality (Muller et al., 2008).

In short, e-maintenance facilitates the bidirectional flow of data and information into the decision-making and planning process at all levels. By so doing, it automates the retrieval of accurate information required by decision-makers to determine which maintenance activities to focus resources on, so that the return on investment (ROI) is optimized (Moore and Starr, 2006).

- *Immediate/online maintenance*: The real-time remote monitoring of equipment status, coupled with programmable alerts, enables the maintenance operator to respond to any situation swiftly and optimally prepare any intervention. In addition, maintenance operators obtain expertise quickly; the feedback reaction in the local loop is accelerated, connecting product, monitoring agent, and maintenance support system. The online guidance based on the results of decision-making and the analysis of asset condition has almost unlimited potential to reduce the complexity of traditional maintenance guidance (Goncharenko and Kimura, 1999).

In this context, potential applications of e-maintenance include the formulation of policies for maintenance scheduling in real time based on up-to-date information of machinery operation history, machine status, anticipated usage, functional dependencies, production flow status, and so on (Muller et al., 2008).

- *Predictive maintenance*: The e-maintenance platform allows any maintenance strategy; the use of a holistic approach combining the tools of predictive maintenance techniques is one of the major advantages of e-maintenance (Lee, 2003).

The potential applications in this area include equipment failure prognosis based on current condition and projected usage, or remaining life prediction of components. In fact, e-maintenance provides companies with predictive intelligence tools (such as a watchdog agent) to monitor their assets (equipment, products, processes, etc.) through Internet wireless communication systems to prevent unexpected breakdown. In addition, these systems can monitor an asset's performance using globally networked monitoring systems; this allows companies to focus on degradation monitoring and prognostics rather than fault detection and diagnostics (Iung et al., 2003).

Prognostic and health management systems that can implement the capabilities mentioned earlier will be able to reduce the overall life cycle costs (LCC) of operating systems and decrease the operations/maintenance logistics footprint (Roemer et al., 2005).

6.2.4.3.2 Potential Improvements in Maintenance Support and Tools Using e-Maintenance

Potential improvements can be summarized as follows:

- *Fault/failure analysis*: The rapid development in sensor technology, signal processing, ICT, and other technologies related to condition monitoring and diagnostics increases the possibility of using data of different types from multiple origins and sources. In addition, by networking remote plants, e-maintenance provides a multisource knowledge and data environment.

These new capabilities allow the maintenance area to improve understanding of causes of failures and system disturbances, to create better monitoring and signal analysis methods, and to improve materials, design, and production techniques. By using failure analysis, companies can move from failure detection to degradation monitoring (Holmberg et al., 2005).

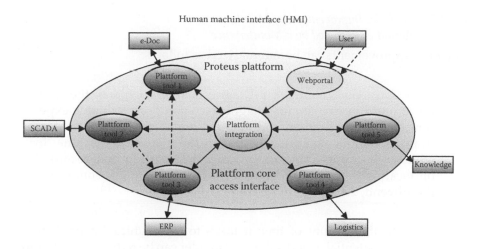

FIGURE 6.9
PROTEUS platform. (Adapted from Bangemann, T. et al., *Comput. Ind.*, 57(6), 539, 2006.)

- *Maintenance documentation/record*: The e-maintenance platform provides a transparent, seamless, and automated information exchange process to access all the documentation in a unified way, independent of its origin, whether from the equipment manufacturer, integrator, or end user. For example, a task completion form can be filled out by users and dispatched to several listeners (software or humans) registered to receive such events (Bangemann et al., 2006). Figure 6.9 shows how this type of e-maintenance platform works, using the PROTEUS platform (plateforme reconfigurable pour l'observation, pour les télécommunications et les usages scientifiques).

At the device level, goods are checked out from stores against a work order or a location, and the transaction is recorded in real time. The massive data bottlenecks between the plant floor and business systems can be eliminated by converting the raw machine health data, product quality data, and process capability data into information and knowledge for dynamic decision-making (Lee, 2003).

- *After-sales services*: With the use of the Internet and web-enabled and wireless communication technology, e-maintenance can transform manufacturing companies into service businesses able to support their customers anywhere anytime (Lee, 2001).

6.2.4.3.3 *Potential Improvements of Maintenance*
              Activities Provided by e-Maintenance

Potential improvements can be summarized as follows:

- *Fault diagnosis/localization*: e-Diagnosis offers experts the ability to perform online fault diagnosis, share their valuable experiences with each other, and suggest remedies to the operators if an anomalous condition occurs in the inspected machine. In addition, lockouts can be performed in isolation and recorded on location thanks to wireless technology and palm computing (Wohlwend, 2005).

Consequently, the amount of time it takes to communicate a production problem to the potential expert solution provider can be reduced, the quality of the information shared can be improved, and the resolution time can be reduced. These factors help to increase the availability of equipment, reduce mean time to repair (MTTR), and significantly reduce field service resources/costs (Muller et al., 2008).

- *Repair/rebuilding*: Using e-connections, remote operators can tap into specialized expertise rapidly without travel and schedule delays. Downtime can conceivably be reduced through direct interaction (troubleshooting) with source designers and specialists. In addition, diagnosis, maintenance work performed, and parts replaced are documented on the spot in structured responses to work steps displayed on a palmtop (Hamel, 2000).

- *Modification/improvement—knowledge capitalization and management*: The multisource knowledge and data environment provided by e-maintenance allows efficient information sharing, leading to important knowledge capitalization and management. With the availability of tools for interacting, handling, and analyzing information about asset state, the development of maintenance engineering for product life cycle (PLC) support, including maintenance and retirement stages (disassembly, recycling, reuse, and disposal), is becoming feasible (Goncharenko and Kimura, 1999).

  The last decade has seen the emergence of many e-maintenance techniques. Together with hardware development, the extensive use of the Internet and web service technology has laid down a new foundation for modern maintenance. Today the use of e-maintenance is still in its infancy, but its development is being encouraged by changes in the strategy of manufacturing industries. Simply stated, many now want to offer services throughout the lifetime of the equipment they have manufactured. This possibility can be supported by hardware improvement but the most significant supporting factor is the availability of new data to facilitate diagnosis and predictive health monitoring. The progress of new signal analysis techniques joined with simulation models can enable a revolution in the prediction of the lifetime of components. The technologies are obtainable today, and their introduction can easily be justified economically, but it will take some time before all available options are used. As innovators tell their success stories, more companies will adopt the new technologies (Jantunen et al., 2011).

6.3 Planned Obsolescence and the End of Traditional Maintenance

The concept of planned obsolescence says a product is designed and produced with the knowledge that it will only be popular, useful, and functional for a limited length of time.

The term was popularized in the 1950s by industrial designer Brooks Stevens, who proposed that corporations should think about a product's entire life cycle during the design phase and anticipate the consumer's natural desire to continually own something a little newer and a little better. Stevens suggested that an iterative approach to product design would not only help keep production costs down, but also inspire the consumer to make a purchase a little sooner than absolutely necessary; this, in turn, would keep the economy growing and create a secondary market for used products.

Over the years, critics have blamed the concept of planned obsolescence for everything from poor product quality to the creation of a disposable society that no longer knows what it is like to live in a world where products are "built to last." Proponents of planned obsolescence point out that technology is changing very quickly, and it is simply common sense for manufacturers to acknowledge the end of a product's life cycle and plan ahead of it.*

Obsolescence is increasingly affecting systems at an early stage of their life cycles. Availability of replacement parts is critical for operational readiness, but the wave of progress in electronics and material innovations in the past 10 years has accelerated parts obsolescence. Traditional support options are no longer effective in minimizing the risk of obsolescence and, thus, have an impact on a system's cost and availability.

---

* http://whatis.techtarget.com/definition/planned-obsolescence-built-in-obsolescence (accessed June 22, 2016).

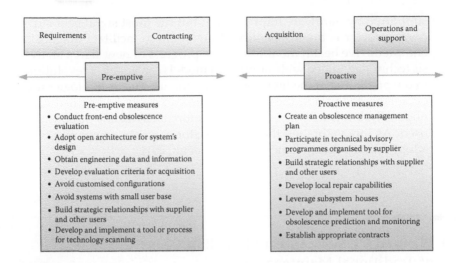

FIGURE 6.10
Measures to manage obsolescence. (Adapted from Lua, Y.A. et al., Comprehensive life cycle approach to obsolescence management, Defence Science and Technology Agency (DSTA) HORIZONS, 2012.)

It has become evident that a more comprehensive approach is needed, with obsolescence management carried out from planning to retirement. During front-end planning, measures can be taken to preempt obsolescence issues and delay their onset in the life cycle of the system. Proactive measures can also be put in place through contracting mechanisms (Lua et al., 2012).

6.3.1 Key Principle and Measures

The key principle of obsolescence management is to manage obsolescence throughout the project or system's life cycle (from front-end planning and acquisition to operation and support) to ensure the most cost-effective strategy. Depending on the project/system phase, preemptive or proactive measures can be adopted (see Figure 6.10).

6.3.1.1 Preemptive Measures

Preemptive measures should be adopted in the early phase of project implementation. Any risk of obsolescence should be identified early to avoid problems downstream. One option is to explore adopting open architecture systems, as these can be modified more easily if the need arises. Due consideration must be given to the selection of the system and the contractor. Conducting market surveys and risk assessments is a suitable method to aid the selection process (Lua et al., 2012).

6.3.1.2 Proactive Measures

Proactive measures should be adopted not only during the contracting phase but also while transiting to the operations and support phase. The project team should engage the contractors constantly to monitor any obsolescence issues. Establishing depot-level maintenance capabilities (i.e., local repair capabilities) will alleviate the impact of obsolescence. Such measures will help to establish through-life support for the system and achieve the maximum benefit for end users (Lua et al., 2012).

6.3.2 Types of Obsolescence and Terminology

Slade's *Made to Break: Technology and Obsolescence in America* looks at the various meanings of obsolescence in America from the late 1800s to the present day. His intent is to describe how America changed global culture from one with limited obsolescence to one economically dependent on it. His emphasis is on deliberate obsolescence and focuses on three major types: "technological, psychological, or planned." He explains how the influence of obsolescence has changed over time and discusses the evolution of the language surrounding it (Slade, 2006).

Table 6.1 uses Slade's book as a foundation and supplements it with other sources to define the different types of obsolescence.

6.3.3 Reasons for Obsolescence

Lynch comments: "The spectrum of obsolescence effects is broad. It encompasses not only advanced microelectronics, but also materials used in legacy systems for which there is no longer a MIL-Spec. nor a supplier; basic electrical piece parts for which the market has dwindled to the re-supply of military hardware; software which is no longer compatible, and is neither supported nor modifiable to meet today's requirements" (Lynch, 2001).

6.3.3.1 Use of Commercial Off-the-Shelf Products

For complex and costly systems like those often developed for defense systems, obsolescence is a significant issue. In 1994, US Defense Secretary William Perry proposed that absolute adherence to military specifications was at times unnecessary and costly. He encouraged, when appropriate, that commercial standards be accepted and military specifications be required only by exception. Since that time, there has been a push to use COTS components in newly developed systems for cost reduction. Of course, the benefits of cost reduction must be weighed against technological concerns occasioned by using COTS components for military applications. Interestingly, an original goal of using COTS was to help mitigate parts obsolescence. Ultimately, because of the speed of technological change in the COTS environment, the use of COTS has actually increased; all new programs have increased levels of parts obsolescence (Devereaux, 2002).

TABLE 6.1

Obsolescence Definitions and Terminology

Term	Definitions	Similar Terminology
Planned obsolescence (1)	"Here, when it is planned, a product breaks down or wears out in a given time, usually not too distant" (Packard, 1960)	Quality obsolescence
Planned obsolescence (2)	When obsolescence of components is expected and planned as part of a system's life cycle	
Psychological obsolescence	"In this situation a product that is still sound in terms of quality or performance becomes 'worn out' in our minds because a styling or other change makes it seem less desirable" (Packard, 1960)	Desirability obsolescence; Progressive obsolescence; dynamic obsolescence; Style obsolescence
Technological obsolescence	"[D]ue to technological innovation" (Slade, 2006) "[i]n this situation an existing product becomes outmoded when a product is introduced that performs the function better" (Packard, 1960)	Functional obsolescence
Human obsolescence	"[H]uman workers could be replaced by machines" (Slade, 2006) When the skill sets of experienced workers become outdated due to specialization in technologically obsolete areas	
Manufacturing and maintenance obsolescence	When there is no need in the current application for a product with increased function; however the market demands do not support a supplier's continued production or support of the legacy component	

Source: Adapted from Slade, G., *Made to Break: Technology and Obsolescence in America*, Harvard University Press, Cambridge, U.K., 2006.

Lynch says: "Somewhat perversely, the rapidity of part unavailability is less for legacy systems designed in the 1970s and 1980s than in their replacements currently being deployed or under development" (Lynch, 2001). This is partially because defense programs no longer have leverage over the development of new components. They lost this luxury when electronics, once very expensive, became commonplace and are now considered a commodity (Devereaux, 2002).

6.3.3.2 Life Cycle Duration

System life cycles are an important reason why obsolescence exists. Unlike a cell phone, which is likely to be discarded in 2–3 years, the life cycle expectations of large, expensive systems, such as defense systems, are significantly longer than most of the critical components that make up the systems themselves (Devereaux, 2002).

Petersen elaborates: "Whereas the life cycles of our weapon systems have become increasingly longer and exceed in many cases 50 years, the introduction cycles of new commercial microelectronics families average approximately 2 to 4 years, for memory devices they are as short as 9 months. And the trend is continuing" (Petersen, 2001).

It is obviously a concern if a system's components will need to be replaced many times over the system's life cycle. But there is a more complex problem associated with uneven update schedules. Such systems are often deployed in unfavorable environments, and upgrade maintenance time periods are correspondingly limited. In addition, if one item requires upgrading before the rest of the system, it can increase the number of configurations (instantiations) of a system. Multiple configurations make design, development, testing, and logistics more complex and more costly (Devereaux, 2002).

6.3.3.3 Increased Use of Electronics

Electronic systems become obsolete faster than mechanical systems. Since many functions can now be executed through software for less initial cost, power, and weight, electronic systems running software have replaced components which executed these functions using analogues and digital hardware. The chart in Figure 6.11 provides a simple example of how electronic typewriters replaced mechanical typewriters over time.

In addition, electrical systems are now able to perform functions that would have been too costly or impractical using mechanical means, increasing the overall capability of a given system.

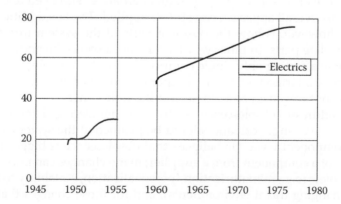

FIGURE 6.11
Electric typewriter sales as a percentage of total typewriter sales. (Adapted from Utterback, J.M., *Mastering the Dynamics of Innovation*, Harvard Business School Press, Cambridge, MA, 1994.)

Despite the advances and gains, obsolescence remains a primary drawback of electronic systems (Devereaux, 2002).

6.3.4 Obsolescence Mitigation

This section is titled "Obsolescence Mitigation" and not "Obsolescence Prevention" because it is impossible to entirely preclude obsolescence in large complex systems. In fact, as multiple sources indicate, for systems using COTS hardware or software, the obsolescence problem is far worse than in the past because the design and support of their components depend on the supplier. In this section, we examine how the literature proposes obsolescence can be mitigated in various phases of the life cycle: systems engineering, software engineering, testing, prognostics, and supply chain (Devereaux, 2002).

6.3.4.1 Systems Engineering

Typically, systems engineering is a front-end activity for a project that peaks early in the project life cycle and then reduces in activity level as the project moves into development. Systems engineering is also important during the integration and test phases and tapers down after system deployment. Obsolescence mitigation can be proactive or reactive, but in either case, to fully leverage a mitigation opportunity, a systems engineering approach should be used. An example of a basic systems engineering process is shown in Figure 6.12.

For proactive obsolescence management, objectives or criteria may include obsolescence considerations in the initial implementation approach. If obsolescence occurs in a legacy system, alternatives still need to be evaluated, and this overall approach can still be applied. The key additional consideration, however, is how the obsolete parts of the system interact with the non-obsolete parts. In other words, can obsolescence be addressed as a local problem affecting only single components or does the impact of obsolescence cross interfaces between components or even interfaces between subsystems or the system as a whole and its environment?

Dowling addresses obsolescence as a form of engineering change and indicates that these changes can and should be planned at the system level, not only the component level. He believes that obsolescence is bigger than the availability of a component from a supplier; many changes can make a given system obsolete (e.g., miniaturization/personalization, scalability to number of users, changing threat environments), and for those changes that can be anticipated, a plan should be in place. Dowling suggests a detailed examination of how the system may be used in the future as well as the current needs of the system (Dowling, 2001).

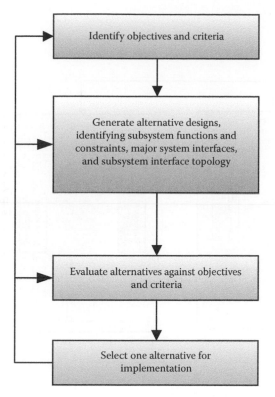

FIGURE 6.12
Leveson's basic systems engineering process. (Adapted from Devereaux, J.E., Obsolescence: A systems engineering and management approach for complex systems, S.B. Aerospace Engineering with Information Technology, Massachusetts Institute of Technology, February 2002.)

Figure 6.14 shows how Dowling anticipates examining solving problems through system design. The figure indicates the importance of evaluating the possible solution options against the anticipated problems and required needs. This step is typically performed in the basic systems engineering process. However, Dowling's process also considers the future problems or requirements of a system and how those might manifest themselves in a solution space. Once those two efforts have been performed, a plan should be developed to evolve the current system into one able to meet the future requirements of the system, thus dealing with functional obsolescence. He proposes adding the following systems engineering artifacts to the typical systems engineering activities for a project: obsolescence management plan, change plan, future systems requirements, future system design, and future system simulations (Figure 6.13) (Dowling, 2001).

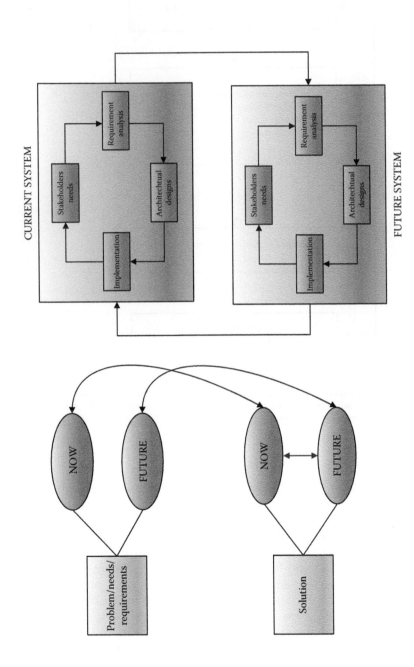

FIGURE 6.13
A systems engineering approach to address obsolescence. (Adapted from Dowling, T., Planning for Change with a Holistic View of the System, NATO Research and Technology Organization Strategies to Mitigate Obsolescence in Defense Systems Using Commercial Components, NATO Research and Technology Organization, Neuilly-Sur-Seine, France, 2001.)

6.3.4.2 Software Engineering

Rothmaier proposes developing modular domain-specific software architectures to mitigate obsolescence. Libraries can be developed to promote reuse and, thus, reduce the qualification and test activities required. Rothmaier also advocates encapsulation and abstraction to help distance the software from potential changes in the hardware (Rothmaier, 2001).

Lane, Beattie, Chita, and Lincoln advocate a model-based architecture development using the UML syntax, object-oriented design, and automatic code generation to limit the impact of platform obsolescence to the code that needs to be developed (Lane et al., 2001.). Theoretically, the model only requires an interface change to the new platform; with this change, the entire code base could be automatically generated from the existing model. They also believe the use of a model tied to the requirements will help bridge the communications gap between the systems engineering and software engineering development groups. Wherever possible, they recommend the reuse and redesign of a model with units that can be easily changed when functionality upgrades are required (Lane et al., 2001). In recent years, this approach has been extended to systems engineering in the model-based systems engineering domain using various languages (e.g., OPM and SYSML) though it is not as mature (Devereaux, 2002).

6.3.4.3 Test Phase

Obsolescence can affect testing in multiple ways. New test methodologies may need to be developed to test the design changes that result from obsolescence. In addition, the product may not be the only area experiencing obsolescence. The test itself may experience obsolescence, and this should be considered early in any activity where obsolescence is being managed (Bach, 2001).

The COTS components' costs are amortized over many users; however, the testing is only focused on the areas of the system most likely to be used by the majority of users. If a new project's application of a COTS component diverges from other projects' applications or contexts, specialized testing is still needed to reduce the likelihood that the difference in application of the COTS component will cause problems during use (Devereaux, 2002).

Weyuker explains this more clearly:

> Even with no explicitly defined operational distribution, testers usually have some information or intuition about how the software will be used and therefore emphasize, at least informally, testing of what they believe to be its central or critical portions. These priorities will likely change, however, if it is decided to incorporate the component into a different software system. The original system may commonly execute portions that the other never will, which makes much of the testing irrelevant to the new user in the new setting. Likewise, the new user may use parts

of the component that correspond to extremely unlikely scenarios in the original system's behavior, and these may have been untested or only lightly tested when the software was developed.

(Weyuker, 1998)

Weyuker stresses the need for various levels of testing, specifically in the software domain. At minimum, unit testing, integration testing and systems testing are essential for a system, and further types of testing, such as feature testing, performance testing, load testing, stability testing, stress testing, and reliability testing, may also be required. She points out that when using COTS systems, the understanding of the lower levels of the system is frequently lost by those trying to integrate and test with COTS products; it is often easier for people to assume that the appropriate levels of testing have been performed by the COTS supplier than to fund and develop a robust testing program specific to a project's application (Weyuker, 1998).

To mitigate some of these concerns with the reuse of COTS components, Weyuker encourages maintaining the software specification whenever new functionality is added and maintaining traceability between the specification and the detailed test cases that have been used to test the software (Weyuker, 1998).

6.3.4.4 Prognostics

Prognostics is a developing field, as mentioned earlier in the chapter, that extends state-of-the-art diagnostics. Where diagnostics tells a maintainer if a failure is happening and, if so, where and what failure, a prognostics system aims to understand the physics of failure models for a system, senses changes that could imply failure in the future, and predicts future failure rates. The latter could be used to predict when obsolescence (i.e., lack of product availability) will have an impact on a system (Smith, 2001). While the technology in this field is still developing, new contracts are being issued with prognostics requirements; these systems should be developed with an eye for how prognostics could mitigate obsolescence issues through improved logistics and supply chain planning (Devereaux, 2002).

6.4 Outsourcing Maintenance: New Frameworks

Maintenance management covers all activities of an organization seeking to preserve the function of different assets, including facilities, IT, or vehicles. All these activities are crucial requirements of any business institution, but as a business develops and competition gets tough, it is not easy to monitor all of them. Keeping up with diversified departments of activities that do not directly affect the business becomes costly, especially in a crisis market, when even the wealthiest companies are reducing costs. Outsourcing maintenance management services ends at least some of the headaches and reduces expenses. While the work could be done in-house, outsourcing allows the business or organization to be more competitive by staying focused on its core competencies (Kurdia et al., 2011). Focusing on primary activities is essential for success (Kurdia et al., 2011).

Alexander defines the outsourcing process as an "act of moving some of a firm's internal activities and decision responsibilities to outside providers" (Alexander and Young, 1996). To this, Lankford and Parsa add: "Outsourcing is the procurement of products or services from sources that are external to the organization" (Lankford and Parsa, 1999).

Outsourcing is contracting out to obtain services or products from an outside provider instead of having them provided by in-house resources. In other words, it is a form of privatization in which one organization contracts with an external organization to provide appropriate functions or services (Ender and Mooney, 1994). The concept is to have the management or day-to-day execution of one or more business functions performed by a third-party service provider who is already in-sourcing these business processes.

Sullivan notes: "When making decisions about the best way to staff, facility executives should keep the focus on finding the right skills, not achieving short-term cost savings" (Sullivan, 2005). The best way to start is by determining the top source for the expertise required.

According to Encon Hui and Albert Tsang (2004), out-tasking (i.e., outsourcing) is a complementary option that may be more suitable in specific situations. While the strategic consequences of out-tasking are generally acclaimed, little is understood about its practice. Two of the key reasons for success are, first, choosing the right outsourcing strategy and, second, implementing it properly. Preferably, the service provider should have the skills to deliver the service with reasonable reliability, certainty, cost-effectiveness, and on-time performance. Salonen concludes: "When companies outsource maintenance, a contrast usually arises between the client's long-term maintenance strategies and the supplier's incentive to provide quality service" (Salonen, 2004).

6.4.1 Maintenance Contracting Strategies

Pakkala et al. (2007) point out that there has been little standardization of the terminology that applies to maintenance contracting. The following are some examples of terms used variously around the world: asset management contracts, asset maintenance contracts, performance-specified maintenance contracts (PSMC), managing agent contracts, performance-based contracts, total maintenance contracting, and other contract methods. These terms basically refer to the outsourcing of routine maintenance, preventive maintenance, both routine and preventive maintenance, or all maintenance services, using some form of outcome-based specification (performance

FIGURE 6.14
Components of a contracting strategy. (Adapted from Menches, C.L. et al., Synthesis of innovative contracting strategies used for routine and preventive maintenance contracts, Project 0-6388, Performing Organization: Center for Transportation Research, The University of Texas at Austin, Austin, TX, 2010.)

levels) or having a certain required "level of service" that must be met over a long duration (often 3–10 years).

Definitions of maintenance processes vary from country to country – and even within countries and organizations – making it difficult to standardize contracted work. To cite one example, the American Association of State Highway and Transportation Officials (AASHTO) Subcommittee on Maintenance provides its own definition of pavement maintenance (routine and preventive) for reference. According to the Subcommittee, routine pavement maintenance "consists of work that is planned and performed on a routine basis to maintain and preserve the condition of the highway system or to respond to specific conditions and events that restore the highway system to an adequate level of service" (Geiger, 2005). However, according to the AASHTO Standing Committee on Highways (1997), preventive pavement maintenance is "a planned strategy of cost-effective treatments to an existing roadway system and its appurtenances that preserves the system, retards future deterioration, and maintains or improves the functional condition of the system (without significantly increasing the structural capacity)" (Geiger, 2005).

A contracting strategy can be defined as a process for allocating the risks and responsibilities for maintaining an existing asset, and consists of three components: (1) a delivery method, (2) a type of contract specification, and (3) a pricing strategy (Figure 6.14).

6.4.1.1 Delivery Methods

The delivery method, as well as the type of contract specification and pricing strategy, must be selected as part of the maintenance outsourcing process. A recent report by Pakkala et al. (2007) investigates and summarizes traditional and nontraditional maintenance delivery models implemented by various countries, including Australia, Canada, England, Estonia, Finland, the Netherlands, New Zealand, Norway, Sweden, and the United States. He characterizes an in-house maintenance model (also referred to as a "traditional model") as one in which in-house personnel carry out nearly all of the maintenance activities. He also identifies seven innovative (or nontraditional) maintenance delivery models:

1. *Activity-based maintenance model*: Specific routine maintenance activities are outsourced by the company. This model is usually based on the lowest price with a unit price payment and its duration is typically for 1 year or season.
2. *Partial competitive maintenance model*: A portion of routine maintenance activities is specifically retained for in-house personnel while the remainder is outsourced. Some companies allow their own workforce to publicly tender against any private sector competitors.
3. *Routine maintenance model*: All routine maintenance activities are outsourced. The duration of this model varies and the present trend is usually between 5 and 10 years. Lump sum or the hybrid of lump sum and unit price is the typical payment of this model.
4. *Integrated maintenance model*: A combination of both routine and preventive maintenance activities is outsourced together as one contract. This model typically uses lump-sum payment but unit price can be implemented if unforeseen conditions require extra work.
5. *Long-term separate maintenance model*: A single maintenance activity is outsourced for a long duration, often because it is unique or risky.
6. *Framework model*: Several contractors are preapproved and receive nominal contracts that make them eligible to be awarded for award of maintenance projects.
7. *Alliance model*: A contractor is selected entirely on qualifications and has the opportunity to gain or lose 15% of the contract value based on performance.

6.4.1.2 Contract Specifications

Once a company has decided to outsource all or a portion of its maintenance activities, and after a delivery method has been chosen, the type of contract specification must be selected. Segal et al. (2003) identifies three primary types of contract specifications used to outsource maintenance work: (1) traditional (i.e., method-based), (2) performance-based, and (3) warranty contract specifications. Hybrid methods combining multiple types are also used.

Traditional contract specifications are often referred to as "method-based" and contractors are typically "paid for the amount of work they do—not on the quality of work that is provided" (Segal et al., 2003). These specifications are usually based on a number of line items that describe the scope of the work to be performed. The company typically specifies the methods, materials, and quantities to be used, and payment is based on amount of output (Stankevich et al., 2005).

Under performance-based contract specifications, the company or customer must define an end outcome goal (e.g., track quality) and the contractor decides how best to achieve the desired outcome. The contract specification identifies clearly defined performance measures, clearly defined outcomes and time-tables, but allows new and innovative methods to be used (Segal et al. 2003). Hence, the customer must establish a minimum performance standard, where payment is based on performance, with options for penalties and rewards.

A warranty contract specification is another form of performance-based contract specification in which the contractor is required to warrant the work for a specified length of time. There is an increasing trend toward the employment of warranty contracts whereby the contractor places a long-term guarantee on his or her work. This further shields the customer company from risk.

6.4.1.3 Pricing Strategies

Typical payment methods for maintenance contracting include unit price, lump sum, cost plus fee, or a hybrid of these methods. Unit price is typically used for method-based contracts because payment is based on the amount of output of a particular line item, such as area of grass mowed during the payment period. However, payment of performance-based contracts is made on a lump-sum basis, normally through 12 equal monthly installments. A hybrid payment method can be used on a performance-based contract that includes line items for emergencies or unknown activities. This allows lump-sum payment for regular monthly maintenance while providing unit price payment for additional line items of work and helping to minimize the unforeseen risks on activities (Pakkala et al., 2007).

6.4.2 Conclusion

Overall, innovative maintenance contracting methods have been largely successful, but the initial implementation of such contracts has often been accompanied by a large learning curve that can only be overcome through patience, persistence, and hard work.

6.4.3 Main Reasons for Outsourcing

The five main reasons to outsource can be listed as (Quélin and Motlow, 1998):

1. Ability to refocus on strategic activities
2. Realization of economies of scale by the service provider
3. Development of reorganization policy
4. Technological reasons
5. Globalization of markets

These are shown in Table 6.2 and explained at greater length below.

TABLE 6.2

Reasons for Maintenance Outsourcing

Reduce cost	Increase the level of service	Enhance risk management
Increase efficiency	Speed project delivery	Overcome a lack of expertise
Improve quality	Spur innovation	Legislative mandate

Source: Adapted from Quélin, B. and Motlow, D., *French J. Commun.*, 6(1), 75, 1998.

6.4.3.1 Refocusing on Strategic Activities

The first argument concerns the strategic analysis of a company's activities. Drawing up a balance sheet of strengths and weaknesses often induces companies to analyze their information systems in terms of their contribution to competitive advantage, looking at the resources they absorb and at the technological state of the art. For example, Kodak Company judged that the functions it sold were not part of its central activities. None of these functions was the result of an accumulation of know-how and skills constituting a source of competitive advantage; therefore, it decided not to spin off development of its applications, although they were at the heart of its international operations.

Making this full appraisal of the contribution of certain business areas to competitive advantage allows a company to adopt a suitable organizational arrangement so as to optimize, for example, customer relations, the development of new products and services, or the entry into new markets (Quélin and Motlow, 1998).

6.4.3.2 Economies of Scale and Costs

The second argument is related to the cost savings devolving from the use of an outside contractor. Costs can be saved on overhead and on hardware upgrades. Outsourcing also enables a firm to reduce the costs associated with hardware and software maintenance.

In some cases, economies of scale are much more easily achieved by the service provider than by the system user. This is why the return on company investment in information systems is often perceived as very low. Another reason is the productivity of in-house services; as these are not always run in competition with external providers, they are not operated in a true market (Quélin and Motlow, 1998).

6.4.3.3 Reorganization Policies

The third argument concerns the reorganization and restructuring policies of companies. Activities and operations that are not a direct part of the company's core business are often sold to refocus and to reduce costs.

Operations of this type are, thus, a source of flexibility for large diversified groups such as General Electric. In other cases, they are needed to meet new demands from customers; banks, for example, have to provide financial software for use on microcomputers, but their information systems are mainly run on large mainframes (Quélin and Motlow, 1998).

6.4.3.4 Swift Technological Change

The fourth argument has more to do with technology. First, a company may consider that it does not have the resources to meet its requirements in terms of expertise and knowledge. In this case, the scarcity of skilled engineers puts strong upward pressure on salaries, especially as there is competition in this job market between companies of different sizes and between client companies and service providers. Second, an increasing number of companies are concerned that technological changes in information and communications systems (ICS) make investment in upgrades risky; this can be avoided by outsourcing (Quélin and Motlow, 1998).

6.4.3.5 Market Globalization

The fifth argument concerns the globalization of markets and the efforts companies make to ensure the technical commercial coverage of their areas of activity. This requires increasingly heavy investment. At the same time, globalization affects the markets of ICS providers. Telecommunications deregulation is diversifying supply at the national level, and the international market is shared by about 10 players (Quélin and Motlow, 1998).

6.4.3.6 Conclusion

This overview of the main arguments suggests there may be some concern about both the intentions and thinking of companies who decide to outsource. Other determinants seem more important than economic ones (Quélin and Motlow, 1998).

6.4.4 Outsourcing: Decision-Making Process

When determining what should be outsourced, the four items listed in Table 6.3 are important. The table lists them by order of importance: the company's core competencies, quality of product or service, cost of internal versus external supply, and need for specialize capability (APICS). Each is discussed more fully in the following sections.

TABLE 6.3

Outsourcing Decision-Making

		Percentage Rated Important or Very Important
1	The company's core competencies	82
2	Quality of product or service	77
3	Cost of internal vs. external supply	72
4	Need for specialized capability	64

Source: Adapted from Quélin, B. and Motlow, D., *French J. Commun.*, 6(1), 75, 1998.

6.4.4.1 Internal Costs versus External Costs

The cost of internal work versus externally supplied work is almost always a major consideration in making the decision to outsource. However, the total costs of the targeted outsourced functions are not well understood. Many companies struggle to identify the actual tasks performed by the functions being outsourced. These unknowns may affect the cost of the outsourcing or the level of satisfaction with the end product or service.

Total costs, including functional interdependencies, must be understood as well, because they often drive costs indirectly related to the outsourced function. These total costs must be included in any quantitative analysis of outsourcing.

In surveys, executives frequently rate the cost of internal work versus external work supply slightly lower than other respondents, as shown in Table 6.4. This is consistent with the results of research on drivers and finds that senior management often views outsourcing as a means to achieve other strategic objectives, such as market share expansion, or to manage risks associated with constrained capacity (APICS).

TABLE 6.4

Costs of Internal Work versus Externally Sourced Work

	SC Leaders (%)	Key Executives (%)
Very important	48	29
Important	28	42
Average importance	15	20

Source: Adapted from APICS, Protiviti independet risk consulting. Managing the risks of outsourcing: A survey of current practices and their effectiveness, APICS community at www.apics.org.

6.4.4.2 Need for Specialized Capability of Suppliers

Companies are realizing that outsourcing decisions should be directly tied to corporate strategy and to an understanding of core competencies. Third-party suppliers can help increase the performance of the entire value chain if each participant in the chain understands and focuses on its respective core competencies. *Note*: core competencies may include capabilities such as technical innovation or rapid response/flexibility.

In surveys, those in the manufacturing industry frequently rate the need for the specialized capability of suppliers slightly lower than those in other industries do. Functional management most often rates the need for specialized capability from suppliers as very important. For companies with revenues of \$500M to \$1B, specialized supplier capabilities are more important than for other companies (APICS).

6.4.5 Framework for a Win-Win Maintenance Outsourcing Relationship

Outsourcing is the process through which a business allocates some activity that it would be capable of doing in-house to an external provider (Byham, 2004). It is a long-term relationship between supplier and beneficiary, with a high degree of risk-sharing, and should not be confused with contracting out, which refers to work assigned to an outside supplier on a job-by-job basis (Embleton and Wright, 1998).

As discussed in previous sections, a few studies from the late 1980s and early 1990s outline the decision criteria that should be used for selecting maintenance activities that should be outsourced.

The following sections propose a set of advisory criteria and a roadmap according to which a company as end user can select the maintenance activities that should be outsourced. However, once the outsourced activities are selected, the company should use a separate set of decision criteria to select a delivery method, type of specification, and pricing strategy (i.e., contract strategy) for outsourcing a single maintenance activity, bundles of activities, or all maintenance activities.

The decision criteria for selecting an appropriate contracting strategy for outsourcing of maintenance activities have not been assembled so far. Therefore, in this section, a set of decisions are described to aid personnel to select an appropriate maintenance contracting strategy, including the delivery method, type of specification (i.e., method-based or performance-based), and pricing strategy (i.e., fixed price, unit price, or cost plus pricing).

The decision aid begins with the maintenance contracting strategy. First, the company decides how many activities it would like to outsource (e.g., nearly all or less than all), following which certain activities will be retained for in-house performance and the rest will be outsourced. Subsequently, the company must select an appropriate maintenance delivery method, specification type, and pricing strategy appropriate for its maintenance outsourcing goals and circumstances (Menches et al., 2010).

6.4.5.1 Selecting the Number of Activities to Outsource

The company should first determine how many activities it intends to outsource under a contract. It can decide to outsource nearly all maintenance activities or it can decide to select a smaller subset of activities. Maintenance directors use several criteria to determine whether nearly all maintenance activities should be outsourced. These criteria include the following:

- To reduce administrative time and cost, all or nearly all maintenance activities are outsourced together.
- To reduce the size of the in-house workforce, all or nearly all maintenance activities are outsourced.

In the absence of one of these criteria, the company may elect to outsource only a portion of the maintenance activities while retaining several activities for performance by in-house personnel (Menches et al., 2010).

6.4.5.2 Selecting Which Activities to Outsource

There are many factors that a company considers when deciding which activities to outsource or retain for in-house performance. Many maintenance managers indicate that the following factors are important:

- Size of the maintenance budget
- Availability of proper equipment
- Availability of in-house workforce
- Availability of in-house expertise
- Availability of contractor expertise
- Need to improve maintenance efficiency
- Quality/experience of contractors
- Need to augment peak workloads
- Required by legislative mandate
- Political reasons or pressure
- Need to encourage innovation
- Need to increase level of service
- Overall risk management strategy
- Need to address weather challenges
- Need to speed up maintenance delivery
- Need to achieve cost savings
- Need to increase responsiveness

- Need to accommodate workload
- Need to accomplish emergency work
- Uniqueness or specialty of the work

Some criteria might specifically compel an activity to be outsourced. For example, lack of equipment or expertise are criteria that would compel a company to outsource an activity. As companies have different in-house and external resources, experience, and maintenance needs, they should base their decisions on what to outsource on the unique characteristics of the end user (Menches et al., 2010).

6.4.5.3 Selecting Which Activities are Outsourced Individually and Which Are Bundled

Once the company has selected one or more maintenance activities to out-source, the next step is to select which activities to outsource individually and which to bundle together. Maintenance managers consider several aspects when deciding how to combine activities:

- Level of control over the work that is desired and would be achieved
- Efficiency that would be achieved by bundling activities together
- Reduction in coordination effort that would be achieved
- Reduction in administrative load that would be achieved
- Available equipment composition
- How similar the bundled activities are and the logic of grouping them
- Whether multiple subcontractors will be needed to complete all of the work
- Time-sensitivity of the bundled work
- Contractor's experience at performing the bundled activities
- Cost-effectiveness of individual versus bundled activities
- Volume of work that would result from individual or bundled activities (Menches et al., 2010)

6.4.5.4 Selecting a Single Outsourced Activity

For a single activity, the following criteria are usually considered:

- Insufficient equipment is available for performing the work.
- The company needs flexibility in when and how to complete the work.

- The duration of the work may be very short or very long.
- The activity is special, unique, or risky.
- There is a need to reduce the amount of time for bidding and awarding projects.
- There is a need to select a contractor quickly for an urgent project.
- There is a need to reduce the overall administrative time, costs, and overhead (Menches et al., 2010).

6.4.5.5 Selecting Bundled Activities

Companies may opt to bundle activities for the following reasons:

- There is a need to reduce the amount of time for bidding and awarding projects.
- There is a need to select contractors quickly for urgent projects.
- There is a need to reduce the overall administrative time, costs, and overheads.
- There is a need to increase the level of competition.
- There is a need to ensure that there is an equal opportunity for in-house employees to get work (Menches et al., 2010).

6.4.5.6 Selecting Nearly All Activities

A company may decide to combine nearly all maintenance activities into one contract, based on the maintenance needs of the customer. In this case, the following are the factors affecting such a decision:

- There is political pressure to outsource nearly all maintenance activities.
- There is a lack of manpower to perform the work in-house.
- There is a need to reduce administrative load.
- The long-term duration of the contract makes it ideal for contracting out nearly all activities.
- There is a need to reduce coordination efforts for different types of maintenance activities.
- The level of control and the desire to shift control are important considerations.
- There is a need to reduce conflicts between owners and contractors.
- There is a need to increase the level of competition among bidders (Menches et al., 2010).

6.4.5.7 Selecting the Type of Contract Specification

Three types of contract specifications are popular. In most cases, each of the three can be selected unless there is a compelling reason to eliminate one from consideration. These three contract specifications are method-based, performance-based, and warranty contracting. The following factors are considered by most managers when selecting a type of contract specification:

- Level of control that the company wants to maintain
- Level of trust in the contractor
- Quality of the contractor
- Political reasons or pressure
- Encouragement of the participation of contractors in the bidding process (Menches et al., 2010)

6.4.5.8 Selecting a Pricing Strategy

Popular pricing strategies include fixed price, unit price, and cost plus pricing. The following factors influence the selection of the pricing strategy:

- Legislative mandate requires or prohibits using a particular pricing strategy.
- The method selected is the most cost-effective for the company.
- Flexibility is needed because of the unique nature of the work (Menches et al., 2010).

6.4.6 Elements in Successful Management of Outsourcing Relationship

Niekerk and Visser (2010) list 22 essential elements for successful management of a win-win maintenance outsourcing relationship (see Table 6.5). They also say five management systems must be in place and properly executed on an ongoing basis:

1. A system to sustain support
2. A system to maintain team unity
3. A system to manage performance
4. A system to manage relationship well-being
5. A system to ensure cohesion

These are shown schematically in Figure 6.15, expanded in Table 6.5 and explained further below.

TABLE 6.5

Essential Systems and Elements of an Outsourced Maintenance Relationship

Management System	Basic Elements
Support retention system	Support from top management
	Clear, agreed roles and responsibilities
	Suitable organization with skilled people and the required resources
	Maintenance strategy aligned with client's business objectives
	Maintenance policies and plants that meet client's operational needs
	Accountability on both sides
Team unity system	Shared vision
	Common goals
	Shared expectations
	Joint planning and review
	Shared responsibility for success
	Shared risk and reward
Performance management system	Objective performance measurement
	Incentives/remuneration that drive improvement
	Effective continuous improvement program
Relationship health management system	Relationship well-being assessment
	Profitability for both partners
	Mechanism to identify and reconcile differences
	Clear contingency plan and exit strategy
Cohesion system	Change tracking and update mechanism
	Regular communication
	Collaboration agreement/commitment

Source: Adapted from Nili, M. et al., *Bus. Manag. Rev.*, 2(5), 20, July 2012.

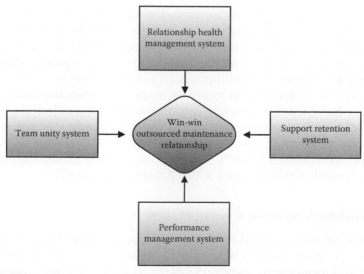

FIGURE 6.15
Outsourced maintenance relationship management framework. (Adapted from Nili, M. et al., *Bus. Manag. Rev.*, 2(5), 20, July 2012.)

- *The support retention system* ensures that all the other aspects of the relationship continue to receive the appropriate amount of support and attention; it becomes a force driving the continued success of the relationship.
- *The team unity system* prevents misalignment and ensures that important but relatively minor details are taken care of.
- *The relationship health management system* proactively identifies and eliminates any unintentional disagreements before the relationship suffers any damage that could have been prevented.
- *The performance management system* regulates the perceived fairness of the engagement. It is also the win-win indicator. It must be as objective as possible and be well aligned with the objectives of both parties.
- *The cohesion system* encircles the other systems and ensures all systems remain operational throughout the life of the relationship. It is the glue binding the systems. It consists of mechanisms to manage change, to ensure regular communication on all levels, and to ensure continued commitment.

For the framework to deliver a sustained win-win relationship, the expected outsourcing benefits (mentioned in Chapter 3) must be in place, sufficiently established, and well and regularly maintained (Nili et al., 2012).

6.4.7 Cost of Selecting a Service Provider

The expense of selecting a service provider can vary from 0.2% to 2% of the total cost of the outsourced service. Selection costs include documentation requirements, sending out requests for proposal (RFPs), evaluating the responses, making a selection, and negotiating a contract. A project leader may be working full time on this, with others chipping in; all of this represents an opportunity cost. There are also legal fees. Some companies hire an outsourcing adviser for about the same cost as doing it themselves. To top it off, the entire process can take from 6 months to a year, depending on the nature of the relationship. At this stage, travel expenses enter the picture as well. Companies may spend an additional 1%–10% on travel costs visiting the service provider. These costs are shown in Table 6.6 (OSF Global Services, 2012).

6.4.8 Benchmarking to the Current Market

If the price for the requested services is not appealing to the client, a typical service supplier reaction is to offer fewer services, reducing the scope. The offer will include only those services for which the supplier can provide the best results, and it will simply leave the rest—by default—to the client (Szatvanyi, 2008).

TABLE 6.6

Cost of Selecting a Supplier

Putting together the RFQ/RFP
Distributing the RFQ/RFP and collecting the feedback
Analyzing and short-listing
Due diligence
Management sponsorship
Decision to outsource

Source: Adapted from OSF Global Services, The real cost of outsourcing, Write paper, 2012, www.osf-global.com.

Note: RFP is a request for a proposal; RFQ is a request for a quotation.

This causes two major problems for the client:

1. *The responsibility issue*: The client must manage the overall service, being responsible by default for everything that will not be done by the supplier.

2. *The cost issue*: The client will incur all costs associated with doing everything that will not be delivered by the supplier.

In sum, a company must make careful note of the following:

- What are the supplier delivery commitments?
- Is cost/quality appropriate for services delivered or is it distorted by supplier issues?
- Does outsourcing compare favorably to in-house service delivery?
- Are internal contract management costs and time appropriate? (Szatvanyi, 2008)

6.5 Warranty Management: Extensions and Claims

As a result of increasing customer expectations, product performance and characteristics are no longer the sole aspects to consider in a competitive global market. Products must perform satisfactorily over their useful life to ensure buyer satisfaction. In this context, the role of post-sale services, particularly during the warranty period, becomes crucial. An efficient warranty program represents a competitive weapon (Gonzalez Díaz et al., 2012b).

The management of the warranty is not a simple issue; it combines technical, administrative, and managerial actions. During the warranty period, an item must be maintained or restored to a state in which it can perform the required function or provide a given service (Gonzalez Díaz et al., 2009). There are different types of warranties; each is suited to a different type of product (consumer, commercial and industrial, standard versus custom-built, etc.) (Lyons and Murthy, 2001; Menezes Melvyn and Quelch, 1990).

For effective warranty management, it is critical to gather proper data and exchange the different types of information between the modules into which a management system can be divided (Lyons and Murthy, 2001). In particular, four important interactions must be considered:

- *Warranty and maintenance*: In many cases, the warranty period is the time when the manufacturer still has strong control over a product and its behavior. The expected warranty costs normally depend not only on warranty requirements, but also on the associated maintenance schedule of the product (Dimitrov et al., 2004).

- *Warranty and outsourcing*: The warranty service or, in general, the after-sales department of a company is usually one of the most likely to be outsourced because of its low risk. In addition, outsourcing provides legal insurance for such assistance services (Gómez et al., 2009).

- *Warranty and quality*: The improvement of the reliability and quality of the product not only has an advantageous and favorable impact on the client, it also greatly reduces the expected warranty cost (Chukova and Hayakawa, 2004). However, in the automobile industry, for example (see Figure 6.16), it is probably an oversimplification to consider a low ratio of client claims to high quality;

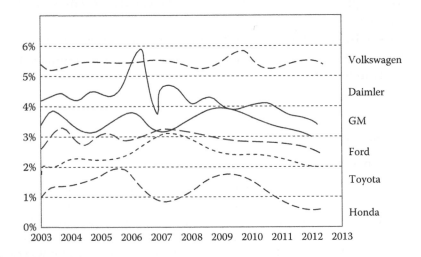

FIGURE 6.16
Ratios of warranty claims from 2003 to 2013. (Adapted from Gonzalez-Prida Díaz, V. and Crespo Márquez, A., *After–Sales Service of Engineering Industrial Assets: A Reference Framework for Warranty Management*, Springer, XXII, 318pp., 2014, ISBN 978-3-319-03710-3, http://www.springer.com/gp/book/9783319037097.)

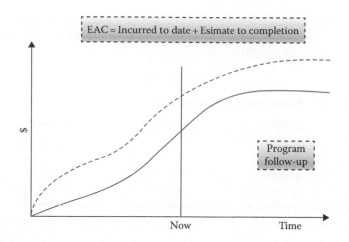

FIGURE 6.17
EAC formula. (Adapted from Christensen, D., *Natl. Contract Manag. J.*, 25, 17, 1993.)

rather, companies may reject customer claims and redirect them to the suppliers. In Figure 6.16, the *y*-axis indicates the percentage of claims dedicated to warranties in vehicle sales (Gonzalez-Prida Diaz and Crespo Marquez, 2014).

- *Warranty and cost analysis*: In reference to costs estimations (see Figure 6.17), and apart from warranty issues, there are several methods to accurately estimate the final cost of a specific acquisition contract. One such method is estimate at completion (EAC).

- In a few words, EAC is a management technique used in a project for cost control. Here, the manager foresees the total cost of the project at completion, combining measurements related to the scope of supply, the delivery schedule, and the costs, using a single integrated system. Warranty costs can be included in the global analysis of the project, providing the estimation costs of the same service generated at the end of contractual responsibility (Gonzalez-Prida Diaz and Crespo Marquez, 2014).

Considering these aspects, problems in warranty management efficiency seem to be correlated to logistic issues. A critical area is spare parts provisioning. Customer satisfaction, strongly related to consumer quality perception, is achieved only when after-sales service is realized in the shortest possible time. This, in turn, is achievable only when the spare parts logistics and repair capacity are adequate. For example, in a warranty program for a complex system, maintenance activities have to be planned, and this necessitates spare parts. Decisions about warehouses and inventory levels must be strategically made to ensure an effective product warranty service at the lowest cost.

Many activities related to warranty, especially logistic ones, are outsourced; their efficient resolution affects both warranty costs and quality perception (González Díaz et al., 2012b).

6.5.1 Warranty Management Framework

Warranty management efficiency, in particular for complex systems, is a real concern (González Díaz et al., 2012b). A framework for warranty management has been suggested; it draws on well-known techniques and methods to increase process efficiency (González Díaz et al., 2009). It consists of four steps, following the plan-do-check-act (PDCA) cycle and principles of quality management systems according to 9001:2008.

The first step of a warranty management process is the definition of generic and specific objectives. This decision is fundamental for the strategic formulation of warranty plans, and it must take different perspectives into account. To avoid contradictions between the warranty program and the overall business strategy, the use of a balanced scorecard (BSC) is suggested. Other useful methods during the planning phase are criticality analysis (CA) and root cause failure analysis (RCFA) to focus actions on specific failures showing rare and high failure frequency (González Díaz et al., 2011).

After establishing the objectives, the warranty capacity (spare parts, warranty tasks schedule, skill levels, etc.) needs to be assessed with the aim of achieving minimum waste or expense. Reliability analysis (RA) and maintenance design tools (MDT) can be adapted to the warranty field. Warranty policy risk–cost–benefit analysis is another possibility, but its results will depend on the information available.

Once designed, planned, and realized, the warranty program must be measured. Performances of warranty tasks have to be evaluated and assessed. Starting from warranty program feedback data, a reliability, availability, maintainability, and safety (RAMS) analysis may help to improve product engineering and manufacturing. Another interesting issue is the warranty contribution to the life cycle cost of the product in terms of repair costs, spare parts, and so on.

The last step of the process is warranty program improvement. Out of a large number of possible approaches, customer relationship management and six sigma seem particularly effective. Other tools that can be used for improvement are related to the implementation of new technologies such as e-warranty strategies, where e-warranty is defined as warranty program support which includes the resources, services, and management necessary to enable proactive decision. This support includes e-technologies and e-warranty activities such as e-monitoring, e-diagnosis, and e-prognosis (González Díaz et al., 2012b).

6.5.2 Importance of a Warranty Management System

Many manufacturing companies spend great amounts of money on their product and service warranties, but this does not always receive much attention. Effective warranty management processes are not yet on the agenda of most executives. In practice, companies often pay attention to their warranty management processes only when high levels of liability are involved with a claim or recall. As a consequence, they miss opportunities to make significant cost savings and to increase the business value through better-quality products and higher levels of customer satisfaction (González Díaz et al., 2010).

Warranty management deals with decision-making at strategic and operational levels. This requires taking into account the interactions between engineering, marketing, and post-sale support elements of manufacturing firms (see Figure 6.18) (Murthy and Djamaludin, 2002). Warranty management needs to be treated not as a cost center but as an asset that can create a higher value of the business performance (González Díaz et al., 2010).

Effective warranty management requires integrating leading practices and identifying critical points in the decision process. These critical points will depend on the already existing service management of the organization and its warranty management capabilities. A company can improve its business

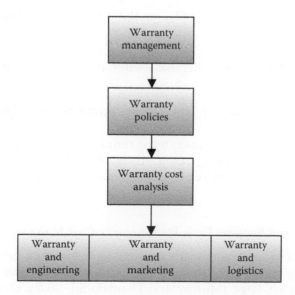

FIGURE 6.18
Warranty interactions. (Adapted from Murthy, D.N.P. and Djamaludin, I., *Int. J. Prod. Econ.*, 79, 231, 2002.)

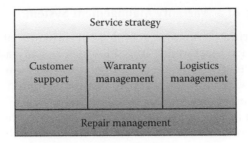

FIGURE 6.19
Services management. (Adapted from González Díaz, V. et al., Warranty cost models State of Art: A practical review to the framework of warranty cost management, in: Briš, R., Guedes Soares, C., and Martorell, S., eds., *Reliability, Risk and Safety: Theory and Applications*, Taylor & Francis Group, London, U.K., 2010.)

performance by maximizing both cost reduction and value creation (see Figure 6.19) in the following areas:

- Claims processing time: improvement by reducing the process
- Claims processing costs: improvement by reducing the procedures through automated processing
- Number of claims processing personnel: improvement by reducing the number of staff to perform claims processing
- Quality: improvement by reducing the customer loss related to poor product quality
- Supplier recovery: improvement by sharing costs; quality also improves because supplier has better information about quality processes

The result is lower warranty costs, better products, and higher levels of customer retention.

Overall, good warranty management requires paying attention to warranty processes and creating better strategies, more effective tools and integrated systems, and better organizational structures (González Díaz et al., 2010).

6.5.3 Models and Support Tools for Warranty Cost Management

A key aspect of strategic warranty management is that decisions must begin at a very early stage in the product life cycle, not as an afterthought just before the launch stage (Murthy and Djamaludin, 2002). Other common problems during the application of warranty services include the following:

- Limited information systems support
- Long cycle time for claims review

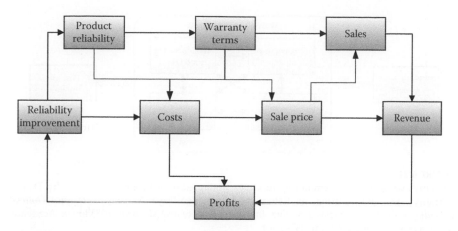

FIGURE 6.20
Services management. (Adapted from Murthy, D.N.P., *Ann. Oper. Res.*, 143, 133, 2006.)

- Excessive numbers of invalid claims
- Inability to distinguish warranty responsibility
- Warranty data not used to improve product quality
- Additional claims due to slow corrective actions in manufacturing, engineering, or product design

An effective management of product reliability must take into account the link between warranty and reliability, as shown in Figure 6.20 (Murthy, 2006).

Basically, a good model of warranty management will help a company achieve its warranty service goals. By reengineering management processes and applying a correct warranty cost model, it is possible to achieve the following:

- Improve sales of products with extended warranties and additional related products
- Improve quality by improving the information flow about product defects and their sources
- Improve customer relationships
- Reduce expenses related to warranty claims and processing
- Improve management and control of warranty costs
- Reduce invalid related expenses and other warranty costs

For effective warranty management, it is critical to collect the proper data and to exchange the different types of information between the modules in which the management system can be divided, as indicated in Figure 6.21 (Lyons and Murthy, 2001).

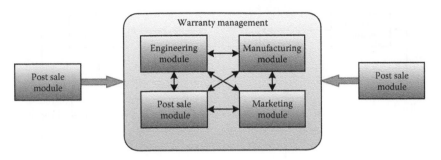

FIGURE 6.21
Warranty management system in four modules. (Adapted from Lyons, K. and Murthy, D.N.P., Warranty and manufacturing, in: Rahim, M.A. and Ben-Daya, M., eds., *Integrated Optimal Modelling in PIQM: Production Planning, Inventory, Quality and Maintenance*, Kluwer Academic Publishers, New York, 2001, pp. 287–324.)

6.5.4 Adapting e-Warranty to Warranty Assistance

Continuous improvement requires the application of emerging techniques and technologies in those areas considered to have more impact. The implementation of new technologies in warranty management includes e-warranty. As shown in Figure 6.22, this concept is a component of e-manufacturing. It applies the advantages of the new and emerging technologies in information and communications to a multiple-user environment in a cooperative and distributed way. Figure 6.22 shows how decisions

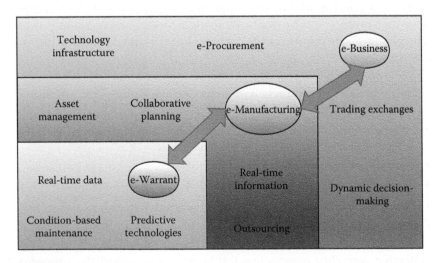

FIGURE 6.22
Integration of e-warranty, e-manufacturing, and e-business systems. (Adapted from Koç, M. et al., Introduction of e-manufacturing, in: *31st North American Manufacturing Research Conference (NAMRC)*, Hamilton, Ontario, Canada, 2003.)

made during presales, fabrication, and operations during aftersales can be integrated in an e-business framework.

e-Warranty can be defined as the support to a warranty program which includes the resources, services, and management needed to enable proactive decision-making. With the application of new technologies in warranty management, technicians in the aftersales service area must work on continuous improvement to reach higher levels of product quality and better technical service assistance effectiveness. This requires higher levels of knowledge, experience, and training (González Díaz et al., 2012a).

e-Warranty basically proposes a new way of managing a warranty program in which the products are controlled via the Internet by the manufacturer. The manufacturer obtains real-time data from the equipment under warranty through digital technologies. The data obtained by sensors preinstalled on the equipment provide information about its status (temperature, run time, pressure, etc.) and allow a continuous diagnostic; this enables the prediction of failure due to the malfunction or bad use of equipment.

e-Warranty incorporates the same principles as those used in the conventional process of warranty management, adding a Web service and the principles of collaboration and/or monitoring. This application may be a useful marketing tool when seeking new business, offering users the possibility of extending the warranty contract by remotely tracking product performance and, consequently, being able to foresee possible failures. In any case, this new advanced model of customer service management breaks down the physical distances between manufacturer and customer through ICT, thereby transforming companies into manufacturing service businesses that provide support to customers anywhere anytime (Gonzalez-Prida Díaz and Crespo Marquez, 2014).

6.6 Insurance: Economic Responsibility of Third Parties

The importance of insurance in modern economies is unquestioned and has been recognized for centuries. Insurance is an essential element in the operation of sophisticated national economies throughout the world today. Without insurance coverage, the private commercial sector would be unable to function.

Insurance enables businesses to operate in a cost-effective manner by providing risk transfer mechanisms whereby risks associated with business activities are assumed by third parties. It allows businesses to take on credit that otherwise would be unavailable from banks and other credit-providers fearful of losing their capital without such protection, and it provides protection against the business risks of expanding into unfamiliar territory—new locations, products, or services. This is critical for encouraging risk-taking and creating and ensuring economic growth.

Insurance is a financial product that legally binds the insurance company to cover the losses of the policyholder when a specific event occurs. The insurer accepts the risk that the event will occur in exchange for a fee, the premium. The insurer, in turn, may pass on some of that risk to other insurers or reinsurers. Insurance makes possible ventures that would otherwise be prohibitively expensive if one party had to absorb all the risk.

In some instances, governments require businesses to purchase insurance. Known as financial responsibility requirements, government-mandated purchases of insurance are intended to ensure that injured parties will be compensated. Businesses also require other businesses to buy insurance. For instance, a retailer may require its suppliers to carry product liability insurance. Similarly, hospitals may require doctors to carry medical malpractice insurance, and mortgage firms often require their clients to insurance the properties used as collateral.

Distribution of insurance is handled in a number of ways. The most common is through the use of insurance intermediaries. Insurance intermediaries serve as the critical link between insurance companies seeking to place insurance policies and consumers seeking to procure insurance coverage. Intermediaries now offer services such as the evaluation and implementation of alternative means of funding for potential losses, risk management strategies, and claims management (WFii).

6.6.1 Breaking Down Third-Party Insurance

Third-party insurance is essentially a form of liability insurance purchased by the insured, the first party, and issued by an insurer, the second party, for protection against the claims of another, the third party. The first party is responsible for its own damages or losses, no matter how they are caused.

In automobile insurance, there are two kinds of third-party liability insurance coverage. Bodily injury liability covers costs to people. These include medical expenses, such as hospital bills, or lost wages, pain, and suffering. Property damage liability covers costs related to property or other material goods.*

6.6.1.1 Importance of Third-Party Liability Insurance

Anyone who drives is required by law to carry at least a minimal amount of both of these types of liability coverage. In the United States, each state has its own minimum requirement for each type of coverage. Even in "no-fault" states, liability coverage is all but essential. No-fault laws were established to clear courtrooms filled with ordinary injury lawsuits with low-dollar price tags, as well as claims for "pain and suffering." Still, no-fault laws do not protect the insured from million-dollar injury lawsuits stemming from seriously

---

* http://www.investopedia.com/terms/t/third-party-insurance.asp.

injured third parties. Both types of third-party insurance are important, spe-cifically for individuals with substantial assets to protect. The more money and other assets an insured individual has, the higher the limit should be for each type of liability coverage.*

6.6.1.2 Other Types of Third-Party Liability Insurance

In most countries, third-party or liability insurance is a compulsory form of insurance for any party who may potentially be sued by a third party. Public liability insurance involves industries or businesses taking part in processes or other activities that affect third parties, such as subcontractors, visitors, or other members of the public. Most companies include public liability insur-ance in their insurance portfolio to protect against damage to property or personal injury.

Product liability insurance is typically mandated by legislation; the scale varies by country and often varies by industry. This type of insurance covers all major product classes and types, including chemicals, agricultural prod-ucts, and recreational equipment, and protects companies against lawsuits over products or components that cause damage or injury.*

6.6.2 Economic View of the Insurance Business

The economic view of the insurance business tracks how and when values are created for owners. In its simplest form, the economic value of earnings is equal to cash flow plus the change in the economic value of assets minus the change in the economic value of liabilities. Economic liabilities are the present value of expected cash flows plus an additional amount that would provide investors a return for placing their economic capital at risk (the "risk margin"). Typically, economic earnings equal the risk margin, and economic earnings divided by economic capital represent the return on capital. This return on capital can be used to measure the value creation from insurance underwriting activities.

Insurance management creates economic value if return on capital equals or exceeds the cost of capital. Because the cash flows associated with insur-ance contracts may not confirm or demonstrate the value of these activities until, perhaps, decades after the policies are sold, companies and investors are looking to answer some basic questions: How much value is created? How was the value created and when was it created (in sales, servicing, or risk management of the contracts)? How and in what manner can investors be convinced that reported "values" are really money and not just a game of numbers (Rubin et al., 2009)?

Recent changes in financial reporting for insurance contracts include a migration away from valuing liabilities and capital based on management

---

* http://www.investopedia.com/terms/t/third-party-insurance.asp.

judgment or regulatory rules to a system incorporating market-based assumptions and risk modeling of the business. Specific issues include the following:

1. Economic capital—and cost of capital as contemplated and implemented today—is not sufficiently market-based to measure whether a company will expect to earn more than its cost of capital.

2. A solvency system that is based solely on past observations may result in inadequate levels of capital and lead to a financial crisis in the insurance industry.

3. There is market evidence of unobserved risks not captured in economic capital modeling.

4. An insurance company can disclose to the market that it expects to earn more than its cost of capital without reporting a gain (Rubin et al., 2009).

6.6.3 Equipment Maintenance Insurance

Equipment maintenance insurance represents an alternative to equipment service contracts with original equipment manufacturers and third-party service providers. It eliminates the inefficiency of multiple service contracts. In addition, such programs can offer significant cost savings, expanded equipment coverage, and complete program management support.

The equipment maintenance insurance provider does not perform the equipment service. Companies continue to choose their preferred service provider. The program reimburses the company or the provider directly for the costs of repairs and preventive maintenance on covered equipment.*

An equipment maintenance insurance program does the following*:

1. Saves money:
   a. In most cases, it can save 15%–50% over present supplier maintenance contract costs.
   b. It budgets an expense that is currently difficult to manage or self-insured.
   c. It offers one policy versus multiple contracts (and hassles).

2. Assists with:
   a. Parts acquisition.
   b. Capital equipment replacement recommendations.
   c. Repair cost review.
   d. Negotiations with service provider.
   e. Competitive service providers.
   f. Program implementation and cost control training.

---

* http://tcim.ca/library/warranty/ (accessed June 27, 2016).

References

AEM (Asociación Española de Mantenimiento), 2005. El mantenimiento en España, encuesta sobre su situación en las empresas españolas españolas, 212 p. España, Barcelona, Spain (in Spanish).

Ahlmann, H., 1984. Maintenance effectiveness and economic models in the terotechnology concept. *Maintenance Management International* 4, 131–139.

Ahlmann, H., 1998. *The Economic Significance of Maintenance in Industrial Enterprises.* Lund University, Lund Institute of Technology, Lund, Sweden.

Alexander, M., Young, D., 1996. Strategic outsourcing. *Long Range Planning* 29(1), 116–119.

Al-Najjar, B., Andersson, D., Jacobsson, M., 2010. A model to describe the relationships man–machine–maintenance–economy. Members, IAENG. University of Jember, Indonesia.

Alsyouf, I., 2004. Cost effective maintenance for competitive advantages. PhD dissertation (terotechnology). Växjö University Press, Acta Wexionensia, Växjö, Sweden.

Anderson, J. C., Narus, J. A., 1999. *Business Market Management: Understanding, Creating and Delivering Value.* Prentice Hall, Upper Saddle River, NJ.

APICS, 2004. Protiviti independet risk consulting. Managing the risks of outsourcing: A survey of current practices and their effectiveness. APICS community at www.apics.org. Accessed June 6, 2004.

Ashton, K., 2009. That 'Internet of things' thing: In the real world, things matter more than ideas. *RFID Journal* 1, 1.

Bach, R., 2001. *A Consideration of Obsolescence Within the Design of Modern Avionics Test Systems.* NATO Research and Technology Organization Strategies to Mitigate Obsolescence in Defense Systems Using Commercial Components. NATO Research and Technology Organization, Neuilly-Sur-Seine, France.

Baldwin, R. C., 2004. *Enabling an E-Maintenance Infrastructure.* Available at http://www.mt-online.com.

Bangemann, T., Reboul, D., Scymanski, J., Thomesse, J. P., Zerhouni, N., 2006. PROTEUS—An integration platform for distributed maintenance systems. *Computers in Industry [Special Issue on e-Maintenance]* 57(6), 539–551.

Berry, L. L., Bendapudi, N., 2007. Health care: A fertile field for services research. *Journal of Service Research* 10, 111.

Bititci, U., Garengo, P., Dörfler, V., Nudurupati, S., 2012. Performance measurement: Challenges for tomorrow. *International Journal of Management Reviews* 14(3), 305–327.

Boyt, T., Harvey, M., 1997. Classification of industrial services. A model with strategic implications. *Industrial Marketing Management* 26, 291–300.

Byham, W. C., 2004. Development Dimensions International: The outsourcing question. www.ddiworld.com/pdf/WPOUTSO.pdf. Accessed: 01/10/2016.

Cameron, I., Duckworth, S., 1995. *Decision Support.* Industrial Development Research Foundation, Norcross, GA.

Candell, O., Karim, R., Söderholm, P., 2009. e-Maintenance—Information logistics for maintenance support. *Journal of Robotics and Computer-Integrated Manufacturing Archive* 25, 937–944.

Cashin, P., McDermott, C. J., 2002. The long-run behavior of commodity prices: Small trends and big variability, IMF Staff Papers. WP/01/68. Vol. 49, No. 2, pp. 175-199. Palgrave Macmillan Journals on behalf of the International Monetary Fund (IMF), Washington, DC.

Chawla, R., Kumar, G., 2013. Condition based maintenance modeling for availability analysis of a repairable mechanical system. *International Journal of Innovations in Engineering and Technology* 2(2), 371–379.

Chen, M., Mao, S., Liu, Y., 2014. Big data: A survey. *Mobile Networks and Applications* 19, 171–209. http://mmlab.snu.ac.kr/~mchen/min_paper/BigDataSurvey2014.pdf. Accessed: 02/10/2016.

Chiang, L. H., Braatz, R. D., Russell, E. L., 2001. *Fault Detection and Diagnosis in Industrial Systems*. Springer Verlag Limited, London, U.K.

Christensen, D., 1993. Determining an accurate estimate at completion. *National Contract Management Journal* 25, 17–25.

Chukova, S., Hayakawa, Y., 2004. Warranty cost analysis: Non-renewing warranty with repair time. *Appllied Stochastic Models in Business and Industry* 20, 59–71.

Cova, B., Salle, R., 2000. Rituals in managing extrabusiness relationships in international project marketing: A conceptual framework. *International Business Review* 9(6), 669–685.

Deardeen, J., Lilien, G., Yoon, E., 1999. Marketing and production capacity strategy for non-differentiated products: Winning and losing at the capacity cycle game. *International Journal of Research in Marketing* 16(1), 57–74.

Devereaux, J. E., February 2002. Obsolescence: A systems engineering and management approach for complex systems. S.B. Aerospace Engineering with Information Technology, Massachusetts Institute of Technology, Cambridge, MA.

Dimitrov, B., Chukova, S., Khalil, Z., 2004. *Warranty Costs: An Age-Dependent Failure/ Repair Model*. Wiley InterScience, Wiley Periodicals, Inc.

Dowling, T., 2001. Planning for change with a holistic view of the system. In: NATO Research and Technology Organization, Strategies to Mitigate Obsolescence in Defense Systems Using Commercial Components. NATO Research and Technology Organization, Neuilly-Sur-Seine, France.

Embleton, P. R., Wright, P. C., 1998. A practical guide to successful outsourcing. *Empowerment in Organizations* 6(3), 94–106.

EN-15341, April 30, 2007. *Maintenance Key Performance Indicators*. BSI. ISBN 978 0 580 50611 6.

Encon Hui, Y. Y., Albert Tsang, H. C., 2004. Sourcing strategies of facilities management. *Journal of Quality in Maintenance Engineering* 10(2), 85–92.

Ender, K. L., Mooney, K. A., 1994. From outsourcing to alliances: Strategies for sharing leadership and exploiting resources at metropolitan universities. *Metropolitan Universities: An International Forum* 5(3), 51–60.

Ferlez, R., Lang, D., 1998. Gear-tooth fault detection and tracking using the wavelet transform. In: *Proceedings of the 52nd Meeting of the Machinery Failure Prevention Technology*, Virginia Beach, VA, March 20–April 2, 1998.

Galar, D., Thaduri, A., Catelani, M., Ciani, L., 2015. Context awareness for maintenance decision making: A diagnosis and prognosis approach. *Measurement* 67, 137–150.

Galar, D., Wandt, K., Karim, R., and Berges, L., 2012. The evolution from e (lectronic) Maintenance to i (ntelligent) maintenance. *Insight-Non-Destructive Testing and Condition Monitoring* 54(8), 446–455.

Galar, D., Kans, M., and Schmidt, B., 2016. Big data in asset management: Knowledge Discovery in asset data by the means of data mining. In: Koskinen, K.T., Kortelainen, H., Aaltonen, J., Uusitalo, T., Komonen, K., Mathew, J., Laitinen, J. (Eds.), *Proceedings of the 10th World Congress on Engineering Asset Management (WCEAM 2015)*. Springer, Cham, Switzerland.

Geiger, D. R., 2005. Pavement preservation definitions. http://www.pavement-preservation.org/PP_Defs_Memo_09_05.pdf.

Gómez, J., Crespo Márquez, A., Moreu, P., Parra, C., González-Prida, V., 2009. Outsourcing maintenance in services providers. In: Martorell, S. et al. (Eds.), *Safety, Reliability and Risk Analysis: Theory, Methods and Applications*. Taylor & Francis Group, London, U.K., pp. 829–837.

Goncharenko, I., Kimura, F., 1999. Remote maintenance for IM [inverse manufacturing]. In: *Proceedings of the First International Symposium on Environmentally Conscious Design and Inverse Manufacturing*, Tokyo, Japan, pp. 862–867.

González Díaz, V., Barberá Martínez, L., Gómez Fernández, J. F., Crespo Márquez, A., 2012a. ICT application on the warranty management process: The "e-Warranty" concept. In: Bérenguer, G., Soares, G. (Eds.), *Advances in Safety, Reliability and Risk Management*. Taylor & Francis Group, London, U.K.

González Díaz, V., Crespo Márquez, A., Pérès, F., De Minicis, M., Tronci, M., 2012b. Logistic support for the improvement of the warranty management. In: Bérenguer, G., Soares, G. (Eds.), *Advances in Safety, Reliability and Risk Management*. Taylor & Francis Group, London, U.K.

González Díaz, V., Gómez Fernández, J. F., Crespo Márquez, A., 2011. Practical applications of AHP for the improvement of waranty management. *Journal of Quality in Maintenance Engineering (JQME)* 17, 163–182.

González Díaz, V., Gómez, J. F., López, M., Crespo, A., Crespo Márquez, A., 2010. Warranty cost models State of Art: A practical review to the framework of warranty cost management. In: Briš, R., Guedes Soares, C., Martorell, S. (Eds.), *Reliability, Risk and Safety: Theory and Applications*. Taylor & Francis Group, London, U.K.

González Díaz, V., Gómez, J. F., López, M., Crespo, A., Moreu de León, P., 2009. Warranty cost models state of art: A practical review to the framework of warranty cost management. In: *Proceedings of the Annual Conference of European Safety and Reliability*, ESREL, Prague, Czech Republic, September 5–9, pp. 2051–2059.

Gonzalez-Prida Díaz, V., Crespo Márquez, A., 2014. *After–Sales Service of Engineering Industrial Assets: A Reference Framework for Warranty Management*. Springer International Publishing, Switzerland, XXII, 318pp. ISBN 978-3-319-03710-3. http://www.springer.com/gp/book/9783319037097.

Goodenow, T., Hardman, W., Karchnak, M., 2000. Acoustic emissions in broadband vibration as an indicator of bearing stress. In: *IEEE Aerospace Conference Proceedings*, Epoch Eng. Inc., Palm City, FL, Vol. 6, pp. 95–122.

Grönroos, C., 2000. *Service Management and Marketing—A Customer Relationships Management Approach*. Wiley, Chichester, U.K.

Grönroos, C., Ojasalo, K., 2004. Service productivity: Toward a conceptualization of the transformation of inputs into customer value in services. *Journal of Business Research* 57(4), 414–423.

Hamel, W., 2000. E-maintenance robotics in hazardous environments. In: *Proceedings of the 2000 IEEE/RSJ International Conference on Intelligent Robots and Systems*, Takamatsu, Japan.

Hansen, R., Hall, D., Kurtz, S., 1994. New approach to the challenge of machinery prognostics. In: *Proceedings of the International Gas Turbine and Aeroengine Congress and Exposition*, June 13–16, 1994, American Society of Mechanical Engineers, pp. 1–8.

Hardman, W., Hess, A., Sheaffer, J., 2000. A helicopter powertrain diagnostics and prognostics demonstration. In: *IEEE Aerospace Conference Proceedings*, Power & Propulsion Dept., Naval Air Warfare Center Aircraft Div., Patuxent River, MD, Vol. 6, pp. 355–366.

Hernandez, A., Galar, D., 2014. Techniques of prognostics for condition-based maintenance in different types of assets. Technical report, Department of Civil, Environmental and Natural Resources Engineering, Operation, Maintenance and Acoustics. Luleå University of Technology, Luleå, Sweden.

Holmberg, K., 2001. New techniques for competitive reliability. *International Journal of COMADEM* 4, 41–46.

Holmberg, K., Adgar, A., Arnaiz, A., Jantunen, E., Mascolo, J., Mekid, S., 2010. Chapter 2: Maintenance today and future trends. In: *e-Maintenance*. Springer, London, U.K., XXV, 511p., Hardcover. http://www.springer.com/978-1-84996-204-9. Accessed: 01/10/2016.

Holmberg, K., Helle, A., 2008. Tribology as basis for machinery condition diagnostics and prognostics. *International Journal of Performability Engineering* 4, 255–269.

Holmberg, K., Helle, A., Halme, J., 2005. Prognostics for industrial machinery availability. In: *POHTO 2005 International Seminar on Maintenance, Condition Monitoring and Diagnostics*, Oulu, Finland.

Iung, B., Morel, G., Leger, J. B., 2003. Proactive maintenance strategy for harbor crane operation improvement. *Robotics [Special Issue on Cost Effective Automation, Erbe, H. (Ed.)]* 21(3), 313–324.

Jääskeläinen, A., 2010. *Productivity Measurement and Management in Large Public Service Organizations. Publication 927*. Tampere University of Technology, Tampere, Finland.

Jantunen, E., Emmanouilidis, C., Arnaiz, A., Gilabert, E., 2011. e-Maintenance: Trends, challenges and opportunities for modern industry. In: *Eighteenth IFAC World Congress*, January 2011, Vol. 44(1), pp. 453–458.

Jardine, A., Lin, D., Banjevic, D., 2006. A review on machinery diagnostics and prognostics implementing condition-based maintenance. *Mechanical Systems and Signal Processing 2006* 20(7), 1483–1510.

Jardine, A., Makis, V., Banjevic, D., Braticevic, D., Ennis, M., 1998. Decision optimization model for condition-based maintenance. *Journal of Quality in Maintenance Engineering* 4, 115–121.

Jonker, R., Haarman, M., 2006. Value driven maintenance: What is the actual added value of maintenance? http://www.reliabilityweb.com/art07/value_driven_maintenance_uptime.pdf. Accessed: 01/10/2016.

Kans, M., 2008. On the utilization of information technology for the management of profitable maintenance. PhD dissertation (terotechnology). Växjö University Press, Acta Wexionensia, Växjö, Sweden.

Kärri, T., Sinkkonen, T., Tynninen, L., Marttonen, S., 2011. Life-cycle thinking for acquisition of machines and preventing lack of maintenance. *Promaint* 25(7), 14–17.

Kemerait, R., 1987. New cepstral approach for prognostic maintenance of cyclic machinery. In: *IEEE SoutheastCon*, pp. 256–262.

Khosrowshahi, F., Ghodous, P., Sarshar, M., 2014. Visualization of the modeled degradation of building flooring systems in building maintenance. *Computer-Aided Civil and Infrastructure Engineering* 29(2014), 18–30.

Koç, M., Ni, J., Lee, J., Bandyopadhyay, P., 2003. Introduction of e-manufacturing. In: *31st North American Manufacturing Research Conference (NAMRC)*, Hamilton, Ontario, Canada.

Koskinen, K. T., Kortelainen, H., Aaltonen, J., Uusitalo, T., Komonen, K., Mathew, J., Laitinen, J., 2016. *Proceedings of the 10th World Congress on Engineering Asset Management (WCEAM 2015)*. Springer International Publishing, Cham, Switzerland 2016. 9783319270647.

Kurdia, M. K., Abdul-Tharim, A. H., Jaffar, N., Azli, M. S., Shuib, M. N., AbWahid, A. M., 2011. Outsourcing in facilities management—A literature review. In: *The Second International Building Control Conference*, Penang, Malaysia.

Lane, C. H., Beattie, E. S., Chita, J. S., Lincoln, S. P., 2001. *Adopting New Software Development Techniques to Reduce Obsolescence. NATO Research and Technology Organization Strategies to Mitigate Obsolescence in Defense Systems Using Commercial Components.* NATO Research and Technology Organization, Neuilly-Sur-Seine, France.

Lankford, W. M., Parsa, F., 1999. Outsourcing: A primer. *Management Decision* 37, 310–316.

Lee, J., 2001. A framework for next-generation e-maintenance system. In: *Proceedings of the Second International Symposium on Environmentally Conscious Design and Inverse Manufacturing*, Tokyo, Japan.

Lee, J., 2003. E-manufacturing: Fundamental, tools, and transformation. *Robotics and Computer Integrated Manufacture* 19(6), 501–507.

Lee, J., Ni, J., Djurdjanovic, D., Qui, H., Liao, H., 2006. Intelligent prognostics tools and e-maintenance. *International Journal of Computers in Industry* 57(6), 476–489.

Lee, J., Wu, F., Zhao, W., Ghaffari, M., Liao, L., Siegel, D., 2014. Prognostics and health management design for rotary machinery systems—Reviews, methodology and applications. *Mechanical Systems and Signal Processing* 42, 314–334.

Lee, J., Yang, S., Lapira, E., Kao, H. A., Yen, N., 2013. Methodology and framework of a cloud-based prognostics and health management system for manufacturing industry. *Chemical Engineering Transcriptions* 33, 205–210.

Lönnqvist, A., Jääskeläinen, A., Kujansivu, P., Käpylä, J., Laihonen, H., Sillanpää, V., Vuolle, M., 2010. *Palvelutuotannon Mittaaminen Johtamisen Välineenä*. Tietosanoma Oy, Helsinki, Finland.

Lua, Y. A., Yu Guang, X., Sow Wai, Z., Jang Wei, L., 2012. Comprehensive life cycle approach to obsolescence management. Defence Science and Technology Agency (DSTA) HORIZONS, DSTA Horizons Editorial Team.

Lynch, D., 2001. Strategies to mitigate obsolescence in defense systems using commercial components. Technical Evaluation Report, pp. T-2–T-3. NATO Research and Technology Organization, Neuilly-Sur-Seine, France.

Lyons, K., Murthy, D. N. P., 2001. Warranty and Manufacturing. In: Rahim, M. A., Ben-Daya, M. (Eds.), *Integrated Optimal Modelling in PIQM: Production Planning, Inventory, Quality and Maintenance*. Kluwer Academic Publishers, New York, pp. 287–324.

Manca, D., Brambilla, S., Colombo, S., 2013. Bridging between virtual reality and accident simulation for training of process-industry operators. *Advances in Engineering Software* 55, 1–9.

Marttonen, S., Viskari, S., Kärri, T., 2013. Appeasing company owners through effective working capital management. *International Journal of Managerial and Financial Accounting* 5(1), 64–78.

McClintic, K., 1998. Feature prediction and tracking for monitoring the condition of complex mechanical systems. MS thesis. Pennsylvania State University, University Park, PA.

McLaughlin, C. P., Coffey, S., 1990. Measuring productivity in services. *International Journal of Service Industry Management* 1(1), 46–64.

Menches, C. L., Khwaja, N., Chen, J., 2010. Synthesis of innovative contracting strategies used for routine and preventive maintenance contracts. Project 0-6388. Performing Organization: Center for Transportation Research. The University of Texas at Austin, Austin, TX.

Menezes Melvyn, A. J., Quelch, J. A., 1990. Leverage your warranty program, *Sloan Management Review* 31(4, Summer), p. 69.

Mitchell, E., Robson, A., Prabhu, V., 2002. The impact of maintenance practices on operational and business performance. *Managerial Auditing Journal* 17(5), 234–240.

Moore, W. J., Starr, A. G., 2006. An intelligent maintenance system for continuous cost-based prioritisation of maintenance activities. *Computers in Industry [special issue on e-maintenance]* 57(6), 595–606.

Muller, A., Marquez, A. C., Iung, B., 2008. On the concept of e-maintenance: Review and current research. *Reliability Engineering and System Safety* 93, 1165–1187.

Murthy, D. N. P., 2006. Product warranty and reliability. *Annals of Operations Research* 143, 133–146.

Murthy, D. N. P., Djamaludin, I., 2002. New product warranty: A literature review. *International Journal of Production Economics* 79, 231–260.

Niekerk, A. J. V., Visser, J. K., 2010. *The Role of Relationship Management in the Successful Outsourcing of Maintenance*. Graduate School of Technology Management, University of Pretoria, Pretoria, South Africa.

Nili, M., Shekarchizadeh, A., Shojaey, R., July 2012. Outsourcing of maintenance activities in oil industry of Iran: Benefits, risks and success factors. *Business and Management Review* 2(5), 20–36.

OSF Global Services, 2012. The real cost of outsourcing. Write paper. www.osf-global. com. Accessed: 03/10/2016

Packard, V., 1960. *The Waste Makers*. David McKay Company, New York.

Pakkala, P. A., De Jong, M., Aijo, J., 2007. *International Overview of Innovative Contracting Practices for Roads*. Finnish Road Administration, Helsinki, Finland.

Pekkarinen, O., Piironen, M., Salminen, R. T., 2012. BOOT Business Model in Industrial Solution Business. *International Journal of Business Innovation and Research* 6(6), 653–673.

Penna, R., Amaral, M., Espýndola, D., 2014. *Visualization Tool for Cyber-Physical Maintenance Systems*. IEEE, Porto Alegre, Brazil. ISBN 978-1-4799-4905-2.

Perera, C., Zaslavsky, A., Christen, P., Georgakopoulos, D., 2014. Context aware computing for the internet of things: A survey. *IEEE Communications Surveys & Tutorials* 16(1), 414–454.

Peters, R. W., 2003. Measuring overall craft effectiveness (OCE). Are you a takeover target for contract maintenance. *Plant Engineering*. 57(10), 39–41.

Petersen, L., 2001. *The Use of Commercial Components in Defense Equipment to Mitigate Obsolescence. A Contradiction in Itself? Strategies to Mitigate Obsolescence in Defense Systems Using Commercial Components*. NATO Research and Technology Organization, Neuilly-Sur-Seine, France, pp. 1-1–1-7.

Price, C., Price, C. J., 1999. Computer-Based Diagnostic Systems. Springer, Heidelberg, Germany, pp. 65–69.

Quélin, B., Motlow, D., 1998. Outsourcing: A transaction cost theory approach. *Réseaux. The French Journal of Communication* 6(1), 75–98.

Reichard, K., Van Dyke, M., Maynard, K., 2000. Application of sensor fusion and signal classification techniques in a distributed machinery condition monitoring system. *Proceedings of SPIE* 4051, 329–336.

Roemer, M., Dzakowic, J., Orsagh, R., Byington, C., Vachtsevanos, G., 2005. An overview of selected prognostic technologies with reference to an integrated PHM architecture. In: *Proceedings of the IEEE Aerospace Conference 2005*, Big Sky, MT.

Roemer, M., Kacprzynski, G., Orsagh, R., 2001. Assessment of data and knowledge fusion strategies for prognostics and health management. In: *IEEE Aerospace Conference Proceedings*, Big Sky, MT, Vol. 6, 2001, pp. 62979–62988.

Rothmaier, M., 2001. *A Modular Signal Processing Architecture to Mitigate Obsolescence in Airborne Systems. NATO Research and Technology Organization Strategies to Mitigate Obsolescence in Defense Systems Using Commercial Components.* NATO Research and Technology Organization, Neuilly-Sur-Seine, France.

Rubin, L., Lockerman, M., Tillis, R., Xiaokai Shi, V., 2009. Economic measurement of insurance liabilities: The risk and capital perspective. Published in the March 2009 issue of the Actuarial Practice Forum. Copyright 2009 by the Society of Actuaries.

Salonen, A., 2004. Managing outsourced support services: Observations from case study. *Facilities* 22(11/12), 317–322.

Schmidt, B., Sandberg, U., Wang, L., 2014. Next generation condition based predictive maintenance. http://conferences.chalmers.se/index.php/SPS/SPS14/paper/viewFile/1742/414. Accessed: 03/10/2016.

Segal, G. F., Moore, A. T., McCarthy, S., 2003. *Contracting for Road and Highway Maintenance. How-to Guide No. 21.* Reason Public Policy Institute, Los Angeles, CA.

Shah, M., Littlefield, M., 2009. Managing risks in asset intensive operations. Aberdeen Group. http://www.aberdeen.com/.

Sinkkonen, T., Kivimäki, H., Marttonen, S., Galar, D., Villarejo, R., Kärri, T., 2016. Using the life-cycle model with value thinking for managing an industrial maintenance network. *International Journal of Industrial and Systems Engineering* 23, 1.

Sinkkonen, T., Kivimäki, H., Marttonen, S., Kärri, T., 2013. A value-based life-cycle framework for networks of industrial maintenance services. *COMDEM*, Helsinki, Finland, June 11–13, 2013.

Slade, G., 2006. *Made to Break: Technology and Obsolescence in America.* Harvard University Press, Cambridge, U.K.

Smith, T., 2001. *Future initiatives for obsolescence mitigation strategies. NATO Research and Technology Organization Strategies to Mitigate Obsolescence in Defense Systems Using Commercial Components.* NATO Research and Technology Organization, Neuilly-Sur-Seine, France.

Stankevich, N., Qureshi, N., Queiroz, C., 2005. Performance-based contracting for preservation and improvement of road assets. Transport Note No. TN-27, The World Bank, Washington, DC.

Stene-erik, B., Baglee, D., Galar, D., Sarbjeet, S., 2013. Maintenance knowledge management with fusion of CMMS and CM. In: *DMIN 2013 International Conference on Data Mining*, Las Vegas, NV, July 22–25, 2013.

Sullivan, E., 2005. *Building Operating Management*, pp. 27–30.

Swanson, D., 2000. Prognostic modelling of crack growth in a tensioned steel band. *Mechanical Systems and Signal Processing* 14(5), 789–803.

Swanson, D., 2001. A general prognostic tracking algorithm for predictive maintenance. In: *Aerospace Conference, 2001, IEEE Proceedings*, vol. 6, March 10–17, 2001, The Applied Research Laboratory, The Pennsylvania State University, State College, PA, pp. 2971–2977.

Szatvanyi, G., September 22, 2008. The real cost of outsourcing. http://ezinearticles.com/?The-Real-Cost-of-Outsourcing&id=1517601.

Takata, S., Kimura, F., Van Houten, F. J. A. M., Westkamper, E., Shpitalni, M., Ceglarek, D., Lee, J., 2004. Maintenance: Changing role in life cycle management. *Annals of the CIRP* 53(2) (2004/8)), 643–655.

Thurston, M., Lebold, M., 2001. Standards developments for condition-based maintenance systems, improving productivity through applications of condition monitoring. In: *55th Meeting of the Society for Machinery Failure Prevention Technology*, Applied Research Laboratory, Penn State University, State College, PA.

Tseng, M., Hu, S., Wang, Y., 2013. *Mass Customization. CIRP Encyclopedia of Production Engineering*. Springer Publications, Berlin, Germany.

Utterback, J. M., 1994. *Mastering the Dynamics of Innovation*. HarvardBusiness School Press, Cambridge, MA.

Vargo, S. L., Lusch, R. F., 2004. Evolving to a new dominant logic for marketing. *Journal of Marketing* 68(1), 1–17.

Wang, L., 2014. Cyber manufacturing: Research and applications. *Proceedings of the 10th International Symposium on Tools and Methods of Competitive Engineering* 1, 39–49.

Weyuker, E. J., 1998. Testing component based software: A cautionary tale. *IEEE Software*, September–October 1998 , 54–59.

WFii. January 1, 2016. The role of insurance intermediaries. https://www.ciab.com/uploadedfiles/resources/roleofinsint.pdf. Accessed: 02/10/2016.

Wilson, B., 1999. Development of a modular in-situ oil analysis prognostic system. In: *Proceedings of the International Society of Logistics (SOLE) 1999 Symposium*, Las Vegas, NV, August 30–September 2, 1999.

Wireman, T., 2003. *Benchmarking Best Practices in Maintenance Management*. Industrial Press Inc., New York. ISBN 0-8311-3168-3 (hardcover). www.TerryWireman.com.

Wohlwend, H., 2005. E-diagnostics Guidebook: Revision 2.1. Technology Transfer # 01084153D-ENG, SEMATECH Manufacturing Initiative, Technology Transfer, Austin, TX.

Woodhouse, J., 2013. *Setting a Good Standard in Asset Management*. The Woodhouse Partnership Ltd, Kingsclere, England, pp. 27–30. Chair of Experts Panel, Institute of Asset Management and Managing Director, TWPL.

Zhang, L., Luo, Y., Tao, F., Li, B. H., Ren, L., Zhang, X., Guo, H., Cheng, Y., Hu, A., Liu, Y., 2014. Cloud manufacturing: A new manufacturing paradigm. *Enterprise Information Systems* 8(2), 167–187.

7

Maintenance Costs Across Sectors

7.1 Maintenance Costs Across the Sectors

As the title suggests, this chapter looks at maintenance costs throughout the industry—not just across industries but within industries. It applies many of the concepts discussed in previous chapters, including the use of maintenance costing policies within life cycle cost analysis (LCCA), among others.

The chapter applies the developed concepts using a series of case studies. It begins with transportation, specifically rail travel, using the Europe–Asia railway system as the main example, and considering the costs involved when the track gauge changes between countries. It also discusses LCC for trams. The next section goes on to consider maintenance costs in infrastructure, while remaining within the railway sector. The third section deals with manufacturing assets, this time using a pump system for illustrative purposes. Section four turns to the US Army for a brief discussion of its use of cost analysis, especially in the acquisition process. Section five looks at the aviation industry, with special focus on how the type and usage of aircraft affect costs. Section six discusses facilities' operation and management (O&M) costs, and section seven looks at life cycle management (LCM) in the energy sector, in this case, the nuclear power industry. Section seven rounds off the chapter with an analysis of mining.

Although we are looking at quite diverse elements of the workplace across the globe, as readers go through the chapter, they will quickly see how concepts apply more or less universally, albeit with certain adaptations to suit a specific sector and its specific cost concerns. We are reminded, once again, that all industries perform maintenance and all want the best bang for their buck.

7.2 Transportation Assets: Rolling Stock Case Study 1

7.2.1 Introduction and Background

A reliable and efficient transport system is the basis for economic development and trade between the East and the West. The railway transport system offers considerable shortening of the duration of cargo transport from Asia to Europe and vice versa. However, there can be problems when the track gauge changes between countries.

The majority of European countries, including Poland, have 1435 mm gauge tracks but the railways of the former USSR, including Lithuania, Latvia, and Estonia, have 1520 mm gauge tracks. Meanwhile, in Spain and Portugal, the tracks are very wide, at 1668 mm. In parts of Asia, trains move on a wide-gauge track (1520 mm), but encounter normal gauge (1435 mm) lines in China and Korea. The map in Figure 7.1 shows these differences (Szkoda et al., 2011).

Such differences seriously impede transportation, as at the points where different gauges meet, the cargo must be either transshipped or the running assemblies of rail vehicles must be exchanged. These operations are costly and time-consuming and require extended infrastructure, along with very expensive storage and transshipment facilities at border-crossing points. They also extend transportation time considerably.

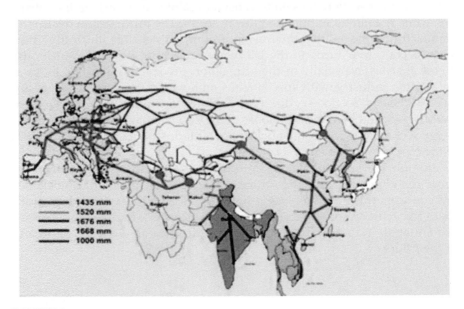

FIGURE 7.1
Variety of track gauge. (Adapted from Szkoda, M., Life cycle cost analysis of Europe-Asia transportation systems, in: *EURO—ZEL 2011, 19th International Symposium*, Žilina, Slovakia, June 8–9, 2011.)

The transport system between Europe and Asia spans 15,000 km and requires a specific type of service because of the changes in rail gauge between countries. Two basic technologies are used to handle the problem:

1. Handling technology
2. Shifting technology

Figure 7.2 suggests some techno-organizational variants of both.

Generally, transshipping technology deals with transporting cargo at the meeting points of different gauge railways from a freight car on one gauge to a car on the other. Depending on the kind of cargo, the following methods may be used:

- Reloading
- Pumping
- Pouring

The shifting technology shifts the means of transport from one gauge to the other in one of two ways:

- Exchanging the vehicle running assemblies
- Using self-adjusting wheel sets (Szkoda et al., 2011)

7.2.2 Decision Models of Transport Systems Evaluation

For effective evaluation of gauge change techniques, the following methods may be applied:

- Techno-economic analysis
- Life cycle cost analysis (LCCA)
- Analytic network process analysis

In this chapter, given the book's focus, the LCC method will be considered even though it may not be the optimum one. LCC includes total costs, comprising of costs of purchase, acquisition, and liquidation. Each element of each of the various datasets used in the life cycle of an asset requires a detailed definition and description of operational and experimental data, as well as data obtained by other means (e.g., expert methods; see Tulecki, 1999). In this particular case, LCCA can be used for the following purposes:

- To make decisions about the organization of transport systems
- To make decisions about system modernization and restructuring
- To create a haulage technology assessment using different haulage technology variants
- To build a basis for costs shaping the transport service

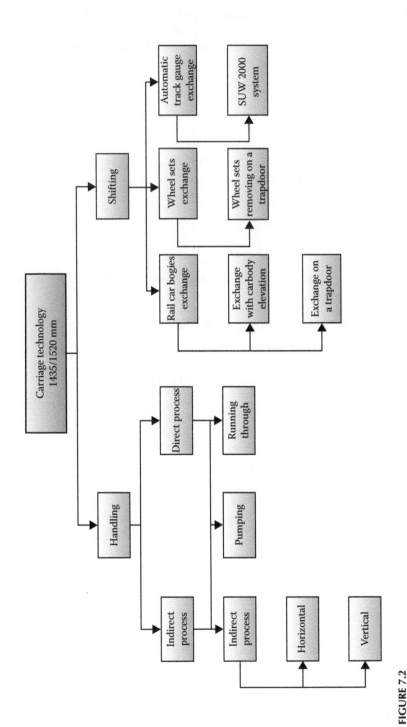

FIGURE 7.2
Techno-organizational variants of railway gauge change. (Adapted from Szkoda, M., Life cycle cost analysis of Europe-Asia transportation systems, in: *EURO—ZEL 2011, 19th International Symposium*, Žilina, Slovakia, June 8–9, 2011.)

In what follows, we discuss a study on the possibility of improving the railway transport of goods by using self-adjusted wheel sets at the border-crossing point 1435/1520 mm to replace the existing wagon bogie exchange. Notably for present purposes, the comparative analysis of the two methods uses a decision-making model based on LCCA (Szkoda et al., 2011).

7.2.2.1 Assumptions and Purpose of LCCA

The LCC study of hazardous material haulage in the East–West transport system uses two variants of track gauge change:

- *Variant 1*: haulage using the currently applied method of wagon bogie exchange
- *Variant 2*: haulage using the system of self-adjusting wheel sets

The study compares the costs generated by the two variants, with the following assumptions accepted for cost structures of the variants:

- *Haulage amount*: 273,000 tons/year.
- *Wagon load capacity*: tank car with a 50 ton load capacity.
- *Haulage distance*: 1100 km; this corresponds to the reality of the East–West hazardous material haulage (Figure 7.3) (Szkoda et al., 2011).

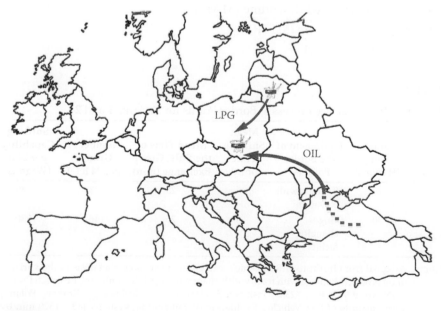

FIGURE 7.3
East–West transport of hazardous materials. (Adapted from Szkoda et al., 2011.)

7.2.2.2 Comparison of Service Process in Analyzed Variants

LCCA must identify the service process at the contact points of different gauge tracks. Table 7.1 presents some parameters of the service process of the two variants used in this study. The parameters were obtained from a techno-organizational evaluation.

Taking into consideration service time and the capability resulting from the time, the table shows that variant 2, i.e., self-adjusted wheel sets, is unrivalled. However, there are some limitations in service universality. Either full train load haulage or initial switching before point 1435/1520 mm is required (Szkoda and Tulecki, 2006; Szkoda 2008).

7.2.2.3 System Breakdown Structure and LCC Model Development

Common elements, which have the same influence in both variants, for example, railway infrastructure and locomotives, are eliminated from the calculations (Table 7.2).

In the study, LCC is preceded by dependability analysis of reliability, availability, and maintainability (RAM) for all elements relevant to both variants. The most important dependability factors are found to be the following:

- Failure intensity $z(t)$
- Mean time between failure (MTBF)
- Mean uptime (MUT)
- Mean accumulated downtime (MADT)
- Mean time to restore (MTTR)
- Mean availability A

TABLE 7.1

Characteristics of Service Process at Border Points for Variants 1 and 2

Variant	Shift Group (Wagons)	Equipment of the Border Point (–)	Mean Shifting Time (min)	Mean Time of the Shift Group Exchange (min)	Number of Groups per 24 h (–)	Shifting Capability per 24 h (Wagons)
1	10	10 stands with elevators	200	25	3	30
2	30	Gauge-changing facility	6	25	46	1380

Sources: Adapted from Institute of Rail Vehicles, Organizing economic analysis of the variants of oil products haulage with track width change using 911Ra tank cars, Research Project KBN No. 9 9454 95 C/2385, Task No. 5, University of Technology, Krakow, Poland, 1995; Institute of Rail Vehicles, Evaluation of rail shifting systems 1435/1520 mm by applying LCC analysis, Research Project No. NB-2/2008, University of Technology, Krakow, Poland, 2008.

TABLE 7.2

Elements of Structure in Analyzed Variants

Analyzed Variant	Element of System Structure (Units)				
	Label	Applied to Rolling Stock	Label	Applied to Point 1435/1520 mm	
Variant 1	1.1	Freight bogies for 1435 mm (106)	1.3	Gantry crane (3)	
	1.2	Freight bogies for 1520 mm (106)	1.4	Staired with elevators (14)	
Variant 2	2.1	Freight bogies with self-adjusted wheel sets of SUW 2000 system (80)	2.2	Track gauge changing stand of the SUW 2000 system (1)	

Source: Adapted from Szkoda, M., Life cycle cost analysis of Europe-Asia transportation systems, in: *EURO—ZEL 2011, 19th International Symposium*, Žilina, Slovakia, June 8–9, 2011.

For its RAM analysis, the study conducts dependability tests to gather and transform the operation information.

The indicated dependability parameters connected with reliability, durability, maintainability, and availability comprise the cost element base in LCC models. The present analysis has a comparative character, so all categories which are the same for both variants are excluded from the cost model. This assumption makes the cost structure much simpler. The LCC model is developed using investments and acquisition costs (7.1). A period of 25 years of operation is assumed for analysis. The determination of LCC can be given as follows:

$$LCC = INC + AQC \qquad (7.1)$$

where
   INC is the investments cost
   AQC is the acquisition cost

Investment costs are the sum of capital investments necessary for the analyzed variants. Acquisition costs constitute both maintenance and operational (O&M) costs (7.2) and can be expressed as follows:

$$AQC = MC + OC = (PMC + CMC) + (POC + UNC) \qquad (7.2)$$

where
   MC is the maintenance cost
   PMC is the preventive maintenance cost
   CMC is the corrective maintenance cost
   OC is the operation cost
   POC is the operation personnel cost
   UNC is the unavailability cost

Cost valuation is based on constant prices tin euros (€) from 2010 (Szkoda, 2011).

7.2.2.4 LCC Assessment and DSS Based on Outcomes

The models are analyzed using CATLOC software. Findings show that carrying hazardous materials 1100 km using the variable-gauge wheel sets is much more effective than using the present bogie exchange. The LCC for variant 2, over 25 years of operation, is €3.62 million or 20.8% lower than for variant 1 (Figure 7.4a). The fundamental difference between variants appears in the acquisition costs: 33.0% lower for variant 2 (Figure 7.4b).

Figure 7.5 presents the share of basic cost categories in the LCC structure for variants 1 and 2. The categories in variant 1 with the biggest share in LCC are investment costs (41.2%), preventive maintenance costs (29.6%), and operation personnel costs (18.8%). The categories with the most significant impact in variant 2 are also investment costs (50.2%) and preventive maintenance costs (28.6%), generated by routine repairs and overhauls of the SUW 2000 bogies.

In variant 1, almost 90% of LCC is generated by handling shifting equipment at the border-crossing point (Figure 7.6a). In variant 2, costs at point 1435/1520 mm comprise only 4.7% of LCC, thanks to the replacement of expensive maintenance bogie exchange facilities with the high availability of a reliable and relatively inexpensive track gauge changing stand (Figure 7.6b).

The study's analysis suggests the application of the self-adjusted wheel sets system in the transport of hazardous materials in certain cases is justified in both technical and economic terms. Ultimately, the effectiveness is determined by the price of the wagon bogie with self-adjusted wheel sets. The current price or initial investment is too high to ensure an economic return for distances of more than 1500 km; therefore, the manufacturer should develop solutions to reduce the wagon bogie price by at least 20%. The efficiency of transport (including the transport of hazardous materials) with the use

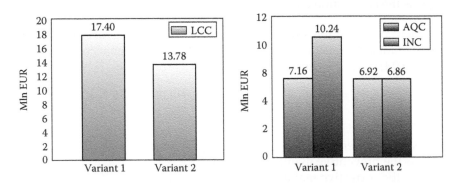

FIGURE 7.4
LCC of variants (a), and investment and acquisition costs Variants (b). (Adapted from Szkoda, M., Life cycle cost analysis of Europe-Asia transportation systems, in: *EURO—ZEL 2011, 19th International Symposium*, Žilina, Slovakia, June 8–9, 2011.)

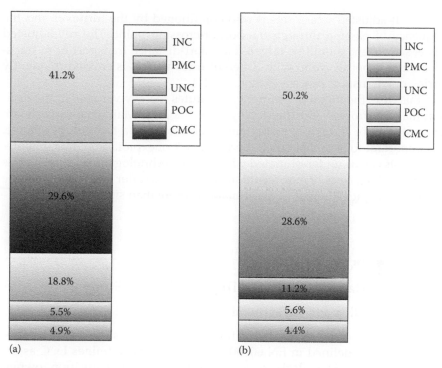

FIGURE 7.5
Share of elementary cost categories in LCC: variant 1 (a), variant 2 (b). INC, investment costs; PMC, preventive maintenance costs; UNC, unavailability costs; POC, operation personnel costs; CMC, corrective maintenance costs. (Adapted from Szkoda, M., Life cycle cost analysis of Europe-Asia transportation systems, in: *EURO—ZEL 2011, 19th International Symposium*, Žilina, Slovakia, June 8–9, 2011.)

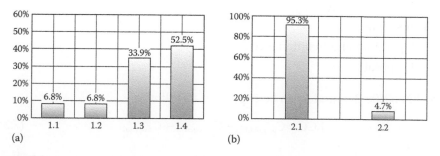

FIGURE 7.6
LCC in system breakdown structure (percentage): variant 1 (a), variant 2 (b). (Adapted from Szkoda, M., Life cycle cost analysis of Europe-Asia transportation systems, in: *EURO—ZEL 2011, 19th International Symposium*, Žilina, Slovakia, June 8–9, 2011.)

of the self-adjusted wheel sets is also conditioned by the turnover and the frequency of passing through the border-crossing point. The self-adjusted wheel set system should be targeted at short-distance transport or at transport where a border point with different track widths is crossed frequently during one transport process (Szkoda, 2011).

In this case, the initial investment constitutes the key variable in the decision-making process and the deployed technology. Maintenance costs are relevant but not enough to justify any decision about a technology change. It is important to note that when there is a lack of exploitation data, maintenance costs can be underestimated when a new technology is selected. In this case, the hidden costs of new techniques popped up after operation for some years, and the deviation of LCC estimates is more than significant.

7.3 Transportation Assets: Rolling Stock Case Study 2

7.3.1 Introduction and Background

The previous chapter discussed different LCC models. One of the most popular ones is defined in EN 60300 3-3. The standard defines LCC as the process of economic analysis to assess the total cost of acquisition, ownership, and disposal of a product. The total costs of the life cycle are similar to the UNIFE (European Society of Railway Industry) model, with some differences (IEC, 2004).

There are six major life cycle phases of a product:

1. Concept and definition
2. Design and development
3. Manufacturing
4. Installation
5. Operation and maintenance
6. Disposal

The EN 60300-3-3 model can be expressed as follows:

$$LCC = C_{acquisition} + C_{ownership} + C_{disposal} \qquad (7.3)$$

Acquisition costs are generally visible and can be readily evaluated before the acquisition decision is made and may or may not include installation cost.

The ownership costs, often a major component of LCC, frequently exceed acquisition costs in the railway domain and are not readily visible.

These costs are difficult to predict and may include the cost of installation. This means the total cost of ownership (TCO) is an important consideration in the railway industry.

Disposal costs may represent a significant proportion of total LCC. Legislation may require certain activities during the disposal phase, especially for major projects (Klyatis, 2012).

In 1997, UNIFE created a working group to study the LCC of the railway. It started by defining the terms and conditions for the LCC of rolling stock. Working group II, led by Pierre Dersin at Alstom, expanded the earlier work by defining the terms and conditions for LCC for total rail systems. Working group III, led by Ulf Kjellsson at Bombardier Transportation, issued a European LCC interface software model, UNIFE-UNILIFE and UNIFE-UNIDATA. More recently, working group IV, led by M. Eberlein at Siemens, issued a document on validation and field data assessment, the UNIFE LCC model (Kjellsson, 1999; Kjellsson and Hagemann, 2000).

The UNIFE LCC costs comprise the following:

- Costs of acquisition
- Costs of operation
- Support costs throughout the following life cycle phases:
  - Concept and definition
  - Design and development
  - Manufacturing
  - Installation
  - Operation and maintenance

Since the UNIFE LCC model is customer-oriented, it emphasizes acquisition costs, operating costs, and support costs throughout the various life stages. It does not consider disposal costs, unlike the EN 60300-3-3 LCC model (Kim et al., 2011).

The UNIFE LCC can be expressed as follows:

$$LCC = C_{\text{acquisition}} + C_{\text{operation}} + C_{\text{support costs through the life phases}} \qquad (7.4)$$

In summary, UNIFE has grouped two major categories—investments and TCOs—to simplify the decision-making process for investors and asset managers. However, this approach may not work if it assigns individual decisions to the different departments.

Railway operators around the world are increasingly applying LCC to assist in their choice of tenders to supply rolling stock and fixed installation equipment. The example in Figure 7.7 illustrates the product breakdown structure (PBS) for rail vehicles used as the basis for an LCC model for a fleet of multiple units to be procured (Ruman and Grencik, 2014).

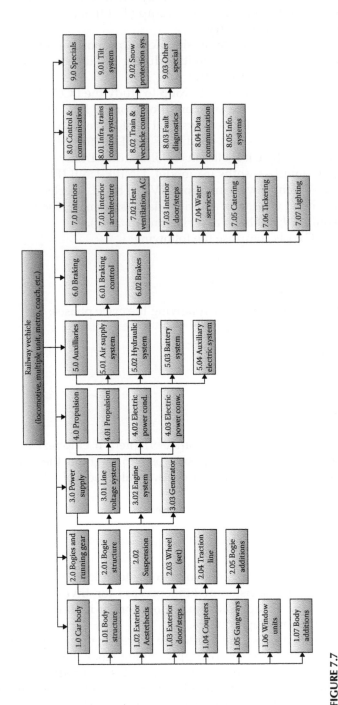

FIGURE 7.7
Vehicle system product breakdown structure. (Adapted from Ruman, F., Grencik, J., *Logistyka*, 3, 5490, 2014.)

7.3.2 Calculation of Cost and Profit Life Cycle of a Tram

To expand the role of LCC, especially TCO in rolling stock, we now consider the life cycle of a tram. The actual purchase cost of a vehicle does not give a picture of the tram's profitability. It is necessary to calculate the other components of the LCC of the vehicle, including infrastructure. In what follows, we present a case study of a tram with a 30-year service life and a capacity of 60 seats.

Despite the long service life, it is important to realize that the renewal of a fleet of trams is a necessity. Nonrenewal leads to internal debt; the business will be passed to another company in a depleted state and will require a large investment. Artificially extending the life of obsolete vehicles is not the solution; it only postpones renewal. The concepts of LCC, TCO, and maintenance costs are the keys to decide when this should happen to maximize the remaining useful life (RUL) of the system (Ruman and Grencik, 2014).

7.3.2.1 Assumptions and Purpose of LCCA

A tram's LCC comprises many different costs. The most important are included here.

7.3.2.1.1 Acquisition Costs

The price of acquisition may or may not include financial costs, transportation, testing, approval, documentation, training, corrective maintenance during the guarantee period, spare parts, tools, and so on. If a loan is required to purchase the tram, it is necessary to calculate the additional cost factor.

For a normal tram, a daily mileage of 164 km is assumed. Annual throughput for preventive and corrective maintenance representing about 90% availability of the vehicle is 55,000 km. Thus, the total mileage over the 30-year service life is 1,650,000 km.

The price of a tram is about €2 million. When financed by a loan, the price is €2,888,888. Acquisition costs calculated per kilometer throughout the lifetime equals €1.74.

7.3.2.1.2 Operational Costs

For operating costs, the greatest impact is the daily mileage of vehicles and the number of passenger seats and daily capacity.

7.3.2.1.3 Preventive Maintenance

The basis for calculating the cost of preventive maintenance is the periodic inspection plan. Table 7.3 shows the cost of preventive maintenance throughout the lifetime of the tram (Ruman and Grencik, 2014).

TABLE 7.3

Preventive Maintenance Costs

Inspection	Cost of One Inspection (€)	Number of Inspections per Period	Costs for the Entire Period (€)
R0	1,481.5	100	148,150.0
R1	9,259.3	16	148,148.8
R2	18,518.5	8	148,148.0
R3	37,037.0	4	148,148.0
R4	166,666.6	2	333,333.2
Total			**925,928.0**

Source: Adapted from Ruman, F., Grencik, J., *Logistyka*, 3, 5490, 2014.

7.3.2.1.4 Corrective Maintenance

Corrective maintenance is not planned but random (stochastic). Its occurrence and extent can be determined based on experience from the operation of similar vehicles and the existing data on failures. The approximate cost of corrective maintenance for the lifetime of the vehicle based on data on failures is €185,185.2. The cost of corrective maintenance over the life cycle of a tram is €0.11/km.

7.3.2.1.5 Conclusion

As shown in Figure 7.8, the total LCC for a 30 m tram with 60 seats and an annual mileage of 55,000 km is €0.07/km/seat.

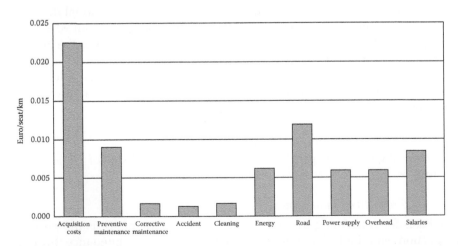

FIGURE 7.8

Costs per kilometer and seat. (Adapted from Ruman, F., Grencik, J., *Logistyka*, 3, 5490, 2014.)

There are a number of additional costs related to the operation of the tram in an urban environment, such as accidents (very common but the damage is light and cost is low), energy (depends on electricity price), operating materials (really small cost which includes windscreen washer fluid, grease for flanges, etc.), salaries (depending on wages, number of staff, and other labor parameters), costs of the transport route (amount paid to the infrastructure manager (IM) for the usage of the track which depends on the deal with the operator), and finally cleaning and overhead costs (Ruman and Grencik, 2014).

7.3.2.2 Decision Models Considering Asset Revenues

Profits in railway transport have several components. To achieve profitability, the minimum tram occupancy rate is 20% for regional transport and 35% for long-distance transport. The proportions of the various sources of revenue for trams are shown in Figure 7.9.

We analyzed the costs and profits using the methodologies suggested earlier, that is, UNIFE and EN 60300-3-3. The calculations suggest the total LCC for a tram with 60 seats and an annual mileage of 55,000 km is €0.07/km/seat. The total approximate revenue of the tram over 30 years with a total mileage of 1,650,000 km is €10,230,000, or €0.10/km/seat. The proportions of costs and profits are shown in Figure 7.10.

These figures provide useful information to managers for fleet renewal and purchasing new vehicles.

TCO is also useful to analyze whether O&M is properly performed or whether the benefit is lost because of problems either in O&M or in the maintenance repair and overhaul (MRO) process.

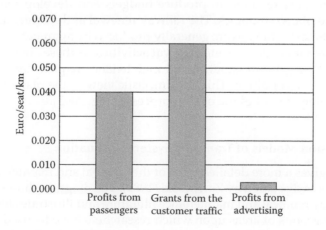

FIGURE 7.9
Tram revenues. (Adapted from Ruman, F., Grencik, J., *Logistyka*, 3, 5490, 2014.)

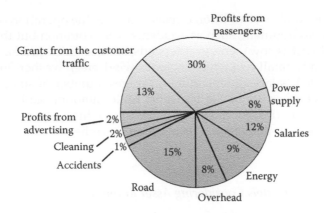

FIGURE 7.10
Cost structure. (Adapted from Ruman, F., Grencik, J., *Logistyka*, 3, 5490, 2014.)

For this business model, the profitability comes when the cost of operation is lower than the revenue from the operation; therefore, the calculation of O&M and MRO figures is relevant to determining the performance achieved and deciding whether the replacement of vehicles is warranted (Ruman and Grencik, 2014).

7.4 Infrastructure: Railway Infrastructure

7.4.1 Introduction and Background

Cost estimates are required to produce budgets and develop renewal and maintenance plans/strategies. The railway renewal and maintenance literature suggests cost estimates are generally produced by aggregating unit costs of the required maintenance and renewal activities, as these are identified by models predicting future renewal and maintenance requirements based on data on the track condition. This is a simplistic view of the planning process but it accurately describes the overall process (Ling, 2005).

7.4.2 Decision Models of Transport Systems Evaluation

Figure 7.11 gives a more detailed view of the renewal and maintenance planning process as discussed by Zarembski (1989). The diagram shows the scope of the railway renewal and maintenance literature and illustrates the individual areas or groups of areas upon which researchers have focused. The bulk of the published work focuses on models fitting into the "analysis, future requirement" area of the diagram; these are concerned with predicting the

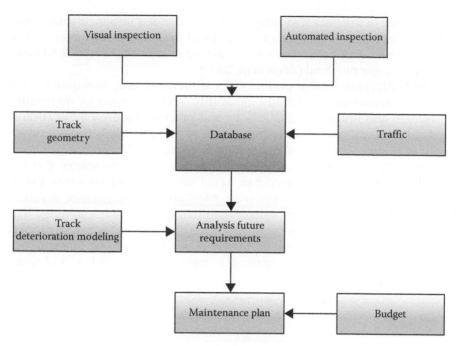

FIGURE 7.11
Renewal and maintenance planning process. (Adapted from Zarembski, A., *On the Development of a Computer Model for the Economic Analysis of Alternate Tie/Fastener Configurations,* American Railway Engineering Association, Chicago, IL, 1989, pp. 129–145.)

future renewal and maintenance requirements. Other models discuss cost–benefit analysis and investigate options (Zarembski, 1989; Zoeteman and Van der Heijden, 2000).

7.4.2.1 Assumptions and Purpose of LCCA with a Focus on Maintenance

Esveld (2001) defines railway infrastructure renewal and maintenance as the necessary process to ensure the track meets safety and quality requirements at minimum cost. Renewal and maintenance are planned considering location conditions and are based on control data from measuring systems, visual inspections, and economic data. Esveld (2001) suggests track maintenance can be divided into six categories:

1. Rail geometry
2. Track geometry
3. Track structures
4. Ballast bed
5. Level crossings
6. Miscellaneous

Maintenance of the track geometry can be subdivided into incidental maintenance (the repair of local irregularities) and systematic maintenance. The latter is done using heavy track maintenance machines (Esveld, 2001) and involves a major overhaul (Zoeteman, 2003).

Esveld (2001) argues that performing maintenance only to requirements, including delivering availability, reliability, and low costs of ownership (Zoeteman and Braaksma, 2001), is a good practice. Figure 7.12 illustrates the track renewal and maintenance process. The process is broken down into four main areas: manual maintenance, mechanical maintenance, manual renewal, and mechanical renewal. The manual maintenance process includes surface welding, switches, level crossing, and structure maintenance, and some spot maintenance. Mechanical maintenance involves tamping, ballast regulating, ballast stabilizing, joint straightening, ballast cleaning, and some spot maintenance. The manual renewal process includes the renewal of certain parts, whereas the mechanical renewal process involves the renewal of continuous track panels, switches, and some structures (Ling, 2005).

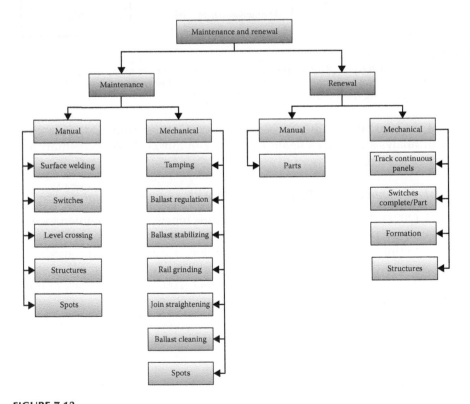

FIGURE 7.12
Track renewal and maintenance process. (Adapted from Esveld, C., *Modern Railway Track*, 2nd edn., Delft University of Technology, MRT Productions, Zaltbommel, Netherlands, 2001.)

7.4.2.2 Availability: The Goal of Infrastructure Mangers

The EU REMAIN Consortium emphasizes that to attract more rail traffic, companies must stress the "availability and reliability of the track infrastructure and of the trains." Stalder (2002) argues that performance-related issues such as "reliability" and "availability" (together with maintainability and safety) are key aspects of infrastructure quality.

Unavailability patterns can be observed by combining different failure rates and different downtimes per failure type. The "penalty cost" of unreliable or unavailable infrastructure is increasingly problematic. These penalty costs include delay costs. Possession costs are another major availability cost driver.

Stalder (2002) claims "unreliability" is a hidden cost driver and suggests that there is a need for a commercial framework to assess the cost of reliability. This would guide decision-making and create a wider whole LCC view from which performance-oriented maintenance strategies could be developed.

Zoeteman and Braaksma (2001) suggest that "availability" is the time the infrastructure is available for operations per calendar period. The "unavailability" of the infrastructure can be attributed to planned possessions (preventive maintenance), infrastructure failures (corrective maintenance), and possession overruns or external factors, such as vandalism or bad weather. They say "reliability" is the time the infrastructure is available for operations during the agreed-upon periods. They only consider unplanned maintenance and repair. Reliability depends on the asset quality and the ease with which it can be maintained, as well as the amount of preventive maintenance, and the failure restore times.

Safety, noise, vibrations, and riding comfort are other areas of concern and are related to maintenance thresholds (e.g., geometry control limits), as well as inspection and failure response strategies (e.g., inspection frequencies and speed restrictions) (Ling, 2005).

7.4.2.3 Prediction of the Infrastructure Downtime as Budget Maintenance Optimizer

Concerns about maintaining and renewing railways with budget constraints have changed little over the past century. The first discussion in the literature of a need to improve maintenance and renewal practice to meet budgets dates back to 1935, when Geyer suggested up to 25% of the annual budget could be considered "dead money." In his view, "dead money" could be attributed to a high turnover of staff. Capability is lost when experienced staff leave, leading to low productivity with new staff. He proposes the need to analyze alternative maintenance and identify which will be the most economical. Although he offers no suggestions on how to perform this analysis, he discusses data collection and proposes the need for a dedicated individual to capture data from time sheets.

Zarembski published research on economic benefit analysis using life cycle techniques 54 years later. It is interesting to observe the similarities between the two researchers. Both suggest the need to analyze alternative options and to collect appropriate data. In fact, the main renewal and maintenance requirements have changed little since 1935. There is still a need to provide safe, reliable, and economic service. Over the last 10 years, however, more pressure has been put on companies to reduce expenditure while still improving performance (Zoeteman and Esveld, 1999).

For this purpose, infrastructure management systems use probabilistic deterioration models. These require accurate data on the condition of the asset.

Studies by Trask and Fraticelli (1991), Mesnick (1991), Esveld (2001), Esor and Zarembski (1992), and Acharya (1991) investigate deterioration mechanisms and the possibilities of controlling deterioration by implementing better maintenance strategies and polices. They argue that developing deterioration models improves the ability to plan track repairs and suggest that the models predicting deterioration are only as good as the deterioration data and the engineering model used.

The idea of using a track degradation model to plan maintenance and renewal activity was first discussed by Trask and Fraticelli (1991). They suggest a model using the current condition as a base; the model can predict the service life of the rail and ties over 5 years. With an understanding of the service life, the user can plan the appropriate renewal or maintenance.

The newest approach to reducing maintenance costs comes from Stirling et al. (2000). They propose an expert system; based on defined rules, it chooses the most appropriate remedial work based on the condition of the asset. They argue that this will reduce maintenance costs because it will provide the most optimized remedial process.

In summary, there is still a need for methodologies providing an understanding of future maintenance and renewal activities to plan budgets. There is also a need to investigate ways to optimize renewal actions and maintenance tasks to reduce costs, while always considering safety requirements. The most economical option over the life of the asset must be known as well. The techniques applied to address these issues involve statistical modeling using historical empirical data (Ling, 2005).

7.4.2.4 Life Cycle Costs of Rail Infrastructure

To estimate LCC, the factors influencing the performance drivers of the railway infrastructure have to be identified, along with their links. The driving factor of maintenance and failure is the degradation of the asset. Figure 7.13 shows a hypothetical degradation pattern for a specific track parameter. With the number of tons carried by the track, measured in million gross

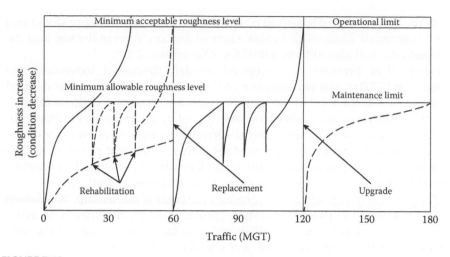

FIGURE 7.13
Hypothetical degradation curve of track geometry. (Adapted from Ebersohn, W. and Ruppert, C.J., Implementing a railway infrastructure maintenance system, in: *CORE98 Conference*, Rockhampton, Queensland, Australia, 1998.)

tons (MGT), the roughness of the track increases (i.e., the track condition decreases) and rehabilitation is needed (Zoeteman, 2001).

The infrastructure managers (IMs) apply maintenance limits (or thresholds) based on their experience that the deterioration will be progressive once the limit is passed. The dotted line shows a replacement of the track which can be postponed by timely maintenance. After some time, the effectiveness of maintenance is too low (decreasing maintenance interval) and a replacement is required. This replacement should be planned before the track condition passes the operational limit, which is set to guarantee the safety and reliability of rail transport.

Another example of a maintenance decision is shown in the same figure: once the limit of 120 MGT is passed, the IM upgrades the track structure. This results in a lower rate of degradation and less maintenance. The improved track quality and reduced amount of maintenance should be traded off against the extra investment to be made.

Track degradation depends on all kinds of factors, such as the initial quality of construction, the quality of the substructure, and the loads on the track. Historical data can provide insight into the actual decline rates and the effectiveness of maintenance activities under specific conditions (Ohtake, 1998; Zaalberg, 1998).

Besides asset degradation, certain other factors influence the LCC, such as the extent of preventive maintenance (rehabilitation), market prices for labor, materials, and machines, and the operational characteristics of the line. The IM can manage some of these factors directly (e.g., maintenance strategy) or with the cooperation of transport operators (e.g., quality of rolling stock) and the government. Exogenous factors, such as the condition of the soil and the interest rate, will also influence the LCC (Zoeteman, 2001).

Figure 7.14 presents a conceptual model, discussed extensively in Zoeteman (2000a). The performance of the railway infrastructure is defined in this figure as the level of safety, riding comfort, noise, vibrations, reliability, availability, and the costs of ownership. Safety and noise standards indirectly influence the LCC, since they determine the tolerances and thresholds for design and maintenance parameters. Other functional parameters, like maximum speed, minimum headway, and maximum axle-load supported, constrain the feasible design or maintenance strategies as well.

The physical design directly determines the costs of ownership. The design also influences the asset degradation (initial quality), together with other conditions, such as traffic intensities and axle-loads, the quality of the substructure, and the effectiveness of performed maintenance.

The quality degradation determines the required volume of maintenance repair and overhaul (MRO). The chosen maintenance strategy influences the amount of preventive and corrective maintenance and renewal over the life span of the track.

The maintenance strategy has a direct impact on the LCC through the costs of restoring failures. A related factor determining the maintenance volume is the annual budget: a backlog in maintenance and renewal can result in a more rapid decline of the quality of the infrastructure. In addition, the performed maintenance and renewal actions cause expenditures and reduced availability due to track possession. Moreover, the performed maintenance, renewal volume, and other factors determine delays which may be blamed on infrastructure failures (reduced reliability). The delay may be converted into penalties for the IM using a performance payment regime.

The performance penalties, the maintenance and renewal costs, and the construction costs make up the TCO. In summary, the TCO for infrastructure is complex. Therefore, models of track degradation need to have a trade-off between maintenance and availability of the infrastructure to satisfy the business goals (Zoeteman, 2001).

7.4.2.5 LCC as Supporting Decision-Making on Design and Maintenance

As seen earlier, railway maintenance is a complex process which depends on a number of variables, some of them exogenous. During each of the steps of the decision-making process, decision-makers need to get quantitative insight into the impact of a specific decision on the LCC.

LCC is the result of a complicated set of partly uncertain conditions. The number of factors affected and the uncertainty involved necessitate an analysis of different scenarios (future conditions). Here, the concept of a decision support system (DSS) can make a significant contribution.

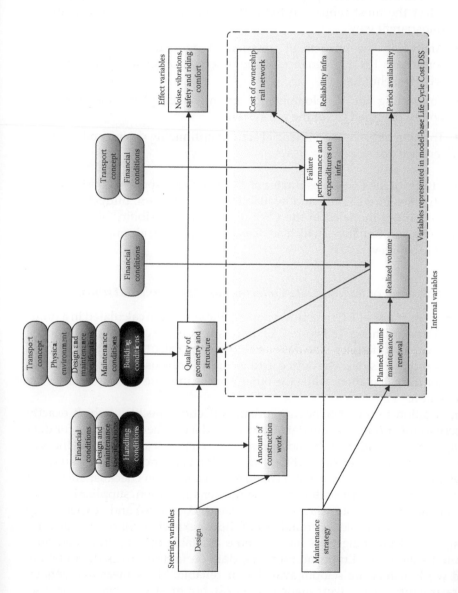

FIGURE 7.14

Factors influencing the performance of rail infrastructure. (Adapted from Zoeteman, A., *Eur. J. Transp. Infrastruct. Res.*, 1(4), 391, 2001.)

With its ability to estimate the LCC of different designs, maintenance strategies, and operational conditions, a DSS can assist IMs in the following:

- Evaluating different physical designs or maintenance strategies to select the most robust, cost-effective solution in a systematic and transparent way
- Analyzing the impacts of (restrictive) operational and financial conditions for maintenance of the assets in order to discuss them with stakeholders
- Supporting the development of maintenance plans that aim at optimizing the LCC of the rail system
- Training technical and financial staff in optimizing design and maintenance decisions

The outputs of the DSS are estimates of TCO during a specified period (LCC) and estimates of reliability and availability of the system. Budget limits are not a constraint modeled in the DSS: it is the responsibility of the user to analyze those alternatives which are feasible, while considering the available budget (Zoeteman, 2001).

7.4.2.6 *Process of Railway Data Collection for Accurate LCC Estimates*

The final outputs—LCC, reliability, and availability—are calculated in a number of steps, as shown in Figure 7.15. In this figure, the calculation processes are shown as rectangles. The data needed for the calculations are shown on the left and right sides. The dotted arrows indicate the use of data from a data table for the calculation, while the other arrows indicate the sequence in the calculation.

In addition to the DSS, having a data collection checklist, which exactly describes the input data and data formats, and a number of sessions for data collection and validation should increase the reliability and the robustness of the analysis and better support the engineering staff and relevant managers.

Many data sources are used to create a data collection checklist, including empirical data (e.g., laboratory tests, computer simulation, supplier information, maintenance history, actual maintenance cost rates) and "expert judgment" on, for example, the number of failures expected. Since maintenance analysis and planning tools have become available only recently and on a limited scale in the European railways, data on failures, track degradation, and work history are seldom available or reliable. Many measurements on infrastructure quality were made in the past, but mostly for operational use, for example, to identify spots requiring urgent maintenance (Esveld, 1989).

Step 5 in the figure estimates the LCC and reliability and availability. If applicable, construction costs are included in the total cash flows. Another choice is to include or exclude specific risks. Finally, the costs of financing are

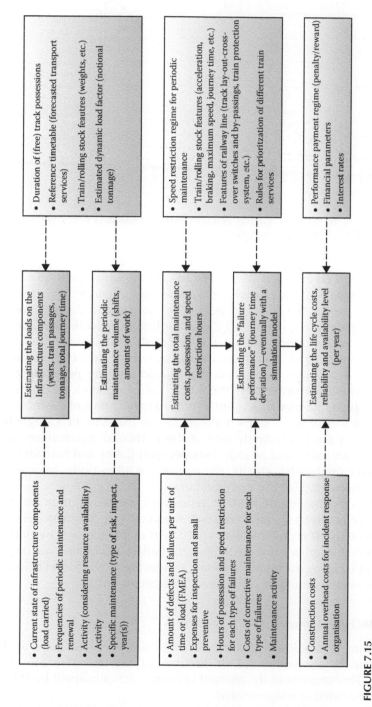

FIGURE 7.15
Structure of the life cycle cost plan decision support system. (Adapted from Zoeteman, A., *Eur. J. Transp. Infrastruct. Res.*, 1(4), 391, 2001.)

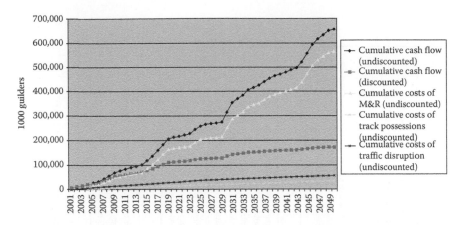

FIGURE 7.16
Cash flows during the period of analysis, and intermediate result after step 5. (Adapted from Zoeteman, A., *Eur. J. Transp. Infrastruct. Res.*, 1(4), 391, 2001.)

calculated based on the interest rate. The LCC of the various design or main-tenance solutions is considered as performance fees. A performance fee is the annuity paid during the analyzed period for interest and depreciation. If a project performance requirement (PPR) is applicable, penalties (and eventu-ally rewards) are included in the fee. Figure 7.16 shows the outcome of step 5.

A model consisting of a set of equations is used for each calculation pro-cess. Table 7.4 provides some formulas to calculate the performance fee to finance periodic maintenance, excluding transport services. Based on a passed threshold, maintenance is initiated for a certain part of the infrastruc-ture. The number of work shifts depends on the duration of a single track possession and the productivity rate of the particular maintenance activ-ity. The costs are calculated as unit costs per shift (labor and machines) and per kilometer (materials). Finally, the annuity of all periodic maintenance expenses is calculated for the analyzed period.

Thresholds can be interdependent: renewals clustered in time and place (clustering of renewals on adjacent track sections or of different infrastruc-ture components). It is possible that a particular maintenance activity harms the condition of another component (e.g., the ballast bed quality can degrade as a result of tamping the track). If these formulas are applied in a DSS model, such relationships should be considered (Zoeteman, 2001).

7.4.2.7 MRO and Infrastructure Expenditures

Infrastructure expenditures can be classified according to the way they enhance the functionality and/or lifetime of the infrastructure (asset approach). This consists of the amount of money actually spent by IMs and can be asset-related or usage-related.

TABLE 7.4

Simplified Formulas for Calculating the Annuity of Renewals

Nr. Formula	Explanation
1. $RQ_{y,a} = Q_a \cdot P_{y,a}(T_f \geq TH_a)$	The quantity of periodic maintenance, RQ, in a particular year (y) for the analyzed activity (a) is determined. The quantity Q is for instance the total track length or number of switches. P is the part of the asset(s) to be renewed given the fact that the notional tonnage (T_f) has passed the threshold (TH). E.g. if P is 0.50 for rail renewal, half of the total number of track kilometers needs to be rerailed.
2. $S_{y,a} = \text{roundup}\left(\dfrac{RQ_{y,a}/PS_a}{TPP_y - L_a}\right)$	The number of hours needed for the periodic maintenance is determined by the production speed (PS). The hours have to be scheduled according to the duration of a track possession period (TPP) provided For each TPP also time is lost due to set up and finishing of the work (L). The total number of shifts S is calculated (whole number).
3. $C_{y,a} = SC_r(TPP_y) \cdot S_{y,a}$ $+ RQ_{y,a} \cdot (MC_a - RV_a)$	The costs for periodic maintenance are calculated by multiplying the number of shifts with the costs per shift (SC) for the given duration of the track possession (TPP), and by adding the material costs (using the unit costs, MC, and the residual value per unit RV).
4. $TPV = \sum_a \sum_{y=0}^{n} \dfrac{C_{y,a}}{(1+i)^y}$	The total present value TPV is the sum of the discounted costs C during all years (y) and for all activities (a) analyzed. Year n is the last year, i.e. the time horizon of the analysis. The interest rate applied is i.
5. $ANN = \dfrac{(1+i)^n \cdot i}{(1+i)^n - 1} \cdot TPV$	This formula is used to convert the total present value of the investment or maintenance strategy into the annuity (ANN) or *performance fee*.

Source: Adapted from Zoeteman, A., *Eur. J. Transp. Infrastruct. Res.*, 1(4), 391, 2001.

- *Asset-related*: expenditures on investment, renewal, and O&M of infrastructure
- *Usage-related*: fixed and variable expenditures on infrastructure

In economics, an investment is the accumulation of some kind of asset expected to have a future return. In the area of transportation infrastructure, the asset is a piece of infrastructure with a certain functionality and lifetime, and the return is an infrastructure service, such as the possibility of traveling between two places in a specific mode at a specific level of comfort.

According to this classification, we can define the following types of expenditures:

- *Investment expenditures*: expenditures on (a) new infrastructure with a specified functionality and lifetime or (b) expansion of existing infrastructure with respect to functionality and/or lifetime

- *Renewal expenditures*: expenditures on replacing existing infrastructure, prolonging the lifetime without adding new functionalities
- *Maintenance expenditures*: expenditures on maintaining the functionality of existing infrastructure within its original lifetime
- *Operational expenditures*: expenditures not related to enhancing or maintaining lifetime and/or functionality of infrastructure
- *Drivers of infrastructure expenditures*: expenditures with the same functionality can be different across countries and among IMs. Table 7.5 shows some drivers.

Social and natural factors influence the level and composition of the capital stock in transport infrastructure as well. Factors such as population density, climate, hydrology, or topography have an impact on the length of a track section, on the need for bridges or tunnels, or on the need for protective structures against floods or avalanches.

In practice, infrastructure expenditures can consist of combined investment, renewal, and/or maintenance activities: IMs plan these activities to be executed in an efficient way to improve functionality, to extend lifetime, and to minimize total expenditures and/or inconvenience for infrastructure users.

To disentangle expenditures, the following should be done:

- Assess whether a project (or expenditure category) is an investment, renewal, or maintenance expenditure.
- Assess what percentage of a project (or expenditure category) should be considered an investment, a renewal, or maintenance.

TABLE 7.5

Main Drivers of Differences in Infrastructure Expenditure

Expenditure Drivers
Construction standards (legal obligations for safety, degree of technical progress applied to infrastructure construction, special standards for mountainous areas or ecologically sensitive areas)
Type of infrastructure: construction and maintenance (motorways/other, high-speed train lines/other, tunnels/bridges, underground system/above-ground system, canals)
Levels of wages and prices per county
Expected traffic mix and occupancy
Weather and climate
Population density (land costs)

Source: Adapted from ECORYS Transport, Infrastructure expenditures and costs, Practical guidelines to calculate total infrastructure costs for five modes of transport, Final Report, Rotterdam, the Netherlands, November 30, 2005.

Basically, both approaches call on expert judgment to make the required distinction (ECORYS Transport, 2005).

7.4.2.8 Infrastructure Costs

A large share of the infrastructure expenditure is related to the creation, renewal, and maintenance of infrastructure assets with an expected lifetime of more than 1 year. This means expenditures in year X do not equal the infrastructure cost for year X, i.e., the yearly value for the use of the infrastructure assets.

Infrastructure costs, the periodic (yearly) value for the use of infrastructure assets, consist of the following:

- *Capital costs*: yearly depreciation costs of investments, renewals, and maintenance of infrastructure assets; yearly interest expenditures
- *Running costs*: yearly recurring (other) O&M expenditures

The drivers of differences in infrastructure expenditures listed in Table 7.5 also influence infrastructure costs. On top of this, we can identify specific drivers that can lead to major differences in the calculation of infrastructure costs between countries and among IMs. These specific cost drivers are listed in Table 7.6 and explained further below (ECORYS Transport, 2005).

1. *Lifetime expectancy*: the lifetime expectancy of infrastructure assets, as well as of the components of specific assets, can be very different (e.g., earthwork, foundation, surface layer). To establish correct depreciation costs, these differences in lifetimes should be accounted for.

2. *Historical costs versus replacement costs*: valuation of assets can be done using historical costs or replacement costs. When calculating on the basis of replacement costs, assumptions are made as to the future value of the asset. These can be different between countries

TABLE 7.6

Main Drivers of Infrastructure Costs

Infrastructure Cost	Cost Drivers
Depreciation	Life expectancy of assets
	Valuation at historical cost versus replacement cost
	Linear versus nonlinear depreciation
	Time span between maintenance expenditures
Interest	Interest rates

Source: Adapted from ECORYS Transport, Infrastructure expenditures and costs, Practical guidelines to calculate total infrastructure costs for five modes of transport, Final Report, Rotterdam, the Netherlands, November 30, 2005.

and across IMs. However, the International Financial Reporting Standards (IFRS) deal with how to establish a "fair value" (IAS 16), leading to more consistency.

3. *Linear versus nonlinear depreciation*: when using historical costs for valuation of assets, depreciation costs can be calculated using types of depreciation functions (linear or nonlinear), resulting in different depreciation estimates. To compare countries or IMs, a single depreciation method should be applied. In principle, either linear or nonlinear depreciation would work.

4. *Time span between maintenance expenditures*: maintenance expenditures are meant to maintain, that is, to restore the original functionality of infrastructure. However, like investments and renewals, these maintenance expenditures—or at least part of them—are not made on an even basis every year, but in "waves." To establish the yearly costs, maintenance expenditures should be capitalized.

5. *Interest rate*: for private IMs, the actual interest expenses are considered the interest costs. For other IMs, there is no "official" interest rate that should be used in all cases to calculate capital costs. To determine a common interest rate, we can look at the interest rate advised in cost–benefit analysis. For European Union (EU) projects, an interest rate of 5% is advised (ECORYS Transport, 2005).

7.4.2.9 *Valorization of Infrastructure Assets*

For some transport modes, capital stock and corresponding capital costs can be derived based on the business accounts of individual IMs. This holds specifically for railways, airports, and harbors.

For other transport modes, other methods should be used to quantify the capital stock. The perpetual inventory method (PIM) is used by most OECD countries (Organization for Economic Cooperation and Development). PIM calculates the asset's value by adding the annual investments and subtracting either the value of those assets that have exceeded their life expectancy or their depreciation. For PIM to be used, a long investment time series must be available (ECORYS Transport, 2005).

For accurate capital stock estimates using PIM, the following information is required:

- *Long investment expenditure time series for each mode (30–40 years)*: investment expenditures comprise expenditures on new construction, extension, reconstruction, and renewals. Non-transport-related capital costs are excluded.
- Life expectancy of the infrastructure as a whole or of infrastructure components (investments per infrastructure component over time must be known as well).

- Depreciation over time (linear, geometric).
- Interest rate (opportunity cost).

Obsolescence or catastrophic loss may occur for infrastructure and should be taken into account when relevant. According to the OECD manual, obsolescence is defined as occurring when an asset is retired before its physical capability is exhausted, and should be included in capital stock data where the asset's owner can be expected to anticipate it.

If any long investment time series does not exist, but a good cross-sectional database for 1 year is available, the synthetic method (discussed below) can be applied for capital valuation. Capital costs can then be calculated by using annuities. If neither the perpetual inventory approach nor the synthetic method can be applied, the use of indicators like capital values per kilometer from other countries is a possible approach.

The synthetic method values the infrastructure network by estimating what it would cost to replace the relevant network with assets of equivalent quality. The method involves measuring the existing physical assets.

Total infrastructure costs consist of capital costs (depreciation of and interest on previous investments, renewals, and non-yearly maintenance) and running costs. The starting point for the calculation of costs is the investment, renewal, and O&M expenditures (see Figure 7.17). Expenditures and costs can be variable (influenced by transport volume) or fixed (not influenced by transport volume) (ECORYS Transport, 2005).

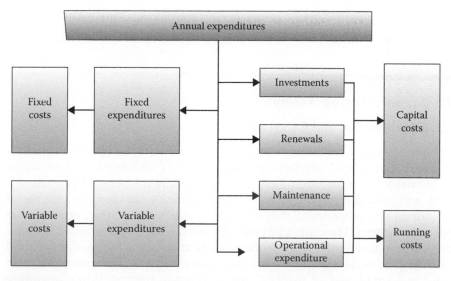

FIGURE 7.17
Components of the total infrastructure costs. (Adapted from ECORYS Transport, Infrastructure expenditures and costs, Practical guidelines to calculate total infrastructure costs for five modes of transport, Final Report, Rotterdam, the Netherlands, November 30, 2005.)

7.4.2.10 Cost of Life Extension of Deteriorating Structures

In many cases, infrastructures created in the 1960s and 1970s (including bridges, for example) are deteriorating quite rapidly and require unavailable financial resources for inspection, maintenance, repair, rehabilitation, and replacement. There is an urgent need to develop cost-effective maintenance strategies for them.

Making decisions on maintenance of existing structures depends on the costs of interventions and the effects of these interventions on the structural safety. Maintenance interventions can be defined as any action whose effects are reliability improvement, reliability deterioration delay, and/or reliability deterioration rate reduction. Traditionally, the cost of a maintenance action is considered as fixed and independent of both the state of the structure and the effect of a maintenance action on the structural performance. However, the cost of maintenance depends not only on the type of maintenance action, but also on the state of the structure before and after its application. As an example, the cost of repairing a corroded steel girder depends on the degree of corrosion and the extension of the repair.

A number of attempts have been made to analyze, predict, or optimize the lifetime maintenance cost of deteriorating structural systems, but most do not explicitly consider the relations between the total cost of maintenance interventions and its effect on the reliability index, that is, reliability improvement, deterioration of reliability, or reduction of deterioration rate of reliability (Frangopol, 2010).

7.5 Manufacturing Assets

7.5.1 Introduction and Background

Pumping systems are widespread; they provide domestic services, commercial and agricultural services, municipal water/wastewater services, and industrial services for food processing, chemical, petrochemical, pharmaceutical, and mechanical industries. Although pumps are typically purchased as individual components, they provide a service only when operating as part of a system. They must be carefully operated and maintained in order to remain so throughout their working lives to ensure the lowest energy and maintenance costs, equipment life, and other benefits. The initial purchase price is a small part of the LCC for high-usage pumps. While operating requirements may sometimes override energy cost considerations, an optimum solution is still possible.

A deeper understanding of all the components that make up the TCO will provide an opportunity to dramatically reduce energy, operational, and maintenance costs. LCCA is a management tool that can help companies minimize waste and maximize energy efficiency for pumping systems (Figures 7.18 and 7.19) (Hydraulic Institute et al., 2001).

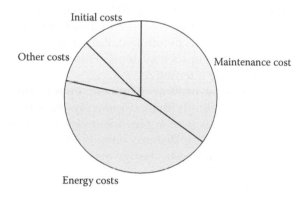

FIGURE 7.18
Typical life cycle costs for a medium-sized industrial pump. (Adapted from Hydraulic Institute et al., Pump life cycle costs: A guide to LCC analysis for pumping systems, Executive summary, DOE/GO-102001-1190, January 2001.)

FIGURE 7.19
Pumping systems.

7.5.2 LCC of Pumping Systems

The LCC of any piece of equipment is the total "lifetime" cost to purchase, install, operate, maintain, and dispose of that equipment. The components of an LCCA typically include initial costs, installation and commissioning costs, energy costs, operation costs, maintenance and repair costs, downtime costs, environmental costs, and decommissioning and disposal costs.

Many organizations only consider the initial purchase and installation cost of a pumping system. It is in the best interests of the asset managers to evaluate the LCC of different solutions before installing major new equipment or carrying out a major overhaul in such systems. This evaluation will identify the most financially attractive alternative.

Existing pumping systems provide a greater opportunity for savings through the use of LCC methods than do new systems for two reasons. First, there are at least 20 times as many installed pump systems as are built each year; second, many existing systems have pumps or controls that are not optimized, as the pumping tasks change over time (Hydraulic Institute et al., 2001).

7.5.2.1 Assumptions and Purpose of LCCA

LCCA, either for new facilities or renovations, requires the evaluation of alternative systems. For a majority of facilities, the lifetime energy and/or maintenance costs will dominate the LCC. It is therefore important to accurately determine the current cost of energy, the expected annual energy price escalation for the estimated life, and the expected maintenance labor and material costs. Other elements, such as the lifetime costs of downtime, decommissioning, and environmental protection, can often be estimated based on historical data. Depending on the process, downtime costs can be more significant than the energy or maintenance elements of the equation. Therefore, careful consideration should be given to productivity losses caused by downtime.

To apply the evaluation process or to select pumps and other equipment, the best information on the output and operation of the plant must be established. The process itself is mathematically sound, but if incorrect or imprecise information is used, an incorrect or imprecise assessment will result. The LCC process is a way to predict the most cost-effective solution; it does not guarantee a particular result, but allows the plant designer or manager to compare alternate solutions within the limits of the available data.

Pumping systems often have a lifespan of 15–20 years. Some costs will be incurred at the outset, and depending on the solution being evaluated, others may be incurred at different times throughout the life of the system. It is therefore practicable, and possibly essential, to calculate a present or discounted value of the LCC to accurately assess the different solutions (Hydraulic Institute et al., 2001).

7.5.2.2 Life Cycle Cost Analysis

The following analysis is concerned with assessments of details of the system design, comparing pump types and control means. Whatever the specifics, designs should be compared on a like-for-like basis. To make a fair comparison, the plant designer/manager may need to consider the measure used. For example, the same process output volume should be considered and, if the two items being examined cannot give the same output volume, it may be appropriate to express the figures in cost per unit of output. The analysis should consider all significant differences between the solutions being evaluated.

Finally, the asset manager might need to consider maintenance or servicing costs, particularly if these are to be subcontracted, or spare parts are to be provided with the initial supply of the equipment for emergency stand-by provision. Once again, whatever is considered must be on a strictly comparable basis. If the plant designer or manager decides to subcontract or carry strategic spares based entirely on convenience, this criterion must be used for all systems being assessed. But if it is the result of maintenance that can be carried out only by a specialist subcontractor, its cost will correctly appear against the evaluation of that system (Hydraulic Institute et al., 2001).

Elements of the LCC equation include the following:

$$LCC = C_{ic} + C_{in} + C_e + C_o + C_m + C_s + C_{env} + C_d \tag{7.5}$$

where

LCC is the life cycle cost

C_{ic} is the initial costs, purchase price (pump, system, pipe, auxiliary services)

C_{in} is the installation and commissioning cost (including training)

C_e is the energy costs (predicted cost for system operation, including pump driver, controls, and any auxiliary services)

C_o is the operation costs (labor cost of normal system supervision)

C_m is the maintenance and repair costs (routine and predicted repairs)

C_s is the downtime costs (loss of production)

C_{env} is the environmental costs (contamination from pumped liquid and auxiliary equipment)

C_d is the decommissioning/disposal costs (including restoration of the local environment and disposal of auxiliary services)

7.5.2.3 Data Collection for LCC and TCO Estimation

7.5.2.3.1 Initial Investment Costs

The plant manager must decide on the pumping system deployed. The smaller the pipe and fitting diameters, the lower will be the cost of acquiring and installing them. However, a smaller-diameter installation requires a more powerful pump, resulting in higher initial and operating costs.

Other choices made during this stage can affect initial investment costs. One important choice is the quality of the equipment selected. There may be options to procure materials with differing wear rates, heavier-duty bearings or seals, or more extensive control packages, all increasing the working life of the pump. These and other choices may incur higher initial costs but reduce LCC costs.

The initial costs also usually include the following items:

- Engineering (e.g., design and drawings, regulatory issues)
- Bid process
- Purchase order
- Testing and inspection
- Inventory of spare parts
- Auxiliary equipment for cooling and sealing water

7.5.2.3.2 Commissioning Costs

Installation costs include the following:

- Foundations—design, preparation, concrete, reinforcing, and so on
- Setting equipment on foundation
- Connection of process piping
- Connection of electrical wiring and instrumentation
- Performance evaluation at start-up

Installation can be performed by an equipment supplier, contractor, or user personnel. This decision depends on several factors, including the skills, tools, and equipment required to complete the installation, contractual procurement requirements, work rules governing the installation site, and the availability of competent installation personnel. Plant or contractor personnel should coordinate site supervision with the supplier. Care should be taken to follow installation instructions carefully. A complete installation includes transfer of equipment and fulfilling O&M requirements by training the personnel responsible for system operation.

Commissioning requires close attention to the equipment manufacturer's instructions for initial start-up and operation. A checklist should be used to ensure the equipment and the system are operating within specified parameters. A final sign-off typically occurs after successful operation is demonstrated.

7.5.2.3.3 Energy Costs

Energy consumption is often one of the larger cost elements and may dominate the LCC, especially if pumps run more than 2000 h per year. Energy consumption is calculated by gathering data on the pattern of the system output. If output is steady, or essentially so, the calculation is simple. If output varies over time, a time-based usage pattern needs to be established.

Performance can be measured in terms of the overall efficiencies of the pump unit or the energies used by the system at the different output levels. Driver selection and application will affect energy consumption. For example, much more electricity is required to drive a pump with an air motor than an electric motor. In addition, some energy use may not be output-dependent. For example, a control system sensing output changes may itself generate a constant energy load, whereas a variable-speed electric motor drive may consume different levels of energy at different operating settings.

The efficiency or levels of energy used should be plotted on the same time base as the usage values to show their relationship to the usage pattern. The area under the curve then represents the total energy absorbed by the system being reviewed over the selected operating cycle. The result will be measured in kilowatt-hours (kWh). If there are differential power costs at different levels of load, the areas must be totaled within these levels.

Once the charge rates are determined for the energy supplied, they can be applied to the total kilowatt-hours for each charge band (rate period). At this point, the total cost of the energy absorbed can be found for each system under review and brought to a common time period.

Finally, the energy and material consumption costs of auxiliary services need to be included. These costs may come from cooling or heating circuits, from liquid flush lines, or liquid/gas barrier arrangements (Figure 7.20) (Hydraulic Institute et al., 2001).

FIGURE 7.20
Pumping system.

7.5.2.3.4 *Operation Costs*

Operation costs are labor costs related to the operation of a pumping system. These vary widely depending on the complexity and duty of the system. For example, a hazardous duty pump may require daily checks for hazardous emissions, operational reliability, and performance within accepted parameters, but a fully automated nonhazardous system may require very limited supervision. Regular observation of how a pumping system is functioning can alert operators to potential losses in system performance. Performance indicators include changes in vibration, shock pulse signature, temperature, noise, power consumption, flow rates, and pressure (Hydraulic Institute et al., 2001).

7.5.2.3.5 *Maintenance and Repair Costs*

Obtaining optimum working life from a pump requires regular and efficient servicing. The manufacturer's recommendations state the frequency and extent of this routine maintenance. Its cost depends on the time and frequency of service and the cost of materials, and also depends on the brand and socio-economic zone. The design can influence these costs through the materials of construction, components chosen, and the ease of access to the parts to be serviced.

The maintenance program includes less frequent but more major attention or more frequent but simpler servicing. Major activities often require removing the pump to a workshop. During the time the unit is unavailable to the process plant, there can be loss of product or a cost incurred by using a temporary replacement. These costs can be minimized by programming major maintenance during annual shutdowns or process changeovers. Major service or overhaul may be summarized as a pump unit which cannot be repaired on site, while routine work is regarding a pump unit that may be repaired on site. The total cost of routine maintenance is found by multiplying the costs per event by the number of events expected during the life cycle of the pump.

Although unexpected failures cannot be predicted precisely, they can be estimated statistically by calculating the mean time between failures (MTBF). This effectiveness indicator can be estimated for components and combined to give a value for the complete machine.

It might be sufficient to simply consider best- and worst-case scenarios where the shortest likely life and the longest likely life are considered. In many cases, historical data are available.

The manufacturer can define and provide the MTBF of those items whose failure will prevent the pump unit from operating or will reduce its life expectancy below the design target. These values can be derived from past experience or from theoretical analyses. Such items include seals, bearings, impeller/valve/port wear, coupling wear, motor features, and other special items that make up the complete system. The MTBF values can be compared with the design working life of the unit and the number of failure events calculated.

Process variations and user practices will almost certainly have a major impact upon the MTBF of a plant and the pumps incorporated in it. Whenever available, historical data are preferable to theoretical data from the equipment supplier. The cost of each unexpected failure and the total costs of these failures can be estimated in the same way as routine maintenance costs.

7.5.2.3.6 *Estimation of Hidden Cost: Failure Cost in Pumping Systems*

The cost of unexpected downtime and lost production is a significant item in the total LCC and can rival the energy cost and replacement parts cost in its impact. Despite the design or target life of a pump and its components, there will be occasions when an unexpected failure occurs. In those cases where the cost of lost production is unacceptably high, a spare pump may be installed to reduce the risk. If a spare pump is used, the initial cost will be greater but the cost of unscheduled maintenance will include only the cost of the repair.

The cost of lost production is dependent on downtime and differs from case to case (Hydraulic Institute et al., 2001).

7.5.2.3.7 *Environmental Costs*

The cost of disposal during the lifetime of the pumping system varies depending on the nature of the pumped product. Certain choices can significantly reduce the amount of contamination, but usually at an increased investment cost. Examples of environmental cost can include cooling water, leakage disposal, used lubricant, and contaminated used parts, such as seals (Hydraulic Institute et al., 2001).

7.5.2.3.8 *Decommissioning Costs*

In most cases, the cost of disposing of a pumping system will vary little across designs. This is certainly true for nonhazardous liquids and generally applies to hazardous liquids as well. Toxic, radioactive, or other hazardous liquids will have legally imposed protection requirements, which will be largely the same for all system designs. There may be differences when one system has the disposal arrangements as part of its operating arrangements (e.g., a hygienic pump designed for cleaning in place) and another does not (e.g., a hygienic pump designed for removal before cleaning). When disposal is very expensive, the LCC becomes much more sensitive to the useful life of the equipment, for example, in nuclear power plants equipment, which is really different from the food industry (Hydraulic Institute et al., 2001).

7.5.2.4 Total Life Cycle Cost

As mentioned in the previous section, certain financial factors must be considered in developing the LCC:

- Present energy prices
- Expected annual energy price increase (inflation) during the pumping system lifetime
- Discount rate
- Interest rate
- Expected equipment life (calculation period)

In addition, the user must decide which costs to include in order to calculate the LCC figures, such as maintenance, downtime, environmental, disposal, and other important costs. In pumping systems, people indeed include downtime costs even though it is not a good practice, since downtime costs are calculated in disparate ways. For harmonization purposes, downtime costs must be calculated in the same way, which in essence is rather difficult. Remember that EN 15341 excludes downtime costs from the maintenance cost calculation.

Proper pumping system design is the most important single element in minimizing the LCC. All pumping systems include a pump, a driver, pipe installation, and operating controls, and each of these elements is considered individually. The characteristics of the piping system must be calculated to determine the required pump performance. This applies to both simple systems and more complex (branched) systems.

Proper design considers the interaction between the pump and the rest of the system and the calculation of the operating duty point(s). In systems with several pumps, the pump workload is divided between the pumps, which together, and in conjunction with the piping system, deliver the required flow. In summary, the degradation, O&M, and MRO costs strongly depend on the duty point and its configuration.

Procurement costs and operational costs must be included, as explained earlier, in the total cost of an installation. A number of installation and operational costs are directly dependent on the piping diameter and the components in the piping system (Hydraulic Institute et al., 2001).

The piping diameter is selected based on the following factors:

- Economy of the whole installation (pumps and system)
- Required lowest flow velocity for the application (e.g., avoid sedimentation)
- Required minimum internal diameter for the application (e.g., solids handling)

- Maximum flow velocity to minimize erosion in piping and fittings
- Plant standard pipe diameter

Decreasing the pipeline diameter has the following effects:

- Piping and component procurement and installation costs will decrease.
- Pump installation procurement costs will increase as a result of increased flow losses with consequent requirements for higher head pumps and larger motors; costs of electrical supply systems will therefore increase.
- Operating costs will increase as a result of higher energy usage because of increased friction losses.

Some costs increase with increasing pipeline size and some decrease. Because of this, an optimum pipeline size may be found, based on minimizing costs over the life of the system.

The duty point of the pump is determined by the intersection of the system curve and the pump curve as shown in Figure 7.21 (Hydraulic Institute et al., 2001).

A pump application might need to cover several duty points, of which the one with the largest flow and/or head will determine the rated duty for the pump. The pump user must carefully consider the duration of operation at the individual duty points to select the correct number of pumps in the

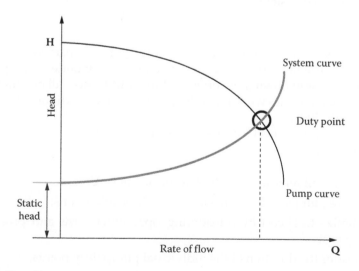

FIGURE 7.21
Duty point on pump and system curves. (Adapted from Hydraulic Institute et al., Pump life cycle costs: A guide to LCC analysis for pumping systems, Executive summary, DOE/ GO-102001-1190, January 2001.)

installation and determine output control (Hydraulic Institute et al., 2001). A wrong selection of the duty point or wrong design will speed up the degradation and ramp up the O&M costs with dramatic increases of MRO expenditures; therefore in pumping systems the design phase is crucial for further rational TCOs.

7.5.2.5 Lessons Learnt in LCC and TCO of Pumping Systems

Observing the operating system allows us to view how the actual system is working, but system operational requirements limit the amount of experimentation plant management will allow. Regardless of the method used, the objective is to gain a clear picture of how the various parts of the system operate and to see where improvements can be made and the system optimized.

The following steps provide guidelines proven successful to improving existing pumping systems:

- Compile a complete document inventory of the items in the pumping system after commissioning.
- Determine the flow rates required for each load in the system; these are clearly mentioned in the specifications.
- Balance the system to meet the required flow rates of each load.
- Minimize system losses needed to balance the flow rates.
- Carry out changes to the pump to minimize excessive pump head in the balanced system.
- Identify pumps with high maintenance cost.

Indeed, the observation of maintenance costs can be considered a lagging indicator of a design problem since the system working under non-optimum conditions will shut down more often and the MRO costs will be much more significant than the ones working under proper design conditions (Hydraulic Institute et al., 2001).

The following is a checklist of some useful ways to reduce the LCC of a pumping system:

- Consider all relevant costs to determine the LCC.
- Procure pumps and systems using LCC considerations.
- Optimize total cost by considering operational costs and procurement costs.
- Consider the duration of the individual pump duty points.
- Match the equipment to the system needs for maximum benefit.
- Match the pump type to the intended duty.
- Don't oversize the pump.

- Match the driver type to the intended duty.
- Specify motors to be high efficiency.
- Match the power transmission equipment to the intended duty.
- Evaluate system effectiveness.
- Monitor and sustain the pump and system to maximize benefit.
- Optimize preventative maintenance (Hydraulic Institute et al., 2001).

7.6 Military Equipment

7.6.1 Introduction and Background

Cost analysis is a critical element in the acquisition process of any army. It supports management decisions by quantifying the resource impact of alternative options. A good analysis includes different acquisition strategies, hardware designs, software designs, personnel requirements, and operating and support concepts.

As a particular program matures and more information becomes available, the cost estimate grows in complexity and detail. In addition, in the army, changes can come about very quickly, so army planners must have reliable and readily available information about the cost consequences of program changes, extensions, or cancellations. Cost analysts must develop models to support these quick turnaround analyses (U.S. Army Cost and Economic Analysis Center, 2002).

Cost analysis plays an ongoing role in the management of base operations. It helps determine base support requirements, develop budgets, conduct cost–benefit analysis, and perform special studies. Cost analysis is also used in the army for the following purposes:

1. To support decisions on program viability, structure, and resource requirements
2. To evaluate the cost implications of alternative materiel system designs
3. To provide credible and auditable cost estimates in support of milestone reviews during the acquisition process
4. To assess the cost implications of new technology, new equipment, new force structures, or new O&M concepts
5. To support the planning, programming, budgeting, and execution system process
6. To determine the funds required for a given level of training or operational activity such as miles driven per year (U.S. Army Cost and Economic Analysis Center, 2002)

7.6.2 Cost Analysis and Uncertainty

Cost analysis applies scientific and statistical methods to evaluate the likely cost of a specific item in a defined scenario. In the real world, there are multiple uncertainties about an item's cost. Some "internal" uncertainties influencing cost are inadequate item definition, poor contract statement of work, overly optimistic proposed solutions, inexperienced management, and success-oriented scheduling. Some "external" uncertainties include funding turbulence, a contractor's underestimation of complexity, a contractor's change of business base, and excessive (or insufficient) government oversight. In spite of uncertainty, the process of cost analysis is the most rigorous approach available to evaluate the costs of alternatives for the decision-maker (U.S. Army Cost and Economic Analysis Center, 2002).

7.6.2.1 Assumptions in LCCA

Cost analysis has its limitations. Analysts develop cost-estimating methodologies with an imperfect understanding of the technical merits and limitations of the item. The applicability of historic data is always subject to interpretation. Because of future uncertainties, there are limitations in determining the degree to which reality varies from the plan. Realistically, the cost analysis process cannot do the following:

1. Be applied with absolute precision, but must be tailored to the problem
2. Produce results that are better than input data
3. Predict political impacts
4. Substitute for sound judgment, management, or control
5. Make the final decisions

Despite these limitations, cost analysis is a powerful tool. Rigorous and systematic analysis leads to a better understanding of the problem. It improves management insight into resource allocation problems. Because the future is uncertain, our best estimate will differ from reality (U.S. Army Cost and Economic Analysis Center, 2002).

7.6.2.2 Economic Analysis

The U.S. Army's Economic Analysis (EA) manual provides a basic framework for implementing the policies of EA concepts, methods, and procedures, and applies to those preparing EAs. The manual describes the EA process, provides information on identifying and quantifying program benefits, identifies methods of comparing alternatives, and gives examples of quantitative techniques. Information for handling sensitivity, risk, and uncertainty is also provided (U.S. Army Cost and Economic Analysis Center, 2002).

7.6.2.3 Cost Analysis Training

Continuing education in cost analysis is crucial to the critical mission of providing U.S. Army decision-makers with quality, timely cost analysis. Department of Defense (DoD) agencies provide several excellent training programs (U.S. Army Cost and Economic Analysis Center, 2002).

7.7 Aviation

7.7.1 Introduction and Background

Aircraft types have different maintenance cost characteristics depending upon their following attributes:

- Technology
- Engines
- Buyer-furnished equipment (BFE)
- Utilization

There are numerous other factors as well. Figure 7.22 reviews how maintenance costs vary between aircraft types and suggests the reason for such variations (Figure 7.23).

Aircraft type and design are equally important to maintenance costs. The following are significant factors to consider:

- Maintainability (efficiency, special tooling) (Figures 7.24 and 7.25)
- Product design and quality
- Engine type
- System integration
- Structural and corrosion control programs
- System and component reliability
- Quality of technical documentation
- Product support capabilities
- Warranty and service life policy of the original equipment manufacturer (OEM)

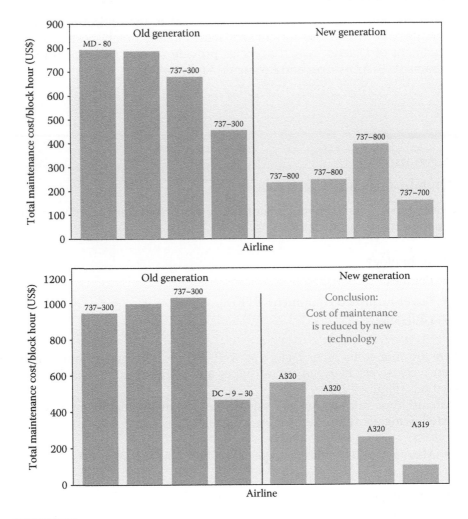

FIGURE 7.22
Maintenance cost variations (Airline Monitor).

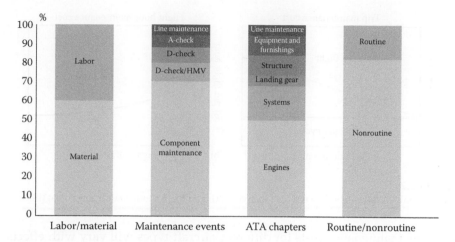

FIGURE 7.23
Maintenance cost distribution (Boeing).

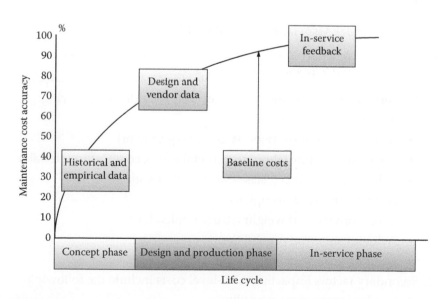

FIGURE 7.24
Maintenance cost prediction.

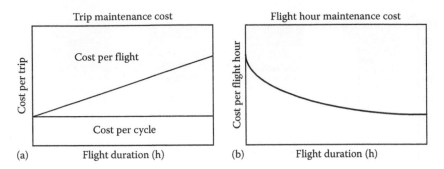

FIGURE 7.25
Aircraft utilization effects. Trip maintenance cost (a), and Flight hour maintenance cost (b).

Baseline maintenance costs for different aircraft types will vary with effects from the following:

- In-service (warranty, service life policy)
- Utilization (F/H–CY ratios)
- Aging (maturity versus increase of scheduled work)
- Modification standard

7.7.2 Air Transport Association (ATA)-Level Costs

Costs at the Air Transport Association (ATA) level can be primary or secondary. Figures 7.26 and 7.27 shows some of these.

1. Primary factors impacting ATA-level costs include the following:
   - Narrow body/wide body
   - Airframe weight (heavier parts and equipment)
   - Airframe structure (aluminum, metal alloys, composite materials)
   - Cabin configuration (classes, seats, galleys, etc.)
   - Number and type of engines
   - Maximum take-off weight (structural loads)
   - Avionics suite and capabilities
   - Landing gear configuration
2. Secondary factors impacting ATA-level costs include the following:
   - System/component reliability
   - Fault tolerance and redundancy
   - Corrosion and fatigue aspects
   - Maintainability aspects
   - Health/condition monitoring

- Built-in test capability
- Accessibility aspects
- Scheduled maintenance program
- Utilization
- Product support (Figures 7.26 and 7.27)

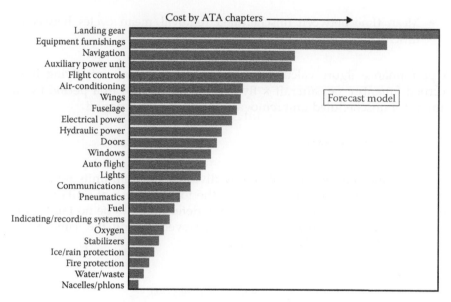

FIGURE 7.26
ATA maintenance costs. (From Boeing.)

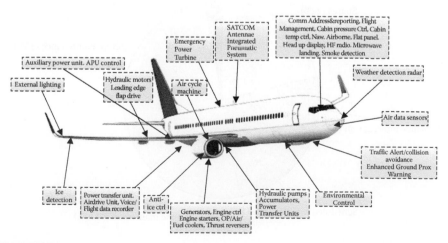

FIGURE 7.27
737NG technology (Honeywell Aerospace Products).

7.7.3 System and Component Reliability

The mean time between unscheduled removals (MTBUR) has a profound effect on airline operations, despite a high MTBF. Airlines also have a high interest in reducing no fault found (NFF) rates, with a goal of less than 20%, since each NFF occurrence represents an unacceptable additional and excessive cost of ownership to the airline.

- Mean time between unscheduled removal/mean cycles between unscheduled removal (MTBUR/MCBUR)

A performance figure calculated by dividing the total unit flying hours accrued (quantity per aircraft × flying hours or cycles) in a period by the number of unscheduled unit removals during the same period.

- No fault found rate

A performance figure is calculated by dividing the total units returned to the shop with no confirmed defect by the total number of units returned to the shop. This is a percentage of units determined to be no fault found out of the whole population of units removed during the same period (Figures 7.28 and 7.29) (Dennehy, 2000).

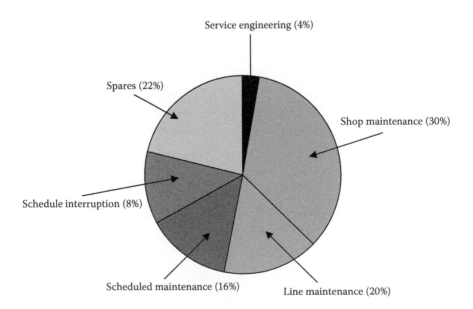

FIGURE 7.28
Direct cost impact.

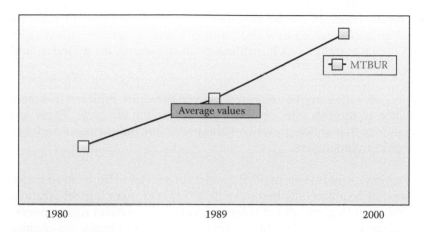

FIGURE 7.29
MTBUR development (Boeing).

7.8 Facilities

O&M typically include the day-to-day activities necessary for a building and its systems and equipment to perform their intended function. These two functions are combined into the common term O&M because a facility cannot operate at peak efficiency without being maintained; therefore, the two are discussed as one (Sapp, 2015).

Facilities O&M can be divided into the following areas:

- *Real property inventory (RPI)*: Provides an overview on the type of system needed to maintain an inventory of an organization's assets and manage those assets.

- *Computerized maintenance management system (CMMS)*: Contains descriptions of procedures and practices used to track the maintenance of an organization's assets and associated costs.

- *Computer-aided facilities management*: Includes the creation and utilization of information technology (IT)-based systems in facility management practice.

- *O&M manuals*: As O&M represent the greatest expense in owning and operating a facility over its life cycle, well-developed, user-friendly, and facility-specific O&M manuals are now required in many companies.

- *Janitorial/cleaning*: Janitorial, custodial, or housekeeping staff and ground maintenance crews are important for an effective overall facility maintenance and cleaning program.

- *Historic buildings O&M*: Includes keeping old equipment running while also installing new and more efficient equipment; this is especially important in such institutions as museums, art galleries, and so on.

O&M includes the activities required to keep the entire built environment in a condition to meet its intended function during the life cycle. These activities include both planned preventive and predictive maintenance and corrective (repair) maintenance:

- Preventive maintenance (PM) consists of a series of time-based maintenance requirements that provide a basis for planning, scheduling, and executing scheduled (planned versus corrective) maintenance. PM includes adjusting, lubricating, cleaning, and replacing components. Time-intensive PM, such as bearing/seal replacement, is typically scheduled for regular (plant or "line") shutdown periods.
- Predictive maintenance attempts to detect the onset of degradation with the goal of correcting it prior to significant deterioration in the component or equipment.
- Corrective maintenance is a repair necessary to return equipment to properly functioning condition or service and may be either planned or unplanned. Some equipment, at the end of its service life, may warrant overhaul.

Requirements will vary from a single facility to several facilities owned by the same company. As the number, variety, and complexity of facilities increase, the organization performing the O&M should adapt in size and complexity to ensure performance is sustained. In all cases, O&M requires a knowledgeable, skilled, and well-trained management and technical staff and a well-planned maintenance program. The philosophy behind the development of a maintenance program is often predicated on the O&M organization's capabilities.

The goals of a comprehensive maintenance program include the following:

- Reduce capital repairs.
- Reduce unscheduled shutdowns and repairs.
- Extend equipment life, thereby extending facility life.
- Realize LCC savings.
- Provide safe, functional systems and facilities that meet the design intent.

Sustainability is an important aspect of the O&M process. A well-run O&M program should conserve energy and water and be resource-efficient, while meeting the comfort, health, and safety requirements of the building occupants. Various laws and standards must be considered in the O&M process as well.

A critical component of an overall facilities O&M program is its proper management. The overall program should contain five distinct functions: operations, maintenance, engineering, training, and administration (OMETA). Beyond establishing and facilitating links across the OMETA functions, O&M managers have the responsibility of interfacing with other department managers (Sapp, 2015).

7.8.1 System-Level O&M Manuals

Organizations requiring a higher level of O&M information beyond the typical vendor equipment documents should ensure sufficient funds are set aside and appropriate scope/content/format requirements are identified during the planning stage. It is important to analyze and evaluate a facility from the system level, then develop procedures to attain the most efficient systems integration.

System-level manuals include information, based on the maintenance program philosophy. O&M procedures at the system level do not replace manufacturers' documentation for specific pieces of equipment, but supplement those publications and guide their use (Sapp, 2015).

7.8.2 Teardowns

Demolishing older or historic buildings and replacing them with new structures that may not be as durable, sustainable, or secure is a problem in both the government and private sectors. Currently there is no single tool available to solve the teardown problem, so a combination of strategies works best. One tool available online is "Teardown Tools on the Web," created as part of the National Trust for Historic Preservation Teardowns Initiative. This tool is intended as an easy-to-share, user-friendly, one-stop shopping, highlighting approximately 30 tools and more than 300 examples of best practices in use in the United States (Sapp, 2015).

7.8.3 Major Resources

7.8.3.1 Planning and Design Phase

O&M activities start with the planning and design of a facility and continue through its life cycle. During the planning and design phases, O&M personnel should be involved and should identify maintenance requirements for inclusion in the design, such as equipment access, built-in condition monitoring, sensor connections, and other O&M requirements that will aid them when the built facility is turned over to the owner/user organization.

The O&M team should be represented on the project development team so they know ahead of time the types of controls, equipment, and systems they will have to maintain once the facility is turned over to them (Sapp, 2015).

7.8.3.2 Construction Phase

System-level and manufacturer manuals of as-installed systems and equipment, including as-built drawings, should be available for review by the owner over the course of the construction phase. To efficiently operate a facility at turnover, O&M information must be available prior to fiscal completion, owner occupancy, and especially before operator/maintainer training. Although obtaining O&M documentation may be overseen by the owner's representative or building commissioning agent, the effort should be coordinated with/overseen by the owner's construction manager to ensure it is being accomplished (Sapp, 2015).

7.8.3.3 O&M Approach

A company's or facility's O&M organization is typically responsible for operating and for maintaining the built environment. To accomplish this, the O&M organization must operate the systems and equipment responsibly and maintain them properly. The utility systems may be simple supply lines/systems or complete production and supply systems. The maintenance work may include planned preventive/predictive maintenance, corrective (repair) maintenance, trouble calls (e.g., a room is too cold), replacement of obsolete items, predictive testing and inspection, overhaul, and grounds care.

O&M organizations may utilize a reliability-centered maintenance (RCM) program that includes a mix of reactive, time- or interval-based, condition-based, and proactive maintenance. The O&M organization is also normally responsible for maintaining records on deferred maintenance (DM), that is, maintenance work that has not been accomplished for some reason—often lack of funds (Sapp, 2015).

7.8.3.4 Life Cycle O&M

According to the International Facilities Management Association (IFMA), the operating LCC of a facility is typically comprised of 2% for design and construction, 6% for O&M, and 92% for occupants' salaries. O&M of the elements included in buildings, structures, and supporting facilities is complex and requires a knowledgeable, well-organized management team and a skilled, well-trained workforce whether the functions are performed in-house or contracted.

The objective of the O&M organization should be to operate, maintain, and improve the facilities to provide reliable, safe, healthful, energy-efficient, and effective performance of the facilities to meet their designated purpose throughout their life cycle. To accomplish these objectives, O&M management must manage, direct, and evaluate day-to-day O&M activities and budget funds to support the organization's requirements (Sapp, 2015).

7.8.3.5 Computerized Maintenance Management Systems

O&M organizations may utilize CMMS to manage their day-to-day operations, track the status of maintenance work, and monitor the associated costs of that work. These systems are vital tools to manage the day-to-day activities and to provide valuable information for preparing facilities' key performance indicators (KPIs)/metrics to use in evaluating the effectiveness of the current operations and to support organizational and personnel decisions. These systems are increasingly integrated with geographic information systems (GIS), building information modeling (BIM) echnologies, and construction operations building information exchange (COBie) to increase/improve a facility's operational functionality (Sapp, 2015).

7.8.3.6 Coordinating Staff Capabilities and Training with Equipment and System Sophistication Levels

Qualified personnel are needed to operate and maintain facilities at peak efficiencies, and to protect significant investments in equipment and systems. Besides posing a potential physical hazard to themselves and others, untrained employees can unknowingly damage equipment and cause unnecessary downtime. Inefficient and improper O&M can also void warranties and reduce expected useful life (EUL) of equipment.

O&M organizations must address the skill level of their staff in light of the O&M systems and components within their facilities. This extends beyond the in-house staff to include any contracted services. If the skills required to support installed systems and equipment are scarce, either training must be provided or less sophisticated equipment systems must be used to provide an economical working arrangement.

As technology advances are incorporated into renovations, major capital repairs, and new building construction, high-tech building systems are being placed in service that current O&M staff are not familiar with and, thus, cannot address problems when they arise. An example is building automation systems (BAS). Often untrained personnel will override programmed settings; over time, these cumulative overrides result in unbalanced system-wide operations.

Regardless of its equipment sophistication levels, every organization should develop training programs and track staff qualifications to ensure they are adequate for existing and planned building systems. This will allow organizations to make improvements to training as needed on an ongoing basis. A recurring training program should consider both the type of skills required and the available labor pool of skills in the geographic area. Topics for consideration include the following:

- Safety regulations and guidelines
- Equipment operational start-up and shutdown procedures
- Normal operating parameters
- Emergency procedures
- Equipment preventative maintenance (PM) plans
- Use of proper tools and materials, including personal protective equipment (PPE)

Training programs should be reviewed at least annually and whenever changes are planned for equipment or new facilities.

In addition to regular assessments of the O&M staff's technical abilities concerning existing equipment, the staff should always be included throughout new project development efforts by design teams. The O&M staff is usually one of the best sources for input on how an existing facility is performing, and they can provide insight into how new equipment will be incorporated into facility maintenance programs. The staff may not always understand the underlying cause of a building problem, but they can identify areas that receive repeated attention and help correct long-standing conditions. O&M staff inputs can guide designers to address these areas in renovation and equipment upgrade projects.

Certifications and proper training of O&M service providers protect the organization, employees, and visitors. Training sources include manufacturers, professional organizations, trade associations, universities and technical schools, commercial education/training courses, and in-house and on-the-job training (OJT) options. Training programs should provide an appropriate mix of these sources to ensure materials are up to date and applicable to the organization's facilities (Sapp, 2015).

7.8.3.7 Non-O&M Work

Most O&M organizations typically perform work that is beyond the definition of O&M but has become a part of their baseline. This work is facilities-related but new in nature and, as such, should not be funded by O&M money but by the requesting organization. Examples include minor facilities work, such as installing an outlet to support a new copier machine, providing a compressed air outlet to a new test bench, day porter services for special event setups and moves. They also include major work, such as construction projects (Sapp, 2015).

7.9 Energy

There is an increasing awareness within the nuclear power industry of the need to perform detailed LCCA to quantify the risks associated with investing in new equipment to improve plant capacity. In undertaking LCCA, three primary factors must be addressed:

- What are the considerations that go into performing LCCA?
- What tools are available for undertaking them?
- Given constrained resources, how does a plant operator determine which improvement or set of improvements will provide the best return on investment? (Hall, 2004)

7.9.1 Nuclear Plant Life Cycle Cost Analysis Considerations

The nuclear industry has seen a substantial rise in average capacity factors (CFs) over the past 15 years, with the average nuclear plant CF rising to well over 90%. For those plants involved in continuous improvement, this trend has resulted in a "good news, bad news" situation. While the plants have vastly improved stakeholder value, it has become increasingly difficult both to justify investment in maintaining and improving plant availability and also to choose between various options when faced with budget constraints. Given the size and nature of the improvement investments being proposed, the undertaking of detailed, quantitative LCCA has become a necessity.

Typically, when LCCA is performed, the cost of the improvement, its expected benefit, and the net present value (NPV) of the proposed change are addressed. Items not addressed include the impact of equipment aging on performance, the variability of replacement power costs, and future spare parts and maintenance costs.

LCCA requires tools that are able to integrate disparate data information. Improvements in computer technology have dramatically increased operating speeds, memory, and storage, making reliability, availability, and maintainability (RAM) simulation an excellent tool for this purpose. Taking into account issues such as equipment aging, overhaul effectiveness, reliability, maintainability, and cost variability, they provide time-based profiles of component and plant performance which, in turn, provide a platform for addressing not only the initial cost of an investment but the spares, operating, and maintenance costs associated with that investment through the remainder of plant life.

Once the means for performing LCCA on individual items is understood, a means for selecting a set of improvements given budgetary constraints must be selected. Typically, decisions on investment are based on the merits of each alternative proposed. Unfortunately, this often results in double-counting the benefits! A way to avoid double-counting and to optimize alternative improvement proposals is presented in the following section (Hall, 2004).

7.9.2 LCCA Defined

LCCA underpins the LCM process, providing a systematic way to address the costs and benefits associated with LCM decisions. LCM in the power industry has been defined as follows (Arey, 2001):

> Life cycle management is the process by which nuclear power plants integrate operations, maintenance, engineering, regulatory, environmental, and economic planning activities in a manner that:
>
> 1. Manages plant material condition (e.g. aging and obsolescence of systems, structures, and components—SSCs),
> 2. Optimizes operating life (including the options of early retirement and license renewal), and
> 3. Maximizes plant value while maintaining plant safety.

In this definition, the LCM process has two parts: technical evaluation and economic evaluation.

Simply stated, LCCA is the process by which future costs for the acquisition, implementation, operation, and maintenance of new equipment or systems are addressed (Hall, 2004).

7.9.2.1 LCCA Considerations

Aspects to be addressed in LCCA can be broken down into the following categories:

- Cost and revenue influences
- Economic factors
- Magnitude of change
- Data uncertainty

Each category is addressed in the following subsections (Hall, 2004).

7.9.2.2 Cost and Revenue Influences

The objective of undertaking an LCCA is to determine whether a proposed action will result in benefit to the plant stakeholders. Figure 7.30, based on the risk-informed asset management (RIAM) concept developed by South Texas Project Nuclear Operating Company (STPNOC) (author/date), gives a view of a comprehensive cost model to address the cost benefit of any proposed

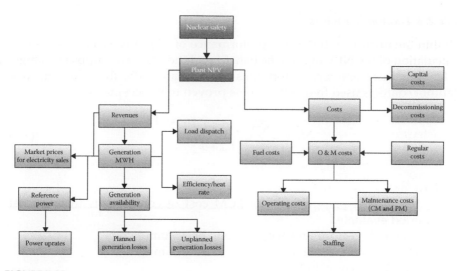

FIGURE 7.30
RIAM value map. (Adapted from Hall, S., Nuclear plant life cycle cost analysis considerations, in: *Proceedings of ICAPP*, Pittsburgh, PA, June 13–17, 2004, Paper 4072.)

option. The goal is to ensure that a proposed action will result in an increase in the plant's NPV (Hall, 2004).

With the premise that the model illustrated in Figure 7.30 is a closed-loop system, the following observations can be made:

- Any suggested improvement likely to reduce safety will probably be rejected. It is assumed that a proposed reduction in safety would either result in relicensing costs or regulatory action.

- The area's greatest susceptibility to volatility or uncertainty (and the primary cost driver) is in the areas of market prices for electricity sales, unplanned generation losses, and maintenance costs.

- The impact of unplanned generation losses and that of corrective maintenance (CM) costs are linked in that the frequency and duration of component and system failure affect the level of generation losses and the frequency and severity of component or system failure affect CM material and labor costs.

- The effect of efficiency and heat rate tends to remain relatively static over a given period. Changes, when they do occur, are relatively predictable because of the time and resources required to address engineering and regulatory issues.

- The remaining revenue and cost areas can be forecast with some predictability and will typically have less influence on the benefit cost of a proposed improvement (Hall, 2004).

7.9.2.3 Economic Factors

Within the nuclear industry, the performance of LCCA typically results in a calculation of the NPV of both the benefit and the cost of a proposed change. Although there are several formulas to calculate NPV, the following one (author/date) is used frequently and is proven to be acceptable:

$$\text{NPV} = \sum_{j\text{first}}^{j\text{last}} C_j * \left[\frac{1+k}{1+d} \right]^{(t_j - t_{\text{NPV}})} \tag{7.6}$$

where

NPV is the net present value of *LCM* cost of components through last period evaluated

j_{first} is the first year for which costs are to be accumulated

j_{last} is the last year for which costs are to be accumulated

C_j is the year j cost in today's dollars

d is the discount rate (cost of money)

k is the inflation rate plus real escalation rate

t_j is the year in which cost is to occur

t_{NPV} is the year for which *NPV* is to be computed

At some power utilities, different values for the discount rate may be used to calculate the NPV of costs and benefits. In addition, different rates may be used to calculate the NPV of hardware or labor costs (Hall, 2004).

7.9.2.4 Determining Magnitude of Change

Performing LCCA relies on determining the expected change in the plant's CF that is attributable to a proposed improvement or set of improvements. Simply stated, the process is as follows (Hall, 2004):

- A change is proposed and the cost of that change ascertained.
- A model is used to determine the change in CF attributable to the proposed change.
- The change in CF is converted to a value using the following (or similar) equation:

$$\text{Benefits}(\$) = \Delta \text{CF} * \left(\frac{\$}{\text{MW-h}} \right)^* \text{Capacity (MW)}^* \text{Period (h)} \tag{7.7}$$

Changes in CF are the result of a change in the equivalent availability (EA) of a plant, a system, or a component, and/or its efficiency.

Reliability or, more appropriately, reliability, availability, and maintainability (RAM) models are used to determine the magnitude of change in the CF for a proposed change in component or system availability. Logic dependency models, typically represented by fault trees or reliability block diagrams (RBD), are coupled with appropriate failure frequency and repair data to calculate the CF of the plant for a given set of conditions. To determine a change in CF, the failure rate and/or the repair time that characterize the proposed change are placed in the model and the CF recalculated.

The process just described appears relatively simple, but its implementation can be complex and resource-intensive. The RAM models developed to evaluate these changes not only require failure and repair data representative of the plant modeled, but they also need to address the following:

- Regulatory requirements (e.g., tech specs)
- Effects of component aging and overhaul effectiveness
- Operation of components and systems (e.g., condensate pumps)
- Operation of standby systems and components
- Availability of maintenance and spare parts resources (Hall, 2004)

7.9.2.5 Data Uncertainty

There is a significant degree of uncertainty associated with certain elements of LCCA. These are the primary uncertainties associated with LCCA:

- *Replacement power cost*: global and local market forces affect this value.
- *Failure and repair data*: typically, failure and repair data used in models reflect average values representing the mean of an underlying distribution. Uncertainty can also apply to the amount of time a component is unavailable because of planned maintenance.
- *Data distributions*: the underlying distribution assumed for failure frequencies (and to a lesser extent, repair times) affects the behavior of RAM models. For example, if an exponential failure distribution is assumed, maintenance will have no effect on average availability. If a Weibull or normal distribution is assumed, maintenance will have an effect.
- *Logistics data*: given a failure, the availability of spares (and the resources to effect the repair) can affect the time a component is inoperable. Should a spare not be available, the time required to order and receive the part can vary from a few hours to weeks, significantly increasing downtime.
- *Values for inflation, escalation, and discount rates used for NPV calculations*: these can vary significantly. During the 1980s and early 1990s, the inflation rate was consistently greater than the discount rate; recently, the reverse has been true. While it is difficult (if not

impossible) to foresee how these factors will change over time, their changing nature should be kept in mind when evaluating the NPV over a 10–20-year period (Hall, 2004).

7.9.3 LCCA Tools

As seen in the previous section, performing LCCA requires the integration of a number of factors. For a typical plant, an LCCA will require the following:

- An availability model reflecting the operation and design of the plant
- Failure and repair data (and underlying distributions) for all systems and components reflected in the model
- Costs for
  - Equipment procurement, installation, and testing (or overhaul/ refurbishment of current equipment)
  - Periodic maintenance (labor and material)
  - Spares
  - Replacement power
  - Fuel, operating, regulatory, and decommissioning
- Refueling and other periodic maintenance schedules
- NPV calculation variables
- Logistics data (spares order and shipping time, labor mobilization time)

The performance of a comprehensive LCCA that integrates all these factors could not be accomplished in a cost-effective and timely manner without high-speed computers and relatively low-cost, proven RAM assessment tools. The question then becomes: which tool best meets the needs of a comprehensive LCCA? Do we want a fault tree or RBD approach? The answer will be driven by the ability of the tool to integrate the information provided here, the desired output, and the cost of developing/modifying models to support the LCCA (Hall, 2004).

7.9.3.1 PRA and RAM Analysis

Some people confuse probabilistic risk assessment (PRA) with RAM analyses. Since they rely on the same logical rules, they are thought to be equivalent, but this is a misconception. The primary use of PRA is to address issues affecting nuclear safety. Because of its nature, PRA is focused on determining the probability of an unlikely event occurring (i.e., core damage). The impact of the unlikely event on the CF is of secondary importance. PRAs are accomplished by applying sophisticated software tools such as the Electric Power Research Institute (EPRI's) computer-aided fault tree analysis (CAFTA). Until recently, an overwhelming portion of the reliability-oriented work has featured PRA.

In contrast to PRA, RAM assessment addresses the frequency of occurrence of more likely events and their impact on the ability of the plant to export power to the grid. Because of the historical focus on safety, relatively little RAM analysis has been done in the nuclear power sector. This is not true, however, in the gas turbine and steam fossil power sectors. These plant types have been assessed using simulation-based RBD-oriented applications that have recently become available (Hall, 2004).

7.9.3.2 LCCA Output

The primary goal of an LCCA is to determine the cost benefit of a proposed plant improvement or change in an operations procedure. Management will want to know the cost benefit of a proposed solution, as well as the uncertainty surrounding that estimate. Table 7.7 is an example of results obtained by the author after performing an LCCA of a hypothetical case involving finding the best alternative for a main generator improvement. The changes in NPV indicate the required net benefit and the uncertainty surrounding the mean value; they also reflect the variables discussed earlier (Hall, 2004).

Figure 7.31 illustrates the distribution of potential NPV changes between case A and case B.

Additional LCCA outputs can support decisions based on LCCA. The output illustrated in Figure 7.32 is a time-based forecast of the costs and benefits involved in replacing Feedwater Heaters. Figure 7.33 is an example of a time-based forecast of the capacity of a Pressurized Water Reactors (PWR) unit with a 2-year refueling cycle. Such outputs provide insight into the expected CF at any particular time in the future, allowing managers to plan outages at times when demand may be less; or if there is a downward trend in forecast performance, they can determine the appropriate time to implement design changes (Hall, 2004).

TABLE 7.7

Improvement Alternatives and Benefit/Cost Results

Parameter ($100,000)	Alternative B-A	Alternative C-A
Most likely NPV change	−1.26	6.23
Mean NPV change	3.67	16.8
5% NPV change	−4.22	−4.32
25% NPV change	−0.75	5.91
50% NPV change	2.23	12.9
75% NPV change	6.29	24.5
95% NPV change	16.3	51.5
Optimum choice		C

Source: Adapted from Hall, S., Nuclear plant life cycle cost analysis considerations, in: *Proceedings of ICAPP*, Pittsburgh, PA, June 13–17, 2004, Paper 4072.

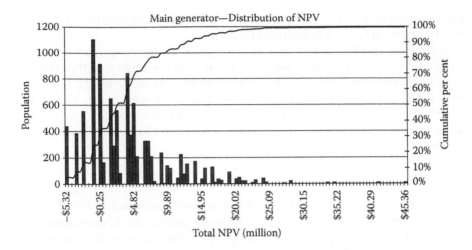

FIGURE 7.31
Distribution of NPV changes between case A and case B. (Adapted from Hall, S., Nuclear plant life cycle cost analysis considerations, in: *Proceedings of ICAPP*, Pittsburgh, PA, June 13–17, 2004, Paper 4072.)

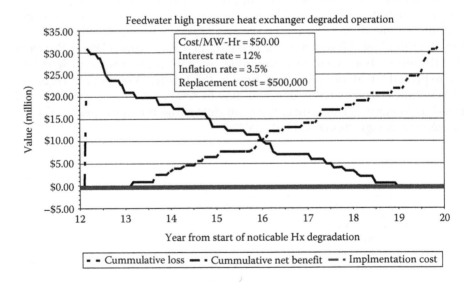

FIGURE 7.32
Feedwater heater operating forecast. (Adapted from Hall, S., Nuclear plant life cycle cost analysis considerations, in: *Proceedings of ICAPP*, Pittsburgh, PA, June 13–17, 2004, Paper 4072.)

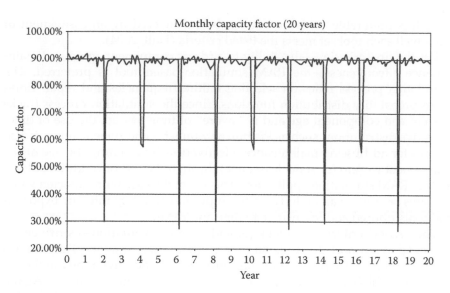

FIGURE 7.33

Time-based CF forecast. (Adapted from Hall, S., Nuclear plant life cycle cost analysis considerations, in: *Proceedings of ICAPP*, Pittsburgh, PA, June 13–17, 2004, Paper 4072.)

7.9.3.3 LCCA Tool Selection

Given the range of inputs and uncertainties to be addressed and the desired output, which tool is preferable for LCCA? There are four criteria on which this decision must be based:

1. The ability to address component aging and maintenance effectiveness
2. Timeliness and ease of evaluation
3. The cost of the software used to support LCCAs
4. The cost to develop the model and maintain it so that it reflects the "as built" condition of the plant

To our knowledge, no single stand-alone application currently fulfills these criteria. To date, LCCAs have been performed using either of two alternatives:

1. A combination of expected value RAM assessment tools (e.g., fault tree applications), electronic spreadsheet applications, and generalized commercial parameter simulation applications
2. A combination of Monte Carlo–based RAM simulation tools and electronic spreadsheets

Which is preferable? This question is best answered by an assessment of how well each tool set meets the listed criteria (Hall, 2004).

For the first criterion—the ability to address component aging and maintenance effectiveness—the RAM simulation-based tool is preferred. The simulation process relies on selecting failure frequencies and repair times from probability distribution functions. Since the simulation process tracks system and component age and is "aware" of time passed, the use of time-sensitive distributions such as Weibull accounts for component aging. It also allows for addressing maintenance effectiveness by resetting the age of the component to zero (or some lesser age) through the use of conditional logic within RAM simulation application. This is not the case with expected value systems because of an underlying assumption that failure rates are exponentially distributed and insensitive to time and repair.

The second criterion—timeliness and ease of evaluation—is driven by the need to support the decision process in a timely manner. Time frames for performing LCCAs are typically compressed and, given the number of variables to be addressed, resources for performing them are limited. Again, the RAM simulation-based approach is preferred. While an expected value application can determine the value of CF for a given set of conditions in seconds, a significant number of these evaluations must be performed to account for each potential plant-operating state and, if so desired, the CF value determined over time. This potentially large number of evaluations imposes a file management and post-processing regime that can slow the evaluation process. The time required to perform an LCCA evaluation using a RAM simulation application can vary from few minutes to a few hours depending on the complexity and size of the model, the number of life histories simulated, and the speed and capacity of the computer on which it is operated. Unlike the expected value applications, the simulation application uses and evaluates a single, unified model that accounts for all potential operating states and determines the value of CF over time. The file management burden is essentially eliminated and post-processing requirements reduced.

When the third criterion—the cost of the application software—is considered, the expected value application tool is preferred. Many expected value applications, such as the EPRI-sponsored CAFTA, are free for the nuclear utilities involved with EPRI's risk and reliability tools. For many other nuclear utilities, the cost of acquisition has already been amortized and is no longer an issue. Published costs for Monte Carlo–based simulation software vary from $10,000 to $75,000. These costs can be offset by the reduced costs involved in performing LCCA; when viewed in the context of the value they deliver, they support decisions that could involve millions of dollars, thus making their use quite reasonable.

In the nuclear industry, when the fourth criterion—the cost to develop and maintain the model—is considered, a non-simulation approach has a slight edge over a simulation-based approach. Most plants have already developed detailed models of the nuclear steam supply system in order to perform PRAs. In addition, the processes for managing quality assurance and configuration control are in place. However, very few plants have developed models for balancing plant systems—the systems most likely to cause generation losses. Recent efforts have been made to allow the automatic incorporation of PRA fault tree elements into RAM simulation models, thereby reducing the cost of model development (Hall, 2004).

7.9.4 Improvement Option Selection

Typically, an LCCA is performed to address the advisability of implementing a specific improvement option or to select the best one from a range of options. The process is focused on evaluating a given component or system, independent of improvements proposed for other components or systems. The selection of the range of improvements optimized to provide the greatest return to the plant, as a whole, is not a straightforward process of selecting alternatives based solely on a single criterion such as expected NPV. A number of different elements must be addressed:

- Is there sufficient budget available to implement a set of projects in a given year?
- Even if sufficient budget is available, is there sufficient outage time available to implement a proposed change?
- Does a proposed change in component availability have a linear effect on unit availability? If we are dealing with redundant components or systems (e.g., if four condensate pumps operating in parallel with three others at any one time are required for 100% plant output), the assumption of linearity is incorrect. In the redundant case, especially with installed spare capacity, no discernible improvement in unit availability will occur unless the availability of more than one pump is improved. Moreover, the improvement in the availability of each additional pump will not result in equal improvements in unit availability.
- Given competition for resources between different units, how is a proposed improvement likely to benefit the company as a whole?

One way to address these complexities, developed by EPRI in the late 1980s, is still valid (Hall and Unkle, 1989). The process, summarized below, was applied to fossil units at two different utilities (Hall and Weiss, 1988). One utility used it to optimize the implementation of approximately 30 different proposed alternatives at three units. The other used it to optimize 180 proposed alternatives at 12 units. In both cases, the process resulted in a reduction in the investment required and an increase in the forecasted return (Hall, 2004).

The improvement LCC optimization process takes a four-step iterative approach, as illustrated in Figure 7.34 and explained as follows:

- The first step is to collect the information and data related to the improvements under evaluation.
- The second step is to apply an economic screening criterion and method to determine which improvement options are potentially cost-beneficial.

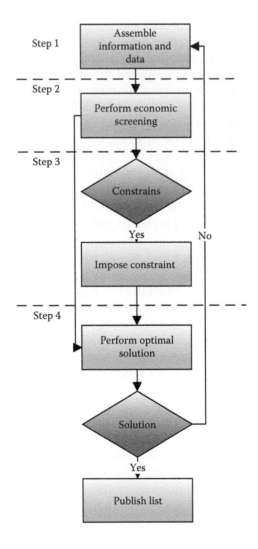

FIGURE 7.34
Cost optimization process. (Adapted from Hall, S., Nuclear plant life cycle cost analysis considerations, in: *Proceedings of ICAPP*, Pittsburgh, PA, June 13–17, 2004, Paper 4072.)

- The third step considers various constraints such as funding limitations, outage schedules, and manpower limitations to further evaluate the candidate improvements.

- The final step is to evaluate surviving candidate improvements through a dynamic program algorithm to arrive at a sequence of improvements that provide the greatest net benefit within established constraints.

As will be shown, LCCA applications like those discussed earlier are used in steps 2 and 4 (Hall, 2004).

7.9.4.1 LCC Optimization Step 1: Data Collection

In order to implement the LCC optimization process, it is necessary to establish a relationship between the cost of implementing an improvement and the expected benefit of that improvement. The relationship is established by determining the cost of the improvement, estimating the expected increase in component availability resulting from it, calculating the effect of the component availability change on the overall unit equivalent availability or CF, and converting the change in unit equivalent availability into a benefit based on an increase in net generation revenue. To do so, the following information is required:

- A listing of the RAM and efficiency improvement options under consideration

- The cost required to implement each improvement option

- The time and resources required to implement each change

- For RAM improvements, the actual or estimated change in event frequency and/or downtime resulting from each improvement option

- For efficiency improvements, the expected percent increase in net revenue from either decreasing the fuel cost or increasing net generation capability

- An LCC simulation or expected value model and associated baseline data for the power plant (or plants) to be evaluated

- The cost relationships between unit availability and costs, such as replacement power, fuel, and O&M expenditures

- Identification of funding, schedule, or other resource constraints

- Economic factors, such as escalation, discount, and interest rates

An LCC model is used to assess changes in unit availability that may occur due to changes in component RAM characteristics so that the relationship between availability and production costs can be studied quickly and accurately.

Finally, information on constraints is required because the cost optimization methodology must be responsive to the possibility of limited capital, outage time, or the labor and engineering resources available for implementing improvements. This is especially true for improvement projects that must compete for funding (Hall, 2004).

7.9.4.2 LCC Optimization Step 2: Economic Screening Analysis

An economic screening analysis is used to identify those candidate improvement options with the potential to produce a positive net benefit. This initial economic screening assumes the proposed improvements are independent.

Before beginning this analysis, an LCCA is performed for the plant (or plants) to evaluate the effect of changes in component availability on unit production. The output of the evaluation is a criticality ranking (C_i). For each component, the ranking indicates the increase in unit productivity to be expected if that component were to achieve "perfect" availability. The forecasted change in component availability for a given proposed improvement (ΔA_c) is then multiplied by the component's criticality ranking ($C_i \times \Delta A_c$) to calculate the approximate change in unit availability that can be expected from implementing the change.

The initial screening relies on the assumption that the relationship between component and unit availability is linear, but as Figure 7.35 illustrates, this relationship can be nonlinear. However, the relationship can be linearly approximated for small changes in component availability. For each

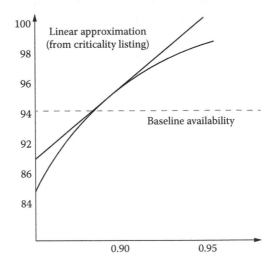

FIGURE 7.35
Component/unit availability relationship. (Adapted from Hall, S., Nuclear plant life cycle cost analysis considerations, in: *Proceedings of ICAPP*, Pittsburgh, PA, June 13–17, 2004, Paper 4072.)

proposed RAM improvement, the expected increase in unit production is used to estimate the increase in annual megawatt hours that may be expected from a specific component improvement. To calculate the change in expected megawatt hours (ΔMW-h), the following equation is used:

$$\Delta MW\text{-}h = \Delta A_u{}^*\left(\text{Unit net capacity}\right)^*\left(\text{Scheduled operating hours}\right) \qquad (7.8)$$

The increase in power production can now be converted to an expected revenue increase and compared to the cost of making the component improvement (Hall, 2004).

Improvements proposed for increasing efficiency are economically screened as follows:

- If the proposed change will result in an increase in net capacity, the benefit is calculated using the following equation:

$$\Delta MW\text{-}h = A_u{}^*\Delta\left(\text{Unit net capacity}\right)^*\left(\text{Scheduled operating hours}\right) \qquad (7.9)$$

- As seen earlier, the increase in power production can be converted to an expected revenue increase compared to the cost of making the component improvement.
- If the proposed change will result in lowering fuel costs, that is, less fuel is required to generate the same amount of power, the following equation is used to calculate the expected benefit:

$$\text{Benefit} = MW\text{-}h's^*\Delta\left(\frac{\text{Cost}}{MW\text{-}h}\right) \qquad (7.10)$$

where

$$MW\text{-}h's = A_u{}^*\Delta\left(\text{Unit net capacity}\right)^*\left(\text{Scheduled operating hours}\right) \quad (7.11)$$

If other cost factors are affected by changes in unit productivity, they can be estimated in a similar manner. Those component improvements likely to yield a cost savings greater than the investment cost become potential economically viable improvement candidates because, as we will see later, they may be dropped from consideration for other reasons. If so desired, the present worth of the costs and benefits can be used in the economic screening process to account for the time value of money over the life of the change.

The output of the economic screening process is a list of potential economically viable improvement projects. These projects, with their costs and benefits, are analyzed considering additional constraints (e.g., minimum cost/benefit ratio, must do for regulatory reasons, negative impact on safety) that may be desired (Hall, 2004).

7.9.4.3 LCC Optimization Step 3: Optimization with Constraint

The third step of the analysis considers any stated constraints on the improvement process, such as funding limitations or manpower resources. If there are no constraints, or the constraints are not exceeded, the optimization process can proceed to the optimal solution process. If the limitations of any constraints are not satisfied, an integer program (IP) algorithm is applied to the economically screened candidate improvement options before taking the last step. The objective of the IP algorithm step is to choose the combination of improvements that provide the optimum benefit while satisfying the limitations of each constraint. The IP step assumes that the benefit resulting from each specific improvement will not affect the benefit of other improvements and that the total benefit is the sum of each individual benefit.

The result of using the IP is a list of candidate component improvements that maximize the net benefits and meet the imposed constraints. If the assumption of independence and linearity reflects the actual relationship between component and unit availability, the IP will provide the final optimum set of improvements. However, the relationship between component availability and unit availability is often nonlinear, and experience with LCC models suggests component improvement effects are not independent (Hall, 2004).

7.9.4.4 Step 4: Optimal Solution Process

The final step in the optimization process is to apply a dynamic programming (DP) algorithm to the set of candidate improvement options. The objective of the DP algorithm is to optimize the solution set, taking into account any nonlinearities between component and unit availability and any interdependency between components. This is done by making a sequence of selections, with the caveat that if the process were prematurely terminated, the changes selected up to that point would still be optimal.

As each component improvement is selected and the baseline design or operation of the unit is changed (via the LCCA model), the ratio of changes in unit availability to changes in component availability of the unmodified components will increase, decrease, or remain the same. Because of these changes, some component improvements that were previously not cost-beneficial may become beneficial. Conversely, it is also possible that some improvements will no longer be beneficial. The unpredictable effect of changes on component criticalities (C_i) is investigated using the DP algorithm. Note that the DP algorithm is dependent on the constraints imposed by the IP algorithm.

The DP algorithm methodically addresses the expected benefit of implementing alternative sets of improvement candidates to ascertain the set that will provide the greatest net benefit. As each improvement candidate is implemented, and the baseline design or operation of the unit is changed (via the LCCA model), the economic screening and imposition of constraint processes are again done on an iterative basis. The economic screening is accomplished with the new baseline design; the imposition of constraints is accomplished with a reduction in the constraint equal to the cost of the candidate improvement(s) implemented.

The result of these optimal analyses is a chronologically ordered list of recommended improvements that should provide maximum return on an improvement investment considering all constraints.

This process has a great deal of relevance to utilities in that it provides a repeatable, verifiable process for selecting LCM projects in the face of limited budgets and other constraints. The nature of the process ensures a rank ordering of projects that will provide the greatest return to the unit or owner (Hall, 2004).

7.10 Mining

Many managers do not know the total cost of maintenance and are unable to calculate or use the ratio of direct maintenance cost to total value-added costs. Figure 7.36 calculates this very important ratio for a number of industries.

As the figure indicates, the ratio of direct maintenance cost to total value-added costs for the mining industry varies between 20% and 50%, towering above the other industries. Clearly, maintenance is a major concern for the mining industry (Barkhuizen, 2002).

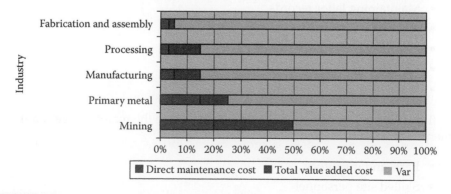

FIGURE 7.36
Ratio of direct maintenance cost to total value-added costs. (Adapted from Campbell, J.D., *Uptime: Strategies for Excellence in Maintenance Management*, Productivity Press, Tampa, FL, 1995.)

In this section, the actual life cycle maintenance data of three front-end loaders, in operation in a South African mine, is used as a case study to elaborate the concepts and ideas discussed elsewhere in the book. Note that some of the data are changed for reasons of confidentiality.

7.10.1 Introduction and Background

Generally speaking, in all industries, a physical asset is put into use to perform a certain function; although the foremost reason for maintenance is to maintain a certain machine condition, the maintenance plan should also preserve the function (Moubray and ALADON, 1999).

As an example in the mining industry, we will consider a front-end loader loading coal into trucks. The foremost reason for maintenance on this machine is to maintain the machine's condition, but maintaining the machine's condition should also be aimed at maintaining the amount of coal loaded into the truck; maintenance procedures should improve machine condition and maintain or improve machine production (Barkhuizen, 2002).

7.10.2 Design of Maintenance Plan

As Chapter 2 suggests, maintenance strategies are much like business strategies and can be analyzed in the same fashion. Simply stated, the maintenance requirements are all the work processes needed to develop the maintenance plan for the equipment, similar to a business unit, which is expected to provide some benefit to the user/owner. These work processes include the following:

- Generating work orders
- Organizing and tracking inventory
- Tracking equipment history
- Scheduling tasks and projecting equipment failure
- Maintaining labor records
- Allocating resources
- Planning and ordering spare parts
- Forecasting
- Managing service exchange units

In this processes, the resources required to achieve the maintenance requirements can be summarized:

- Parts and inventory
- Skilled site personnel
- Tools and equipment
- Stores and offices

Finally, the last element to be considered is the maintenance control systems. These systems ensure all the information collected is used to improve the maintenance plan (Barkhuizen, 2002).

7.10.3 Analysis of Maintenance Data

Maintenance information and data are gathered to predict future events based on past events. When the maintenance planner has access to maintenance information, every maintenance activity can be treated as a single project. The work breakdown and scope of work can be based on history, and the project will eventually become a recurring event (Gouws, 1995).

Every time a project or maintenance task is completed, the new information is fed into the information system, helping the maintenance planner make a more realistic estimate when the project recurs. The data captured must be correct, complete, and sufficient to be of any use during the analysis or to build a maintenance model. Using a standardized format ensures correct data are captured.

Information is a valuable tool that can be used to sharpen the maintenance program and improve the maintenance approach (Barkhuizen, 2002).

7.10.3.1 Machine Hours

This section was compiled using data from the case study to illustrate what information can be derived from the data gathered. Note that machine hours are mostly used as the time reference for equipment. The machine hour for each machine is illustrated in Figure 7.37. As the figure shows, loader no. 1 has the most machine hours and is, therefore, the oldest of the three

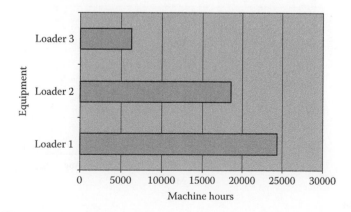

FIGURE 7.37
Machine running hours. (Adapted from Barkhuizen, W.F., Life cycle management for mining machinery, The Rand Afrikaans University, Johannesburg, South Africa, November 2002.)

machines. As Chapters 1 and 2 have pointed out, the age of equipment is a very important KPI and influences the maintenance strategy significantly (Barkhuizen, 2002).

7.10.3.2 Machine Availability

Availability is calculated using Equations 7.12 and 7.13. The machine hours use Equation 7.2 and the calendar hours use Equation 7.12, resulting in the difference illustrated in Figures 7.38 through 7.40 for loaders 1, 2, and 3, respectively (Barkhuizen, 2002).

$$\text{Availability} = \frac{\text{Scheduled production} - \text{All unplanned delays}}{\text{Scheduled production time}} \quad (7.12)$$

$$\text{Availability} = \frac{\text{MTBF}}{\text{MTBF} + \text{MTTR}} \quad (7.13)$$

7.10.3.3 Mean Time between Failures and Mean Time to Repair

The mean time between failures (MTBF indicates the reliability of the machine, and the mean time to repair (MTTR indicates the maintainability of the machine (Barkhuizen, 2002).

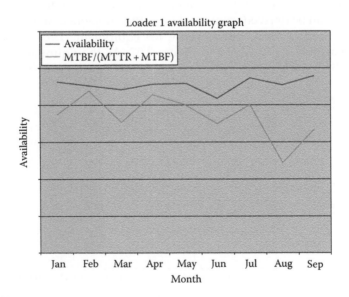

FIGURE 7.38
Loader 1 availability graph. (Adapted from Barkhuizen, W.F., Life cycle management for mining machinery, The Rand Afrikaans University, Johannesburg, South Africa, November 2002.)

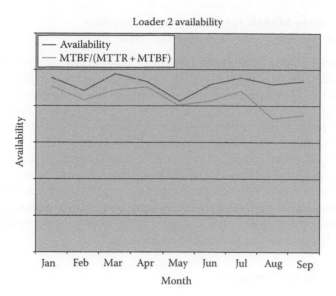

FIGURE 7.39
Loader 2 availability graph. (Adapted from Barkhuizen, W.F., Life cycle management for mining machinery, The Rand Afrikaans University, Johannesburg, South Africa, November 2002.)

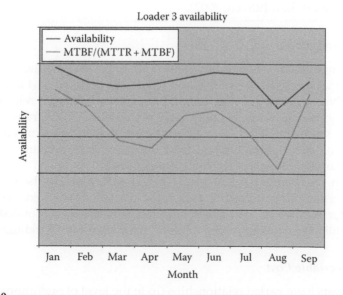

FIGURE 7.40
Loader 3 availability graph. (Adapted from Barkhuizen, W.F., Life cycle management for mining machinery, The Rand Afrikaans University, Johannesburg, South Africa, November 2002.)

7.10.4 Decision Models for Maintenance Decision and Optimum Life Cycle Management

As the previous section suggests, reliability is generally concerned with the failures during the life of a product. Thus, RCM is an essential aspect of LCM. RCM is part of LCM when the failure of equipment in its operating context is addressed. Therefore, the connection between RAM and LCM is clear and implies the translation of effectiveness into costs where RCM plays a key role for rationalization of mentioned maintenance expenditures.

LCM is typically an agreement between a company (in this case, a mine) and a contractor (a service supplier). The contractor agrees to perform certain maintenance functions, and in return for these maintenance functions, the company agrees to reward the contractor for the maintenance service delivered.

A very important aspect of this agreement between the company and the contractor is the maintenance cost (Barkhuizen, 2002).

7.10.5 Cost of Maintenance

Today's production facilities are focused on availability, cost-effectiveness, production output, and flexibility. These are the key issues in any maintenance or LCM contract (Campbell, 1995).

Maintenance is one of the costliest parts of operation, especially if the costs of production losses due to downtime caused by failure and/or maintenance activities are added to the equation. Maintenance has a huge impact on the performance of any operation; in fact, it determines the profitability of the whole operation (Barkhuizen, 2002).

7.10.5.1 Fixed Costs

Life cycle maintenance agreements, similar to those used by the mining case study, normally consist of variable and fixed costs. A fixed cost is a group of costs involved in an ongoing activity whose total will remain relatively constant throughout the range of operational activity (ECORYS Transport 2005). In the case study, the fixed costs are made up of such cost items as salaries to site technicians, equipment and tools, infrastructure, and any other costs incurred that are not part of the maintenance cost.

Although fixed costs remain relatively constant, they can be expected to increase in a stepped pattern with an increase in activity. When extra equipment is added, for example, an extra site technician will have to be appointed to maintain this equipment, causing a step increase in the fixed rate (Barkhuizen, 2002).

7.10.5.2 Variable Cost

Variable costs have varied relationships up to the level of operational activity (Thuesen and Fabrycky, 1993). The variable cost is expected to increase with an increase in required availability.

In the case study, the variable cost consists of a rate per machine hour or a rate per ton. Thus, the uptime of the machine has direct financial implications for the company maintaining the machine, as well as for the mine; more uptime results in more production (Barkhuizen, 2002).

7.10.6 Life Cycle Maintenance Cost

Most products are brought into being and utilized over a life cycle. The maintenance life cycle starts when a machine is at zero running hours and ends with phase out and disposal or selling the machine at salvage value (Barkhuizen, 2002).

The variable cost (i.e., the rate per machine hour or the rate per ton) varies during the life cycle of the machine, in this case the loader. This variance can be linked to the preventative maintenance strategy and coincides with the planned parts change-out (International Society of Parametric Analysts (ISPA), 1995).

For comparative purposes, Figure 7.41 shows a typical variable maintenance cost for a drill and a front-end loader; it illustrates that the maintenance on each type of equipment is unique. In the first three maintenance periods, the variable rate curve of the loader and the drill is similar. In the fourth period,

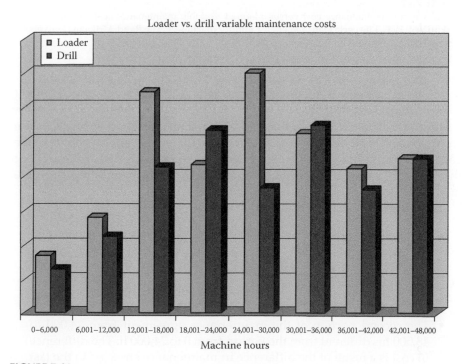

FIGURE 7.41
Variable maintenance cost: loader versus drill. (Adapted from Barkhuizen, W.F., Life cycle management for mining machinery, The Rand Afrikaans University, Johannesburg, South Africa, November 2002.)

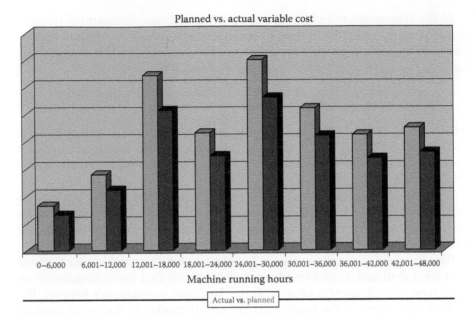

Planned vs. actual variable cost

Machine running hours

0–6,000   6,001–12,000   12,001–18,000   18,001–24,000   24,001–30,000   30,001–36,000   36,001–42,000   42,001–48,000

Actual vs. planned

FIGURE 7.42
Actual variable maintenance cost versus planned variable maintenance cost. (Adapted from Barkhuizen, W.F., Life cycle management for mining machinery, The Rand Afrikaans University, Johannesburg, South Africa, November 2002.)

the comparative variable rate starts to vary; this can be attributed to the different life expectancy and replacement of parts on the two types of equipment (Barkhuizen, 2002).

Figure 7.42 illustrates the actual and planned maintenance cost for the case study loaders. The difference between the actual and planned maintenance cost is the profit margin. The profit margin should be constant throughout the life cycle of the machine (Mather, 2002), but it is not, suggesting improvements need to be made.

The background details for the figure are the following:

- A major overhaul is scheduled for the period 18,000–24,000 h to improve reliability; some components are rebuilt at 18,000 h, others at 24,000 h.
- The life cycle of the maintenance agreement is 48,000 h although major components are replaced during a midlife service.
- The second portion of the maintenance agreement from 24,000 to 48,000 his different from the period from 0 to 24,000 h. The difference in cost is a result of the difference in maintenance strategy. In the last period, maintenance costs are kept low to lower the salvage value of the machine at the end of the agreement. During this period the guaranteed availability will also be lower (Barkhuizen, 2002).

7.10.7 Outcomes of LCCA: Added Value

Value in this context can be defined as a measure of the worth that a person ascribes to a good service. Thus, the value of an object is inherent not in the object but in the regard a person has for it (Thuesen and Fabrycky, 1993). The value of LCM to the customer can be represented by the following relationship (Meredith and Mantel, 2000):

$$\text{Value} = \frac{\text{Quality}*\text{Service}}{\text{Cost}*\text{Time}*\text{Risk}} \qquad (7.14)$$

In this case, value to the customer is gained by delivering the best quality and service at the lowest cost and in the shortest time (Barkhuizen, 2002).

References

Acharya, D., Mishalani, R., Martland, C., Eshelby, J., 1991. *REPOMAN: An Overall Computer-Aided Decision Support System for Planning Rail Replacements.* International Heavy Haul Railway Conference, Vancouver, British Columbia, Canada, pp. 113–123.

Arey, M., 2001. Demonstration of life cycle management planning for systems, structures, and components—With pilot applications at Oconee and Prairie Island nuclear stations. EPRI Report 1000806, Electric Power Research Institute, Palo Alto, CA.

Barkhuizen, W. F., November 2002. Life cycle management for mining machinery. The Rand Afrikaans University, Johannesburg, South Africa.

Campbell, J. D., 1995. *Uptime: Strategies for Excellence in Maintenance Management.* Productivity Press, Tampa, FL. ISBN: 1-56327-053-6.

Dennehy, P., 2000. *Improved Performance Through the Use of Defined Metrics.* The Boeing Company, Chicago, IL.

Dersin, P., Valenzuela, R. C., December 2012. Application of non-Markovian stochastic Petri nets to the modeling of rail system maintenance and availability. In: *Proceedings of the 2012 Winter Simulation Conference*, Berlin, Germany.

Ebersohn, W., Ruppert, C. J., 1998. Implementing a railway infrastructure maintenance system. In: *CORE98 Conference*, Rockhampton, Queensland, Australia, 1998.

ECORYS Transport, 2005. Infrastructure expenditures and costs. Practical guidelines to calculate total infrastructure costs for five modes of transport. Final Report, Rotterdam, the Netherlands. November 30, 2005.

Esor, R., Zarembski, A., 1992. Development of track maintenance planning models for rail rapid transit system. *Computers in Railways III: Proceedings of the Third International Conference, Management.* Computational Mechanics Publications, Boston, MA.

Esveld, C., 1989. *Modern Railway Track.* MRT Publications, Zaltbommel, the Netherlands.

Esveld, C., 2001. *Modern Railway Track*, 2nd edn. Delft University of Technology, MRT Productions, Zaltbommel, Netherlands.

EU REMAIN Consortium, 2000. Modular system for reliability and maintainability management in European rail transport. Final Report, IITB.

Frangopol, D. M., 2010. Reliability deterioration and lifetime maintenance cost optimization. Department of Civil, Environmental and Architectural Engineereing, University of Colorado, Boulder, CO.

Gouws, L. E., November, 1995. *Maintenance Management with Emphasis on Condition Monitoring of Excavation Machines.* Rand Afrikaans University, Johannesburg, South Africa.

Hall, S., 2004. Nuclear plant life cycle cost analysis considerations. In: *Proceedings of ICAPP*, Pittsburgh, PA, June 13–17, 2004, Paper 4072.

Hall, S., Unkle, R., 1989. Demonstration of an availability optimization methodology. EPRI Report GS-2462-1, Electric Power Research Institute, Palo Alto, CA.

Hall, S., Weiss, J., 1988. Life extension—A strategy for cost optimization. In: *Proceedings, Life Assessment and Extension*, Vol. III. Nederlands Instituut voor Lastechniek, The Hague, the Netherlands, p. 190.

Hydraulic Institute, Europump, US Department of Energy's Office of Industrial Technologies (OIT), January 2001. Pump life cycle costs: A guide to LCC analysis for pumping systems. Executive summary. DOE/GO-102001-1190, U.S. Department of Energy, Washington, DC.

Institute of Rail Vehicles, 1995. Organizing economic analysis of the variants of oil products haulage with track width change using 911Ra tank cars. Research Project KBN No. 9 9454 95 C/2385, Task No. 5. University of Technology, Krakow, Poland.

Institute of Rail Vehicles, 2008. Evaluation of rail shifting systems 1435/1520 mm by applying LCC analysis. Research Project No. NB-2/2008. University of Technology, Krakow, Poland.

International Society of Parametric Analysts (ISPA), 1995. Parametric cost estimating handbook. Department of Defence, Washington, DC, Fall 1995.

International Standard IEC 60300-3-3, 2004. Dependability management, Part 3-3: Application guide—Life cycle costing, 2nd edn. IEC 2004.

Kim, J., Park, J., Jeong, D., 2011. A study on the life cycle cost calculation of the Maglev vehicle based on the maintenance information. In: *The 21st International Conference on Magnetically Levitated Systems and Linear Drives*, Daejeon, Korea, October 10–13, 2011.

Kjellsson, U., 1999. From X2000 to Crusaris Regina: Development of LCC technology. In *Proceedings of World Congress Railroad Research*. Tokyo, Japan. 1999.

Kjellsson, U., Hagemann, O., 2000. Unife-unilife and unife-unidata-the first european life cycle cost interface software model. Research Report.

Klyatis, L. M., February 3, 2012. *Accelerated Reliability and Durability Testing Technology*. John Wiley & Sons, Hoboken, NJ, p. 430.

Ling, D., 2005. Railway renewal and maintenance cost estimating. PhD thesis. School of Applied Sciences, Cranfield University, Cranfield, U.K.

Mather, D., 2002. Calculating the savings from Implementation of the CMMS. Plant Maintenance Resource Center, Cosmo, Western Australia, Australia, July 2002. Accessed November 2011.

Meredith, J. R., Mantel Jr., S. J., 2000. *Project Management—A Managerial Approach*, 4th edn. Wiley, Hoboken, NJ. ISBN: 0-471-29829-8.

Mesnick, D., 1991. Implementation of a track maintenance planning and component degradation modelling program at a transit property. *International Heavy Haul Railway Conference*, Vancouver, British Columbia, Canada, pp. 267–284.

Moubray, J., 1999. The responsible custodianship of physical assets. In: *The Tenth Annual Canadian Maintenance Management Conference*, Toronto, Ontario, Canada.

Neves, L. C., Frangopol, D. M., Cruz, P. S., 2004. Cost of life extension of deteriorating structures under reliability-based maintenance. *Computers and Structures* 82, 1077–1089.

Ohtake, T., 1998. Practical use of TOSMA on track maintenance work. In: *Conference on Cost Effectiveness and Safety Aspects of Railway Track*, ERRI and UIC, Paris, France.

Ruman, F., Grencik, J., 2014. LCC and LCP analysis of rail vehicles. *Logistyka* 3, 5490–5498.

Sapp, D., 2015. Facilities operations & maintenance. Plexus Scientific. Updated by the Facilities O&M Committee. https://www.wbdg.org/om/om.php. Accessed August 2016.

Stalder, O., 2002. The cost of railway infrastructure. *Innovations for a Cost Effective Railway Track* 2, 32–37.

Stirling, A., Roberts, C., Chan, A., Madelin, K., Booking, A., 2000. Trail of an expert system for the maintenance of plain line track in the UK. In: *Railway Engineering 2000*, London, U.K., July 5–6, 2000.

Szkoda, M., 2008. Method of assessing the durability and reliability of rail gauge-changing systems. Doctoral dissertation. University of Technology, Institute of Rail Vehicles, Krakow, Poland.

Szkoda, M., 2011. Life cycle cost analysis of Europe-Asia transportation systems. In: *EURO—ZEL 2011, 19th International Symposium*, Žilina, Slovakia, June 8–9, 2011.

Szkoda, M., Tulecki, A., 2006. Life cycle cost in effectiveness evaluation of rail vehicles' modernization. In: *Materials of XVII Science Conference "Rail Vehicles"*, Kazimierz Dolny, Poland, September, 2006.

Thuesen, G. J., Fabrycky, W. J., 1993. *Engineering Economy*, 8th edn. Prentice Hall, Englewood Cliffs, NJ. ISBN: 0-13-279928-6.

Trask, E., Fraticelli, C., 1991. The track degradation model. In: *International Heavy Haul Railway Conference*, Vancouver, British Columbia, Canada, pp. 54–60.

Tułecki, A., 1999. Life Cycle Cost (LCC) as a measure of effectiveness railway means of transport. Instytut Pojazdów Szynowych TABOR, Pojazdy Szynowe No. 10/1999.

UNIFE LCC Group, 1997. Guidelines for life cycle cost. Union of European Railway Industry. Brussels, BE.

U.S. Army Cost and Economic Analysis Center, 2002. Cost analysis manual. ATTN: SFFM-CA-CP, Department of the Army, May, 2002, Washington, DC.

Zaalberg, H., 1998. Economising track renewal and maintenance with ECOTRACK. In: *Conference on Cost Effectiveness and Safety Aspects of Railway Track*, ERRI and UIC, Paris, France.

Zarembski, A., 1989. *On the Development of a Computer Model for the Economic Analysis of Alternate Tie/Fastener Configurations*. American Railway Engineering Association, Chicago, IL, pp. 129–145.

Zoeteman, A., 2001. Life cycle cost analysis for managing rail infrastructure. Concept of a decision support system for railway design and maintenance. *European Journal of Transport and Infrastructure Research* 1(4), 391–413.

Zoeteman, A., 2003. Whole life costing: The Dutch experience. *Journal and Report of Proceedings* 121(4), 355–370.

Zoeteman, A., Braaksma, E., 2001. An approach to improving the performance of rail systems in a design phase. In: *World Conference on Railway Research*, Cologne, Germany, pp. 1–9.

Zoeteman, A., 2000. Bepalen waarvoor te betalen: een instrument voor de analyse van de prijs/prestatieverhouding van railinfrabeheer (Knowing what to pay for: An instrument for the analysis of the price and performance of railway infrastructure), *CVS Transport Conference*, Amsterdam, the Netherlands.

Zoeteman, A., Esveld, C., 1999. Evaluating track structures: Life cycle cost analysis as a structured approach. In: *World Congress on Railway Research*, Tokyo, Japan.

Zoeteman, A., Van der Heijden, R., 2000. Planning the infrastructure performance of railways? Decision support for track design of a Dutch high speed link. In: *Developing Rail Policy for People and Freight: Proceedings of Seminar H held at the European Transport Conference*. Homerton College Association for European Transport, Cambridge, U.K.

Index

A

Aa, *see* Achievable availability
AAS, *see* Asset automation systems
ABC, *see* Activity-based costing
Achievable availability (Aa)
    capital improvements, 54, 56
    definition, 46
    design, 48, 50
ACI, *see* Asset condition index
Acquisition costs, 401, 404
Activity-based costing (ABC), 274
Activity-based maintenance
    model, 363
Affordability studies, of LCC, 69
AFI, *see* Asset functionality index
After-sales services, 349
Aircraft maintenance cost
    aircraft type and design, 441, 444
    ATA-level costs, 444–445
    attributes, 441
    cost distribution, 441, 443
    cost variations, 441–442
    maintainability prediction, 441, 443
    utilization effects, 441, 444
Air Transport Association (ATA) level
    costs, 444–445
Alliance model, 363
Alternate method impact (AMI) costs
    complexity *vs.* accuracy, 316
    cost procedure, 313–315
    definition, 312–313
    graph, 313–314
    timeline, 313–314
Al-Zahrani audit programs, 263
American Association of State Highway
    and Transportation Officials
    (AASHTO) Subcommittee on
    Maintenance, 362
AMI costs, *see* Alternate method
    impact costs
Analytic hierarchy process (AHP), 262
Annually recurring costs, 72–73
Annual maintenance budget

affecting factors, 144–145
categories, 128–129
composition, 128–129
decision considerations, 135
development, 129–133
estimating formula, 145–146
execution, 136–138
funding calculation, 132
importance, 128
maintenance cost control, 141–142
vs. maintenance costs, 142–144
maintenance strategy, 136–137
manager's role, 127–128
preparation, 135–136
proactive use, 40
review, 138–140
summary, 134
workload, 137–139
worksheet, 133
Ao, *see* Operational availability
API, *see* Asset performance index
ARI costs, *see* Associated resource
    impact costs
Army cost analysis
    EA process, 440–441
    internal and external
        uncertainties, 440
    limitations, 440
    purposes, 439
    training, 441
Asset automation systems (AAS), 118
Asset condition index (ACI), 99
Asset functionality index (AFI), 100
Asset life cycle model, 89
Asset maintenance program
    (AMP), 198
Asset management (AM)
    business benefits, 40
    condition/performance, 24–25
    creation/acquisition, 23–24
    definition, 21–22
    disposal/rationalization, 25–26
    financial services sector, 21

Printed and bound by CPI Group (UK) Ltd, Croydon, CR0 4YY

24/10/2024

01778301-0016